Lecture Notes in Electrical Engineering

Volume 780

The book series *Lecture Notes in Electrical Engineering* (LNEE) publishes the latest developments in Electrical Engineering - quickly, informally and in high quality. While original research reported in proceedings and monographs has traditionally formed the core of LNEE, we also encourage authors to submit books devoted to supporting student education and professional training in the various fields and applications areas of electrical engineering. The series cover classical and emerging topics concerning:

- Communication Engineering, Information Theory and Networks
- Electronics Engineering and Microelectronics
- Signal, Image and Speech Processing
- Wireless and Mobile Communication
- Circuits and Systems
- Energy Systems, Power Electronics and Electrical Machines
- Electro-optical Engineering
- Instrumentation Engineering
- Avionics Engineering
- Control Systems
- Internet-of-Things and Cybersecurity
- Biomedical Devices, MEMS and NEMS

For general information about this book series, comments or suggestions, please contact leontina.dicecco@springer.com.

To submit a proposal or request further information, please contact the Publishing Editor in your country:

China

Jasmine Dou, Editor (jasmine.dou@springer.com)

India, Japan, Rest of Asia

Swati Meherishi, Editorial Director (Swati.Meherishi@springer.com)

Southeast Asia, Australia, New Zealand

Ramesh Nath Premnath, Editor (ramesh.premnath@springernature.com)

USA, Canada:

Michael Luby, Senior Editor (michael.luby@springer.com)

All other Countries:

Leontina Di Cecco, Senior Editor (leontina.dicecco@springer.com)

**** This series is indexed by EI Compendex and Scopus databases. ****

More information about this series at http://www.springer.com/series/7818

Chandima Gomes
Editor

Lightning

Science, Engineering, and Economic Implications for Developing Countries

Springer

Editor
Chandima Gomes
School of Electrical and Information Engineering
Faculty of Engineering and the Built Environment
University of the Witwatersrand
Johannesburg, South Africa

ISSN 1876-1100 ISSN 1876-1119 (electronic)
Lecture Notes in Electrical Engineering
ISBN 978-981-16-3439-0 ISBN 978-981-16-3440-6 (eBook)
https://doi.org/10.1007/978-981-16-3440-6

This Springer imprint is published by the registered company Springer Nature Singapore Pte Ltd.
The registered company address is: 152 Beach Road, #21-01/04 Gateway East, Singapore 189721, Singapore

Foreword

I feel very privileged to write this foreword for this book on lightning, which has been the subject of predilection of my scientific research for almost 35 years. My relationship with the book Editor, Prof. Chandima Gomes, goes back to the late 1990s, when I met him during a visit to Uppsala University. At that time, Chandima was a young graduate student working towards a Ph.D. degree under the supervision of my dear friend and colleague, Prof. Vernon Cooray, who is one of the world's most eminent lightning scientists. I remember an interesting discussion I had with Chandima on return stroke modelling, which, among many others, was one of the topics in his thesis. I was very much impressed by his knowledge and maturity, despite his young age. Later, in 2000, he published a paper in the IEEE Transactions on Electromagnetic Compatibility[1] in which he presented a very general categorization of the return stroke models, showing that all of the models can be described by two general sets of mathematical equations. This paper is considered today as a classic in return stroke modelling.

After finishing brilliantly his Ph.D. thesis, Chandima established a successful international academic career starting at the University of Colombo in Sri Lanka. Over the following years and during the past 20 years or so, I was in close contact with him and watched him develop into a first-class researcher.

Beyond his outstanding scientific achievements, especially in the field of lightning and lightning protection, Chandima always felt concerned about the ethical aspects of scientific work, in general, and, in particular, about how to reduce deaths, injuries and property damage from lightning in the developing countries, hence this book. Who else but Chandima could have designed and edited it?

The book comprises a logically organized sequence of 11 chapters that are, at the same time, self-contained and can therefore also be read separately. It starts with three fundamental and general chapters describing the physics of lightning, basic approaches for lightning detection and warning systems, and risk assessment, making the book also very accessible to the technical non-expert. The ensuing chapters are focused on the protection of different types of systems and infrastructure, with

[1] C. Gomes and V. Cooray, Concepts of Lightning Return Stroke Models, IEEE Trans. EMC, Volume 42, Issue 1, Feb 2000.

special attention to the protection of renewable energy systems. The final chapters are concerned with economic, human and technical aspects of lightning protection, with a strong emphasis on developing countries.

One of the particularly remarkable aspects of the book is that, except for two chapters, Chandima has solicited authors who are lightning scientists from developing countries in Asia, Africa and South America.

In 2015, the United Nations adopted Agenda 2030, which includes 17 Sustainable Development Goals (SDGs) which aim at fighting poverty in all forms and at promoting sustainable development. This book presents, in a comprehensive way, methods and technologies that can greatly contribute to the attainment of these 17 SDGs.

October 2020

Farhad Rachidi, Ph.D.
Swiss Federal Institute of Technology
Lausanne, Switzerland

Preface

Lightning is one of the most revered extreme natural events revered by many civilizations for a few millennia. Started nearly three hundred years ago, lightning was explored as per the modern scientific methodologies in many countries by a number of scientists, so far. These researches shed light on many phenomena related to lightning; however, the bulk of the physics of lightning is still far being fully understood. On the other hand, technologies of lightning protection measures of structures and systems, and methods of lightning safety of human beings and animals have been deeply studied. The outcomes of these researches have been documented as standards and guidelines, and consequently put into practice, however, mostly in developed countries. As such, the temporal records of many developed countries clearly delineate that lightning-related losses of both property and life in these countries have significantly reduced over the last century.

The lightning ground flash density maps reveal that the occurrence of lightning is much higher in equatorial and oceanic tropical regions in the world than that in temperate regions. Unfortunately, most of the countries in this region belong to the developing world where the literacy rate, level of education, quality of sheltering, medical facilities, infrastructure, etc., are below par compared to the same in the developed world. In turn, these countries experience the highest fatality rates, property damage, service interruption and down time-related losses due to lightning. The scientists and researchers, engineers, educators and social activists in these countries are at a distinct disadvantage of accessing the latest knowledge and know-how due to both economic and strategic reasons. These lapses further suppress the development of their own technology or methodologies in the developing world for curbing lightning accidents.

This book has been developed in the above backdrop; thus, its major goal is to channel the concurrent knowledge from the leading scientists and engineers in the field to the scientific public in the developing countries. The contents of the book will also be a very valuable source of information for academics, researchers, funding agencies and business communities in developed countries who interact with fellow professionals, grantees and clients/customers of developing countries. The will also be a very useful handbook for early career researchers and engineers involved with lightning-related research and consultancy, irrespective of their location.

The concept of developing such a book has been proposed by Dr. Amitava Bandopadhyay, Director General, NAM S&T Centre, New Delhi. Since 2007, NAM S&T Centre has generously supported organizing international symposia on lightning sciences in Sri Lanka, Nepal, Uganda and Zambia, during which several authors of this book gathered invaluable facts and figures on lightning-related incidents in developing countries. They could also acquire first-hand information on the level of awareness on lightning safety and protection among both educated masses and the general public of these less-privileged communities. I may confidently state that this wealth of knowledge has made the platform for several authors of this book for understanding the needs of the potential readers.

This book contains eleven chapters contributed by 19 authors from Brazil, China, Colombia, France, Germany, India, Mexico, South Africa, Sweden and the USA. The contents basically cover the physics of lightning, lightning detection, lightning protection of structures, low-voltage systems and high-risk installations, lightning accidents, injury mechanisms and safety guidelines, economic implications and future research opportunities.

I would like to express my heartfelt gratitude to all the authors, Prof. Vladimir Rakov, Dr. Anirban Guha, Dr. Yakun Liu, Prof. Earle Williams, Dr. Carina Schumann, Dr. Hugh Hunt, Mr. Alain Rousseau, Mr. Alexis Barwise, Prof. Ruy Alberto Corrêa Altafim, Mr. Sergio Roberto Silva dos Santos, Prof. Hélio Eiji Sueta, Dr. Arturo Galván Diego, Prof. Michael Rock, Mr. Ronald L. Holle, Prof. Emerita, Dr. Mary Ann Cooper, Dr. Norberto Navarrete-Aldana, Mr. Ashen Gomes and Prof. Adônis F. R. Leal, who kindly accepted my invitation to write the chapters and providing the complete contents well in time.

Thanks are also due to Dr. Amitava Bandopadhyay and his team at NAM S&T Centre, for all the support rendered; Prof. Farhad Rachidi, Swiss Federal Institute of Technology (EPFL), Lausanne, Switzerland, for writing the forward; and Dr. Loyola D'Silva, Senior Editor, Springer Nature, Singapore, and his team for making this endeavour a success.

Chandima Gomes, Ph.D.
University of the Witwatersrand
Johannesburg, South Africa

Introduction

Lightning and thunderstorms are of concern for the human beings for many centuries. Even today, many communities and tribes believe that lightning is a God itself or it is a weapon hurled by the divine powers. As people started analysing lightning as a physical phenomenon, around 300 years back, the attention was mainly on understanding the nature of lightning and the ways of avoiding the lightning strikes damaging their properties. With time, this focus was shifted to safeguarding of human beings and livestock, protecting power lines that stretch over hundreds of kilometres over highly exposed land, protection of defence systems and electronics & communication system.

In the 1970s, scientists realized that similar to other atmospheric phenomena, lightning may also be forecasted in advance. Such forecasting could play a significant role in planning outdoor activities and preparation of emergency crews for power system breakdowns. However, the enormous number of dynamic variables that determines the spatio-temporal location of lightning flashes prevented them from making these predictions with reasonable accuracy, at least by few minutes in advance. Such limitations gave birth to a new term, *Lightning Nowcasting*, where the detection system could reconstruct the lightning strike point as soon as it is attached to a ground point. More specifically, *Lightning Nowcasting* refers to the prediction of areas and times where lightning activity may occur within 0–1 hour based on intelligent software. Even such systems could provide very useful information on planning outdoor activities in the event of an approaching thunderstorm, developing lightning density maps which are essential in the risk assessment of buildings and properties and flight control management at airports.

In the last decade, as super-fast computers and smart algorithms started emerging, scientists were empowered with the ability of processing millions of data with real-time feedback loops. Thus, gradually prediction of lightning and even the movement of thunderstorms became a possibility with high accuracy. In parallel, several research groups found that there could be some strong correlations between electric discharge activities in thunderclouds and extreme weather events to take place subsequently, such as tornadoes, micro-bursts, cloud bursts and flash floods. At present, many researchers focus on their research in this direction, so that highly expensive and complex weather radar systems could be replaced by an accurate lightning detection system.

The frontiers of lightning protection of buildings are rather stagnating for the last couple of centuries, although there were moderate successes in developing more accurate risk assessments and documentation of protection measures. Some developments emerged as "Modern Technologies" in the 1970s onwards still struggle to justify the validity of their abilities. In the safety measures of equipment, known as surge protection, there is a significant development as the protection industry has to meet the demands of safeguarding electronic systems that become highly sophisticated, vulnerable and miniaturized at a rapid pace.

Research has also been underway for the last three decades to trigger atmospheric lightning artificially, mainly for the purpose of testing and evolving the protection methods and understanding the physics of lightning. So far, rocket-triggered lightning has been done very successfully, now with a success rate of over 80%, in USA, China and few other countries. The interesting phenomena of laser-triggered lightning have been started and abandoned few times in several countries during this period. In the last few years, this topic came back to the fore due to the latest development of high-power pulsed lasers. There are speculations that if laser-triggered lightning is successful, one could avoid undesired lightning by discharging the charged clouds into predetermined destinations.

For the last few decades, scientific world has also been working on several other concepts of academic interests, such as ionospheric lightning (red sprites, blue jets and whistlers), lightning in volcanic eruptions and discharge phenomena in the atmospheres of other planets. However, now several researchers, specifically those who emerged from the developing world, work intensively on developing low-cost structural protection measures and personal protection gears targeting at safeguarding the underprivileged communities in the world, where a clear majority of lightning mishaps occur annually.

In this connection, I am proud to mention that in the past, the NAM S&T Centre has made significant contributions in capacity building and exchange of knowledge for its member countries on the subject of lightning and other extreme natural events in partnership with various S&T institutions and agencies by organizing international workshops, roundtables, symposiums and training programmes. These programmes of the centre also led to the establishment of the *African Centre for Lightning and Electromagnetics Network (ACLENet)* based in Kampala, Uganda and *Zambian Centre for Lightning Information and Research (ZaCLIR)*. Recently, similar network has been established in South Asia (SALNet) of which the present headquarters is in Kathmandu, Nepal. Several other African countries such as South Africa, Kenya and Zimbabwe are now in the process of establishing their own national centres.

I am extremely delighted that the NAM S&T Centre has reached another milestone by publishing its first scientific monograph titled *Lightning—Science, Engineering and Economic Implications for Developing Countries*. I gratefully acknowledge the contributions made by eminent experts from various countries with their papers on different themes on lightning including lightning science research; protection of buildings, structures and power systems; lightning detection and warning, etc.

I express my sincere gratitude to Prof. Chandima Gomes, Professor of High Voltage Engineering, University of the Witwatersrand, Johannesburg, South Africa, for accepting our request to take charge of our first monograph project as the *Chief Editor*. He has conceived and planned the entire monograph, and we are thankful to him for his guidance and support to the centre in bringing out this valuable publication. We are also thankful to Prof. Farhad Rachidi, Professor and Head of the EMC Laboratory at the Swiss Federal Institute of Technology (EPFL), Lausanne, Switzerland, for kindly agreeing to write the *Foreword* of the monograph in spite of his extremely busy schedule.

I am thankful to Dr. Loyola D'Silva, Senior Editor, Springer Nature, Singapore, for considering this book worthy for publication under the reputed name of *Springer Nature*. I am also thankful to Dr. Suvira Srivastav, Associate Editorial Director, Springer, New Delhi, for her initial guidance to the NAM S&T Centre and helping us to get in touch with Dr. D'Silva. I am confident that our first ever association with "Springer" would be highly successful and would lead to many more such collaborative endeavours in future.

My sincere thanks are also due to the entire team of the NAM S&T Centre, especially to Mr. M. Bandyopadhyay, Senior Adviser, and Ms. Jasmeet Kaur, Programme Officer, for facilitating this new initiative of the centre.

I am sure that this book would be a valuable reference material for the scientists, researchers and other professionals working in the area of "lightning" and other relevant fields.

Amitava Bandopadhyay, Ph.D.
Director General
NAM S&T Centre
New Delhi, India

Contents

Lightning, the Science

Vladimir A. Rakov

Abstract Lightning can be defined as a transient, high-current (typically tens of kiloamperes) electric discharge in air whose length is measured in kilometers. As for any discharge in air, lightning channel is composed of ionized gas, that is, of plasma, whose peak temperature is typically 30,000 K, about five times higher than the temperature of the surface of the Sun. Lightning was present on Earth long before human life evolved and it may even have played a crucial role in the evolution of life on our planet. The global lightning flash rate is some tens to a hundred per second or so. Each year, some 25 million cloud-to-ground lightning discharges occur in the United States, and this number is expected to increase by about 50% due to global warming over the twenty-first century. Lightning initiates many forest fires, and over 30% of all electric power line failures are lightning related. Each commercial aircraft is struck by lightning on average once a year. A lightning strike to an unprotected object or system can be catastrophic. In this chapter, an overview of thunderclouds and their charge structure is given, basic lightning terminology is introduced, and different types of lightning (including the so-called rocket-triggered lightning) are described. For the most common negative cloud-to-ground lightning, main lightning processes are identified and the existing hypotheses of lightning initiation in thunderclouds are reviewed. Additionally, current and electromagnetic field signatures of lightning are characterized and the techniques to measure lightning electric and magnetic fields are discussed.

Keywords Thunderstorm · Return stroke · Ground flash · Stepped leader · Positive lightning

---

V. A. Rakov (✉)
Department of Electrical and Computer Engineering, University of Florida, Gainesville, Florida, USA
e-mail: rakov@ece.ufl.edu

C. Gomes (ed.), *Lightning*, Lecture Notes in Electrical Engineering 780,
https://doi.org/10.1007/978-981-16-3440-6_1

1 Thunderclouds and Their Charge Structure

The primary source of lightning is the cloud type termed cumulonimbus, commonly referred to as the thundercloud. Sometimes the term "thunderstorm" is used as a synonym for thundercloud, although thunderstorm is usually a system of thunderclouds rather than a single thundercloud. Lightning-like electrical discharges can also be generated in the ejected material above volcanoes, in sandstorms, and in nuclear explosions.

Before reviewing the electrical structure of thunderclouds it is worth outlining their meteorological characteristics. In effect, thunderclouds are large atmospheric heat engines with the input energy coming from the Sun and with water vapor as the primary heat-transfer agent [17]. The principal outputs of such an engine include (but are not limited to) (1) the mechanical work of the vertical and horizontal winds produced by the storm, (2) an outflow of condensate in the form of rain and hail from the bottom of the cloud and of small ice crystals from the top of the cloud, and (3) electrical discharges inside, below, and above the cloud, including corona, lightning, sprites, halos, elves, blue starters, blue jets, and gigantic jets. The processes that operate in a thundercloud to produce these actions are many and complex, most of them being poorly understood. A thundercloud develops from a small, fair-weather cloud called a cumulus which is formed when parcels of warm, moist air rise and cool by adiabatic expansion; that is, without the transfer of heat or mass across the boundaries of the air parcels. When the relative humidity in the rising and cooling parcel exceeds saturation, moisture condenses on airborne particulate matter to form the many small water particles that constitute the visible cloud. The height of the condensation level, which determines the height of the visible cloud base, increases with decreasing relative humidity at ground. This is why cloud bases in Florida are generally lower than in arid locations, such as New Mexico or Arizona. Parcels of warm, moist air can only continue to rise to form a cumulus and eventually a cumulonimbus if the atmospheric temperature lapse rate, the decrease in the temperature with increasing height, is larger than the moist-adiabatic lapse rate of about 0.6 °C per 100 m. The atmosphere is then referred to as unstable since rising moist parcels remain warmer than the air around them and thus remain buoyant. When a parcel rises above the 0 °C isotherm, some of the water particles begin to freeze, but others (typically smaller particles) remain liquid at temperatures lower than 0 °C. These are called supercooled water particles. At temperatures lower than about −40 °C all water particles will be frozen. In the temperature range from 0 to −40 °C liquid water and ice particles coexist forming a mixed phase region where most electrification is thought to occur.

Convection of buoyant moist air is usually confined to the troposphere, the layer of the atmosphere that extends from the Earth's surface to the tropopause. The latter is the boundary between the troposphere and the stratosphere, the layer which extends from the tropopause to a height of approximately 50 km. In the troposphere the temperature decreases with increasing altitude, while in the stratosphere the temperature at first becomes roughly independent of altitude and then increases with altitude. A zero or positive temperature gradient in the stratosphere serves to suppress convection and, therefore, hampers the penetration of cloud tops into the stratosphere. The height of the tropopause varies from approximately 18 km in the tropics in the summer to 8 km or so in high latitudes in the winter. In the case of vigorous updrafts, cloud vertical growth continues into the lower portion of the stratosphere. Convective surges can overshoot the tropopause up to 5 km in severe storms.

Although the primary thunderstorm activity occurs in the lower latitudes, thunderclouds are occasionally observed in the polar regions. Thunderstorms commonly occur over warm coastal regions when breezes from the water are induced to flow inland after sunrise when the land surface is warmed by solar radiation to a temperature higher than that of the water. Similarly, because mountains are heated before valleys, they often aid the onset of convection in unstable air. Further, horizontal wind blowing against a mountain will be directed upward and can aid in the vertical convection of air parcels, a process which is referred to as the "orographic effect." While relatively-small-scale convective thunderstorms (also called air-mass thunderstorms) develop in the spring and summer months when the potential for convection is usually the greatest and an adequate water vapor is available, larger-scale storms associated with frontal activity are observed in temperate latitudes at all times throughout the year.

Lightning is usually associated with convective cloud systems ranging from 3 to 20 km in vertical extent. The horizontal dimensions of active air-mass thunderstorms range from about 3 km to greater than 50 km. Seemingly merged thunderstorms may occur in lines along cold fronts extending for hundreds of kilometers. Ordinary thunderstorms are composed of units of convection, typically some kilometers in diameter, characterized by relatively strong updrafts ($\geq$10 m/s). These units of convection are referred to as cells. The lifetime of an individual cell is of the order of 1 h. Thunderstorms can include a single isolated cell, several cells, or a long-lived cell with a rotating updraft, called a supercell. At any given time, a typical multicell storm consists of a succession of cells at different stages of evolution. Large frontal systems have been observed to persist for more than 48 h and to move more than 2000 km. Thunderstorms over flat terrain tend to move at an average speed of 20–30 km/h. Further information on thunderstorm morphology and evolution can be found in the book by MacGorman and Rust [16] and in the book chapter by Williams [34].

The distribution and motion of thunderstorm electric charges, most residing on hydrometeors (various liquid or frozen water particles in the atmosphere) but with some free ions, is complex and changes continuously as the cloud evolves. Hydrometeors whose motion is predominantly influenced by gravity (fall speed $\geq$ 0.3 m/s) are called precipitation. All other hydrometeors are called cloud particles. In calculating remote electric fields due to cloud charges and lightning-caused field changes, the typical gross thundercloud charge structure is often approximated by an idealized model in which three vertically stacked point charges (or spherically symmetrical charged volumes): main positive at the top, main negative in the middle, and lower positive at the bottom (see Fig. 1). This charge configuration is assumed to be located in an insulating atmosphere above a perfectly conducting ground. The magnitudes of the main positive and negative charges are typically some tens of coulombs, while the lower positive charge is probably about 10 C or less.

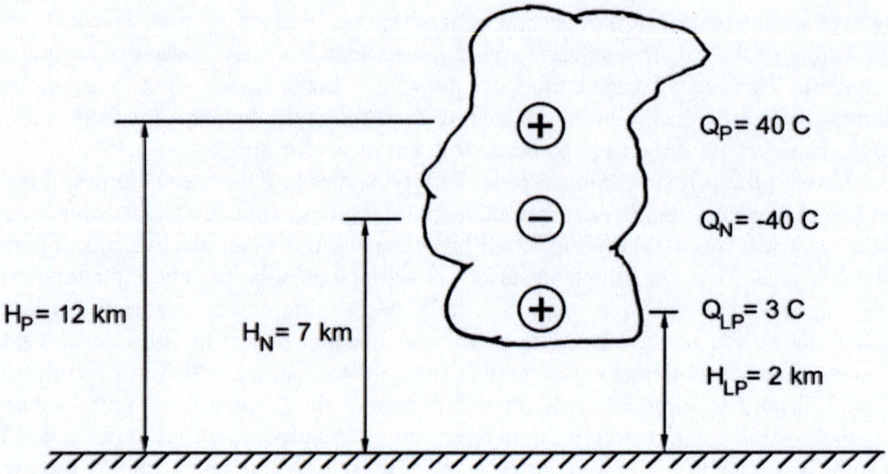

Fig. 1 A vertical tripole representing the idealized gross charge structure of a thundercloud

It has been inferred from a combination of remote and in-situ measurements that, regardless of the stage of storm development, location, and season, negative charge is typically found in the same relatively narrow temperature range, roughly −10 to −25 °C, where the clouds contain both supercooled water and ice. This feature has important implication for the dominant cloud electrification mechanism and is illustrated in Fig. 2.

Many cloud electrification theories have been proposed. There is growing consensus that the so-called graupel-ice mechanism is the dominant one, at least at the initial stages of cloud electrification. In this mechanism, the electric charges are produced by collisions between graupel (millimeter-size soft hail or snow pellet) particles and small ice crystals in the presence of water droplets, and the large-scale separation of charged particles is provided by the action of gravity. The presence of water droplets is necessary for significant charge transfer, as shown by the laboratory experiments. A simplified illustration of this mechanism is given in Fig. 3. The heavy graupel particles (two of which are shown in Fig. 3) fall through a suspension of smaller ice crystals (hexagons) and supercooled water droplets (dots). The droplets remain in a supercooled liquid state until they contact an ice/grouped surface, whereupon they freeze and stick to the surface in a process called riming. Laboratory experiments (e.g., Jayaratne et al. [11]) show that when the temperature is below a critical value called the reversal temperature, T_R, the falling graupel particles acquire a negative charge in collisions with the ice crystals. At temperatures above T_R they acquire a positive charge. The charge sign reversal temperature T_R is generally thought to be between −10 and −20 °C, the temperature range characteristic of the main negative charge region found in thunderclouds. The graupel which picks up positive charge when it falls below the altitude of T_R could explain the existence of the lower positive charge region in the cloud. It is believed that the polarity of the charge that is

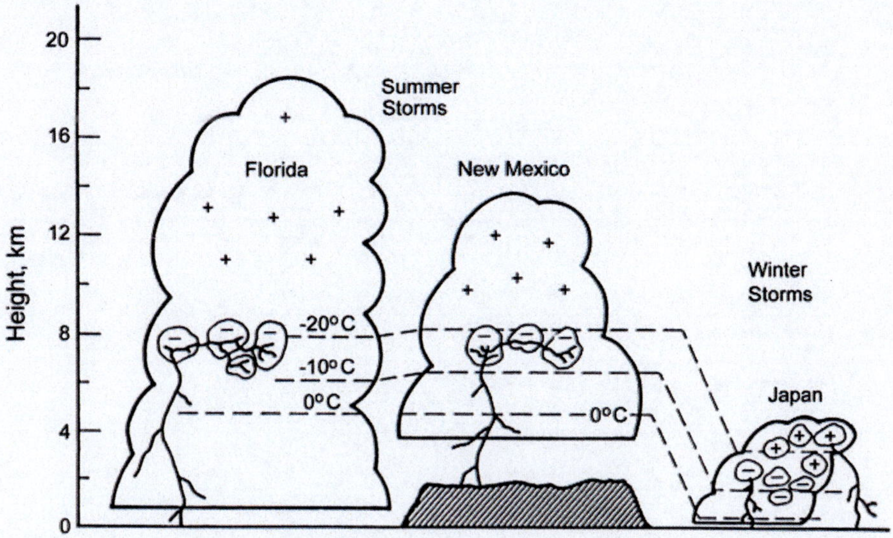

Fig. 2 The locations, shown by the small irregular contours inside the cloud boundaries, of ground-flash charge sources observed in summer thunderstorms in Florida and New Mexico and in winter thunderstorms in Japan using simultaneous measurements of electric field at a number of ground stations. Adapted from Krehbiel [12]

separated in ice-graupel collisions is determined by the rates at which the ice and graupel surfaces are growing. The surface that is growing faster acquires a positive charge.

It appears that the graupel-ice mechanism is capable of explaining the "classical" tripolar cloud charge structure shown in Fig. 1 and the location of the negative charge region in the −10 to −20 °C temperature range seen in Fig. 2.

2 Initiation, Propagation, and Attachment

The lightning discharge in its entirety, whether it strikes ground or not, is usually termed a "lightning flash" or just a "flash." A lightning discharge that involves an object on ground or in the atmosphere is referred to as a "lightning strike." A commonly used nontechnical term for a lightning discharge is a "lightning bolt." About three-quarters of lightning discharges do not involve ground. They include intracloud, intercloud, and cloud-to-air discharges and are collectively referred to as cloud discharges or cloud flashes (see Fig. 4) and sometimes as ICs. Lightning discharges between cloud and Earth are termed cloud-to-ground (or just ground) discharges and sometimes referred to as CGs. The latter constitute about 25% of global lightning activity.

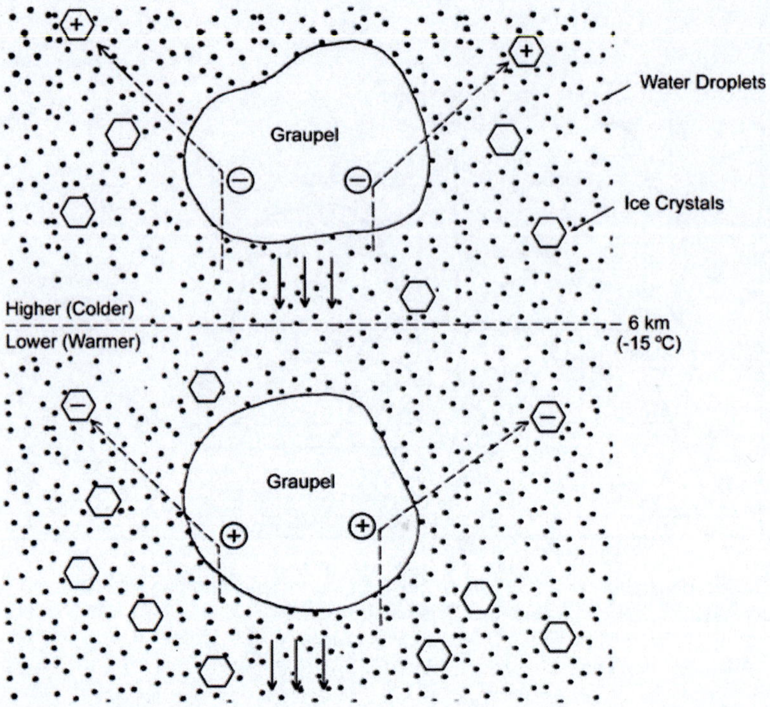

Fig. 3 Schematic representation of the graupel-ice mechanism of cloud electrification, in which the charge transfer occurs via collision of graupel with small ice crystals in the presence of supercooled water droplets. It is assumed that the reversal temperature T_R is -15 °C, and that it occurs at a height of 6 km

About 90% or more of global cloud-to-ground lightning is accounted for by downward (the initial process begins in the cloud and develops in the downward direction) negative (negative charge is effectively transported to the ground) lightning. The term "effectively" is used to indicate that individual charges are not transported all the way from the cloud to ground during the lightning processes. Rather the flow of electrons (the primary charge carriers) in one part of the lightning channel results in the flow of other electrons in other parts of the channel. Other types of cloud-to-ground lightning include downward positive, upward negative, and upward positive discharges (see Fig. 5). Downward flashes exhibit downward branching, while upward flashes are branched upward. Upward lightning discharges [types (b) and (d) in Fig. 5] are thought to occur only from tall objects (higher than 100 m or so) or from objects of moderate height located on mountain tops. There are also bipolar lightning discharges (not shown in Fig. 5) sequentially transferring both positive and negative charges during the same flash. Bipolar lightning discharges are usually initiated from tall objects (are of upward type). Downward bipolar lightning discharges do exist, but appear to be rare.

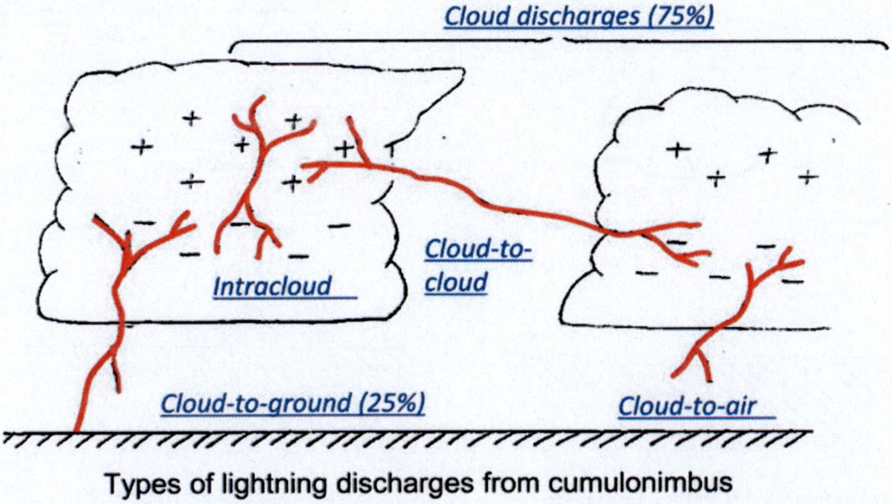

Fig. 4 General classification of lightning discharges from cumulonimbus (thunderstorm clouds). Cloud discharges constitute 75% and cloud-to-ground discharges 25% of global lightning activity

We first introduce, referring to Fig. 6a, b, the basic elements of the negative downward lightning discharge, termed component strokes or just strokes. Then, we will introduce, referring to Fig. 7, the two major lightning processes comprising a stroke, the leader and the return stroke, which occur as a sequence with the leader preceding the return stroke.

Two photographs of the same negative cloud-to-ground discharge are shown in Fig. 6a and b. The image in Fig. 6a was obtained using a stationary camera, while the image in Fig. 6b was captured with a separate camera that was moved horizontally during the time of the flash. As a result, the latter image is time resolved showing several distinct luminous channels between the cloud and ground separated by dark gaps. The distinct channels are associated with individual strokes, and the time intervals corresponding to the dark gaps are typically of the order of tens of milliseconds. These dark time intervals between strokes explain why lightning often appears to the human eye to "flicker." In Fig. 6b, time advances from right to left, so that the first stroke is on the far right. The first two strokes are branched, and the downward direction of branches indicates that this is a downward lightning flash.

Now, we consider sketches of still and time-resolved (much better than in Fig. 6b) optical images of the three-stroke lightning flash shown in Fig. 7a, b, respectively. A sketch of the corresponding current at the channel base is shown in Fig. 7c. In Fig. 7b, time advances from left to right, and the time scale is not continuous. Each of the three strokes in Fig. 7b, represented by its luminosity as a function of height above ground and time, is composed of a downward-moving process, termed a leader, and an upward-moving process, termed a return stroke. The leader creates a conducting path between the cloud charge source region and ground and distributes negative charge from the cloud source region along this path, and the return stroke traverses

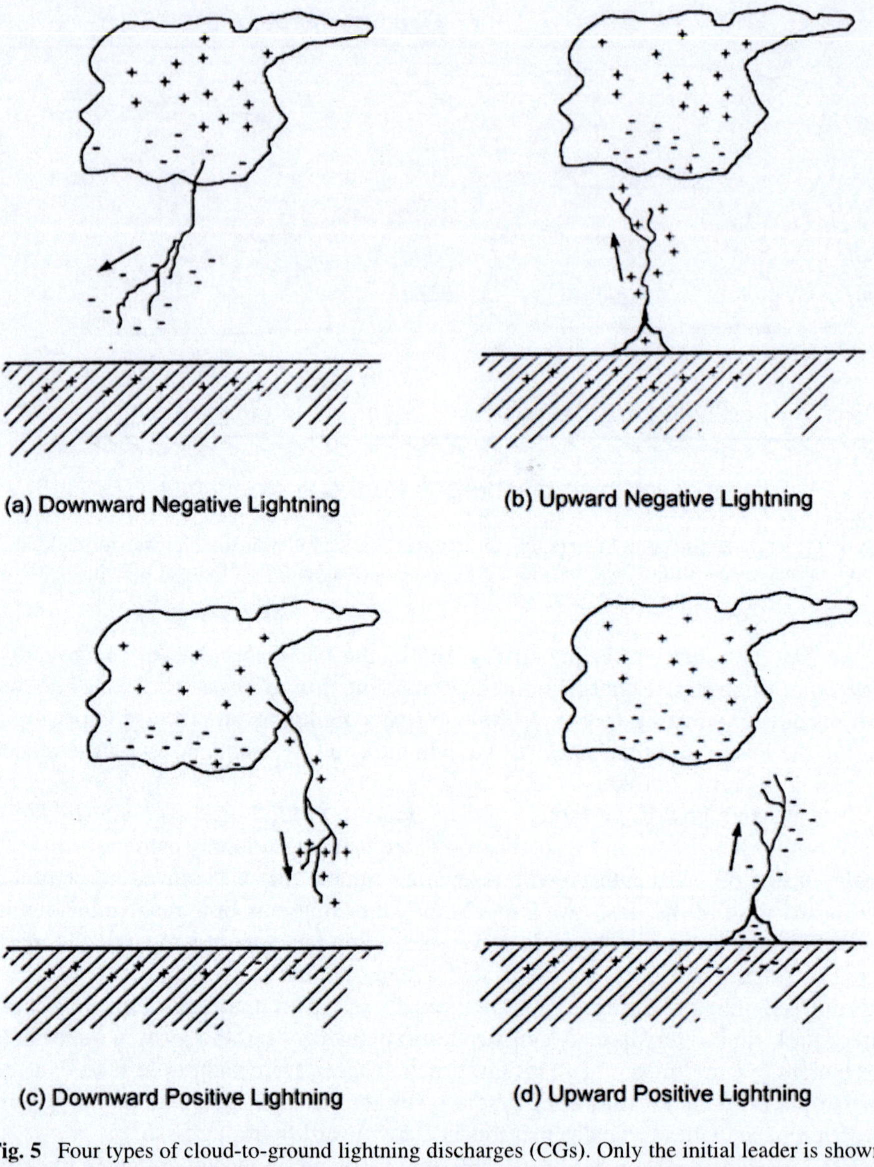

Fig. 5 Four types of cloud-to-ground lightning discharges (CGs). Only the initial leader is shown for each type. For each lightning-type name given below the sketch, direction (downward or upward) indicates the direction of propagation of the initial leader and polarity (negative or positive) refers to the polarity of the cloud charge effectively lowered to ground. In **a**, **c**, the polarity of charge lowered to ground is the same as the leader polarity, while in **b**, **d** those polarities are opposite

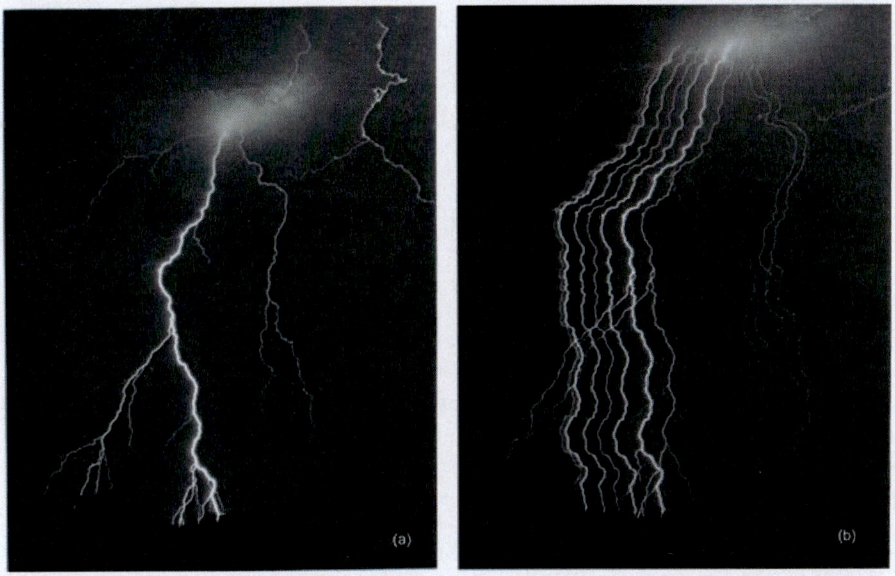

Fig. 6 Lightning flash which appears to have at least 7 (perhaps as many as 10) separate ground strike points: **a** still-camera photograph, **b** moving-camera photograph. Some of the strike points are associated with the same stroke having separate branches touching ground, while others are associated with different strokes taking different paths to ground. Adapted from Hendry [9]

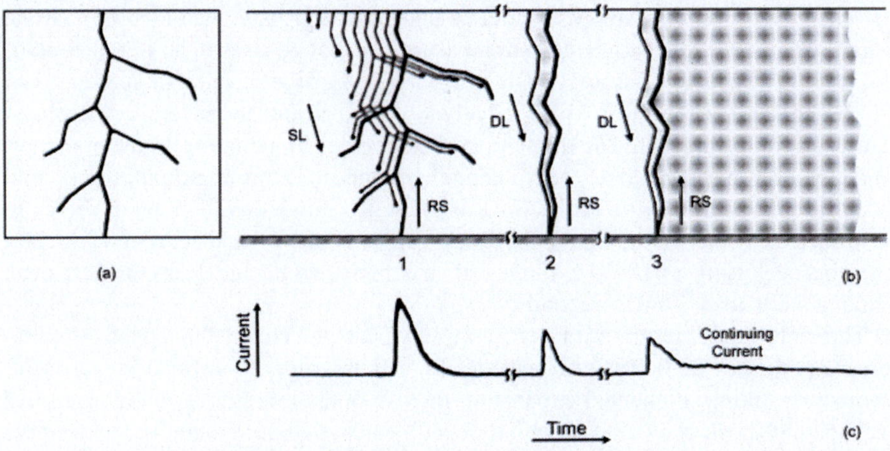

Fig. 7 Schematic diagram showing the luminosity of a three-stroke downward negative flash and the corresponding current at the channel base: **a** still-camera image, **b** streak-camera image, and **c** channel-base current. Stroke 3 is followed by continuing current whose luminosity in (**b**) is represented by an array of dark spots

that path moving from ground toward the cloud charge source region and neutralizes the negative leader charge. Thus, both leader and return-stroke processes serve to effectively transport negative charge from the cloud to ground. As seen in Fig. 7b, the leader initiating the first return stroke differs from the leaders initiating the two subsequent return strokes (all strokes other than first are termed subsequent strokes). In particular, the first-stroke leader appears optically to be an intermittent process, hence the term stepped leader, while the tip of a subsequent-stroke leader appears to move continuously. The continuously moving subsequent-stroke leader tip appears on streak photographs as a downward-moving "dart," hence the term dart leader. The apparent difference between the two types of leaders is related to the fact that the stepped leader develops in virgin air, while the dart leader follows the "pre-conditioned" path of the preceding stroke or strokes.

The electric potential difference between a downward-moving stepped-leader tip and ground is probably some tens of megavolts, comparable to or a considerable fraction of that between the cloud charge source and ground. The magnitude of the potential difference between two points, one at the cloud charge source and the other on ground, is the line integral of electric field intensity between those points. The upper and lower limits for the potential difference between the lower boundary of the main negative charge region and ground can be estimated by multiplying, respectively, the typical observed electric field in the cloud, 10^5 V/m, and the expected electric field at ground under a thundercloud immediately prior to the initiation of lightning, 10^4 V/m, by the height of the lower boundary of the negative charge region above ground. The resultant range is 50–500 MV, if the height is assumed to be 5 km.

When the descending stepped leader attaches to the ground, the first return stroke begins. The first return-stroke current measured at ground rises to an initial peak of about 30 kA in some microseconds and decays to half-peak value in some tens of microseconds. The return stroke effectively lowers to ground the several coulombs of charge originally deposited on the stepped-leader channel including all the branches. Once the bottom of the dart leader channel is connected to the ground, the second (or any subsequent) return-stroke wave is launched upward, which again serves to neutralize the leader charge. The subsequent return-stroke current at ground typically rises to a peak value of 10–15 kA in less than a microsecond and decays to half-peak value in a few tens of microseconds.

The high-current return-stroke wave rapidly heats the channel to a peak temperature near or above 30,000 K and creates a channel pressure of 10 atm (1 megapascal) or more, resulting in channel expansion, intense optical radiation, and an outward propagating shock wave that eventually becomes the thunder (sound wave) we hear at a distance. Each cloud-to-ground lightning flash involves an energy of roughly 10^9–10^{10} J (one to ten gigajoules). Lightning energy is approximately equal to the energy required to operate five 100 W light bulbs continuously for one month. Note that not all the lightning energy is available at the strike point, only 10^{-2}–10^{-3} of the total energy, since most of the energy is spent for producing thunder, hot air, light, and radio waves.

Given above is only basic information about downward negative lightning. In the following, referring to Fig. 8, we present a more complete sequence of processes involved in a typical downward negative lightning flash.

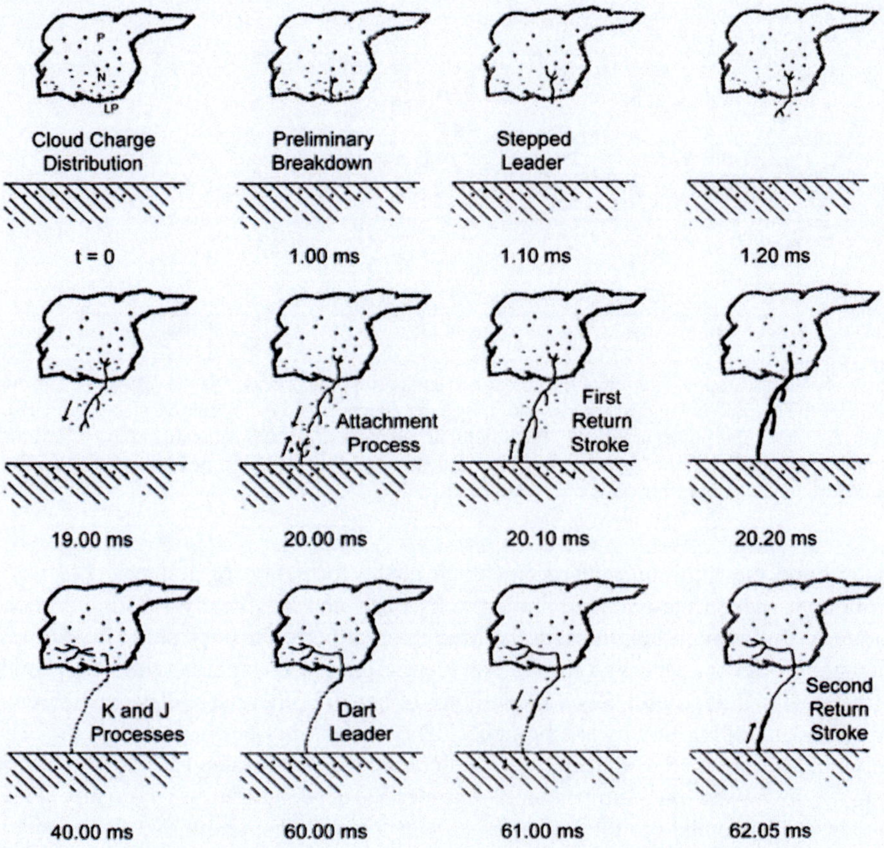

Fig. 8 Various processes comprising a two-stroke negative cloud-to-ground lightning flash. Time labels below the sketches can be used to roughly estimate typical durations of the processes and time intervals between them ($t = 0$ corresponds to the beginning of preliminary breakdown process which ends at $t = 1$ ms). Continuing current and M-components are not illustrated in this figure. Adapted from Uman [29, 30]

The source of lightning is usually a cumulonimbus (see Sect. 1), whose idealized charge structure is shown in Fig. 8 at t = 0 as three vertically stacked regions labeled "P" and "LP" for main positive and lower positive charge regions, respectively, and "N" for main negative charge region. The stepped leader is preceded by an in-cloud process called the preliminary or initial breakdown. It may be a discharge bridging the main negative and the lower positive charge regions, as shown in Fig. 8. The initial breakdown serves to provide conditions for the formation of the stepped leader. The latter is a negatively charged plasma channel extending toward the ground at an average speed of 2×10^5 m/s in a series of discrete steps. Each step produces a current pulse that originates at the tip of the downward-extending leader channel and propagates upward, like a mini return stroke, as schematically shown in Fig. 9. Krider et al. [13] inferred that the peak step current is at least 2–8 kA close to the

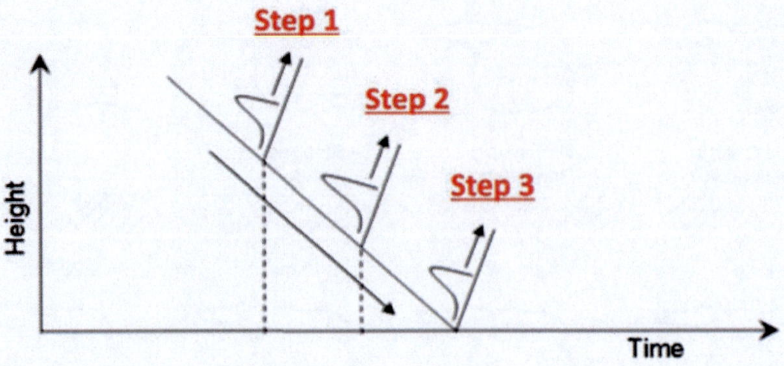

Fig. 9 Schematic representation of the downward leader stepping process. Negatively-sloped arrow indicates the overall downward extension of the leader channel. Three consecutive steps giving rise to current (and light) pulses are shown. Each step-current pulse originates at the tip of the downward-extending channel and propagates upward (as indicated by a positively-sloped arrow), like a mini return stroke. Adapted from Nag and Rakov [18]

ground and the minimum charge involved in the formation of a step is 1–4 mC. From high-speed time-resolved photographs, each step is typically 1 μs in duration and tens of meters in length, with the time interval between steps being 20–50 μs. The stepped leader serves to form a conducting path or channel between the cloud charge region and ground. Several coulombs of negative charge are distributed along this path, including downward branches. The leader may be viewed as a process removing negative charge from the cloud charge region and depositing this charge onto the downward extending channel. The stepped-leader duration is typically some tens of milliseconds, the total charge is about 5 coulombs, and the average leader current is some hundreds of amperes.

As the leader approaches ground, the electric field at the ground surface, particularly at objects or relief features protruding above the surrounding terrain, increases until it exceeds the critical value for the initiation of one or more upward leaders, one of which will become the upward connecting leader. It is usually assumed that the initiation of upward connecting leader (UCL) from ground in response to the descending stepped leader marks the beginning of the attachment process. The attachment process includes the so-called breakthrough phase that is assumed to begin when the relatively low conductivity streamer zones developing ahead of the two propagating leader tips meet to form a common streamer zone. The subsequent accelerated extension of the two relatively high conductivity plasma channels toward each other takes place inside the common streamer zone (see Rakov and Tran [22] and references therein). The attachment process ends when contact is made between the hot channels of the downward and upward moving leaders, probably some tens of meters above ground (more above a tall structure), where the first return stroke begins (see Fig. 10 where the attachment process is schematically shown). The return stroke serves to neutralize the leader charge or, equivalently, to transport the negative charges stored on the leader channel to the ground. It is worth noting that the

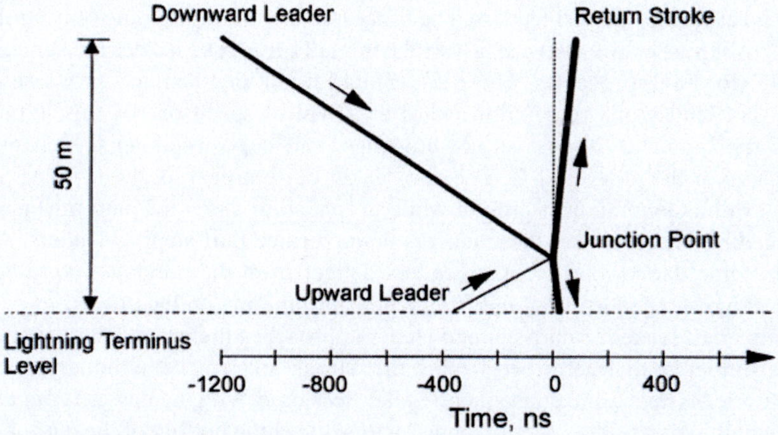

Fig. 10 Schematic representation of the attachment process followed by the bidirectional return-stroke process observed in a rocket-triggered lightning stroke. Rocket-triggered lightning strokes are similar to subsequent strokes in natural lightning. Adapted from Wang et al. [32]

return-stroke process may not neutralize all the leader charge and is likely to deposit some excess positive charge onto the upper part of the leader channel and into the cloud charge source region. The speed of the return stroke, averaged over the visible channel, is typically between one-third and one-half of the speed of light. There is no consensus about whether or how the first return-stroke speed changes over the bottom 100 m or so, but over the entire channel, the speed decreases with increasing height, dropping abruptly after passing each major branch. At the same time, a transient enhancement of the main channel luminosity below the branching point, termed a branch component, is often observed.

When the first return stroke, including any associated in-cloud discharge activity (discussed later), ceases the flash may end. In this case, the lightning is called a single-stroke flash. However, more often the residual first-stroke channel is traversed by a leader that appears to move continuously, a dart leader. During the time interval between the end of the first return stroke and the initiation of a dart leader, J (for junction) and K processes occur in the cloud. K processes can be viewed as transients occurring during the slower J process. The J processes amount to a redistribution of cloud charge on a tens of milliseconds time scale in response to the preceding return stroke. There is controversy as to whether these processes, which apparently act to extend the return-stroke channel further into the cloud, are necessarily related to the initiation of a following dart leader. The J process is often viewed as a relatively slow positive leader extending from the flash origin into the negative charge region, with the K process being a relatively fast "recoil process" that begins at the tip of the positive leader or in a decayed positive leader branch and propagates toward the flash origin. Both the J processes and the K processes in cloud-to-ground discharges serve to transport additional negative charge into and along the existing channel (or its remnants), although not all the way to the ground. In this respect, K processes may be viewed as attempted dart leaders. The processes that occur after the only stroke in single-stroke flashes and after the last stroke in multiple-stroke flashes are sometimes termed F (for final) processes. These are similar, if not identical, to J processes.

The dart leader progresses downward at a typical speed of 10^7 m/s, typically ignores the remnants of first stroke branches, and deposits along the channel a total charge of the order of 1 C. The dart-leader current peak is about 1 kA. Some leaders exhibit stepping near ground while propagating along the path traversed by the preceding return stroke; these leaders being termed dart-stepped leaders. Additionally, some dart or dart-stepped leaders deflect from the previous return-stroke path, become stepped leaders, and form a new termination on the ground.

When a dart leader or dart-stepped leader approaches the ground, an attachment process similar to that described for the first stroke takes place, although it occurs over a shorter distance and consequently takes less time, with the upward connecting leader length being of the order of some meters. Once the bottom of the dart or dart-stepped leader channel is connected to the ground, the second (or any subsequent) return-stroke wave is launched upward, which again serves to neutralize the leader charge. The upward propagation speed of subsequent return strokes is similar to that of first return strokes, although due to the absence of branches the speed variation along the channel does not exhibit abrupt drops.

Wang et al. [32], using the digital optical imaging system ALPS with 3.6-m spatial and 100-ns time resolution, observed the attachment process in one rocket-triggered-lightning stroke (see Sect. 5 of this chapter). A sketch of the time-resolved image for that event is shown in Fig. 10, in which the return stroke begins at $t = 0$. Note that the return stroke was initially a bidirectional process that involved both upward- and downward-moving waves which originated from the junction point of the downward negative dart leader and the upward positive connecting leader.

The impulsive component of the current in a subsequent return stroke is often followed by a continuing current which has a magnitude of tens to hundreds of amperes and a duration up to hundreds of milliseconds (median duration is 6 ms). Continuing currents with a duration in excess of 40 ms are traditionally termed long continuing currents. Between 30 and 50% of all negative cloud-to-ground flashes contain long continuing currents. The source for continuing current is the cloud charge, as opposed to the charge distributed along the leader channel, the latter charge contributing to at least the initial few hundred microseconds of the return-stroke current observed at ground. Continuing current typically exhibits a number of superimposed surges that rise to peak and fall off to the background current level in some hundreds of microseconds, with the peak being generally in the hundreds of amperes range but occasionally in the kiloamperes range. These current surges are associated with enhancements in the relatively faint luminosity of the continuing-current channel and are called M components. Note that continuing current and M-component processes are not shown in Fig. 8.

The time interval between successive return strokes in a flash is usually several tens of milliseconds, although it can be as large as many hundreds of milliseconds if a long continuing current is involved and as small as one millisecond or less. Note that interstroke intervals are usually measured between the peaks of current or electromagnetic field pulses. The total duration of a flash is typically some hundreds of milliseconds, and the total charge lowered to ground is some tens of coulombs. The average number of strokes per flash is 3 to 5. The overwhelming majority (typically about 80%) of negative cloud-to-ground flashes contain more than one stroke.

One-third to one-half of all lightning discharges to earth, both single- and multiple-stroke flashes, strike ground at more than one point with the spatial separation between the channel terminations being up to many kilometers. The average number of channels per flash is 1.5 to 1.7. In most cases, multiple ground terminations within a given flash are associated not with an individual multi-grounded leader but rather with the deflection of a subsequent leader from the previously formed channel.

The salient properties of downward negative lightning discharges are summarized in Table 1.

Lightning initiation in thunderclouds remains a mystery. Indeed, maximum electric fields typically measured in thunderclouds (see Table 3.2 of Rakov and Uman [23] and references therein) are 1–2 $\times$ 10^5 V/m (the highest measured value is 4 $\times$ 10^5 V/m), which is lower than the expected conventional breakdown field, of the order of 10^6 V/m. Two general mechanisms of lightning initiation have been suggested. One relies on the emission of positive streamers from hydrometeors when the electric field exceeds 2.5–9.5 $\times$ 10^5 V/m, and the other involves high-energy cosmic ray particles and the so-called runaway breakdown that occurs in a critical field, calculated to be about 10^5 V/m at an altitude of 6 km. Either of these two mechanisms permits, in principle, creation of an ionized region ("lightning seed") in the cloud that is capable of locally enhancing the electric field at its extremities. Such field enhancement is likely to be the main process leading to the formation (via conventional breakdown) of a hot, self-propagating lightning channel.

3 Current and Electromagnetic Field Signatures

The most complete characterization of the return stroke in negative downward flashes is due to Karl Berger and co-workers (e.g., Berger [3], Berger et al. [4]). The data of Berger were derived from oscillograms of current measured using resistive shunts installed at the tops of two 70 m high towers on the summit of Monte San Salvatore in Lugano, Switzerland. The summit of the mountain is 915 m above sea level and 640 m above the level of Lake Lugano, located at the base of the mountain. The towers are of moderate height, but because the mountain contributed to the electric field enhancement near the tower tops, the effective height of each tower was a few hundred meters. As a result, the majority of lightning strikes to the towers were of the upward type. Here we only consider return strokes in negative downward flashes. A total of 101 are included in the summary by Berger et al. [4]. Berger's data were additionally analyzed by Anderson and Eriksson (1980).

The results of Berger et al. [4] are still used to a large extent as the primary reference source for both lightning protection and lightning research. These results are presented in Figs. 11 and 12 and in Table 2.

Table 1 Characterization of negative cloud-to-ground lightning

Parameter	Typical value[a]
Stepped leader	
Step length, m	50
Time interval between steps, μs	20–50
Step current, kA	>1
Step charge, mC	>1
Average propagation speed, m s^{-1}	2×10^5
Overall duration, ms	35
Average current, A	100–200
Total charge, C	5
Electric potential, MV	~50
Channel temperature, K	~10,000
First return stroke[b]	
Peak current, kA	30
Maximum current rate of rise, kA μs^{-1}	10–20
Current risetime (10–90%), μs	5
Current duration to half-peak value, μs	70–80
Charge transfer, C	5
Propagation speed, m s^{-1}	$(1–2) \times 10^8$
Channel radius, cm	~1–2
Channel temperature, K	~30,000
Dart leader	
Speed, m s^{-1}	$(1–2) \times 10^7$
Duration, ms	1–2
Charge, C	1
Current, kA	1
Electric potential, MV	~15
Channel temperature, K	~20,000
Dart-stepped leader	
Step length, m	10
Time interval between steps, μs	5–10
Average propagation speed, m s^{-1}	$(1–2) \times 10^6$
Subsequent return stroke[b]	
Peak current, kA	10–15
Maximum current rate of rise, kA μs^{-1}	100
10–90% current rate of rise, kA μs^{-1}	30–50
Current risetime (10–90%), μs	0.3–0.6
Current duration to half-peak value, μs	30–40

(continued)

Table 1 (continued)

Parameter	Typical value[a]
Charge transfer, C	1
Propagation speed, m s^{-1}	(1–2) × 10^8
Channel radius, cm	~1–2
Channel temperature, K	~30,000
Continuing current (longer than 40 ms or so)[c]	
Magnitude, A	100–200
Duration, ms	~100
Charge transfer, C	10–20
M component[b]	
Peak current, A	100–200
Current risetime (10–90%), μs	300–500
Charge transfer, C	0.1–0.2
Overall flash	
Duration, ms	200–300
Number of strokes per flash[d]	3–5
Interstroke interval, ms	60
Charge transfer, C	20
Energy, J	10^9–10^{10}

Adapted from Rakov and Uman [23]
[a]Typical values are based on a comprehensive literature search and unpublished experimental data acquired by the University of Florida Lightning Research Group
[b]All current characteristics for return strokes and M components are based on measurements at the lightning channel base
[c]About 30–50% of lightning flashes contain continuing currents longer than 40 ms or so
[d] About 15–20% of lightning flashes are composed of a single stroke

Figure 11 shows, on two time scales, A and B, the average current wave shapes for negative first and subsequent strokes. The averaging procedure involved the normalization of waveforms from many strokes to their respective peak currents (so that all have peaks equal to unity) and subsequent alignment using the 0.5 peak point on the initial rising portion of the waveforms. The overall duration of the current waveforms is some hundreds of microseconds. The rising portion of the first-stroke waveform has a characteristic concave shape. The averaging procedure masked secondary maxima typically observed in first-stroke waveforms and generally attributed to major branches.

Figure 12 shows the cumulative statistical distributions (solid-line curves) of return-stroke peak currents for (1) negative first strokes, (2) negative subsequent strokes, and (3) positive strokes (each was the only stroke in a flash). These empirical distributions are approximated by log-normal distributions (dashed lines) and shown on cumulative probability distribution graph paper, on which a Gaussian (normal)

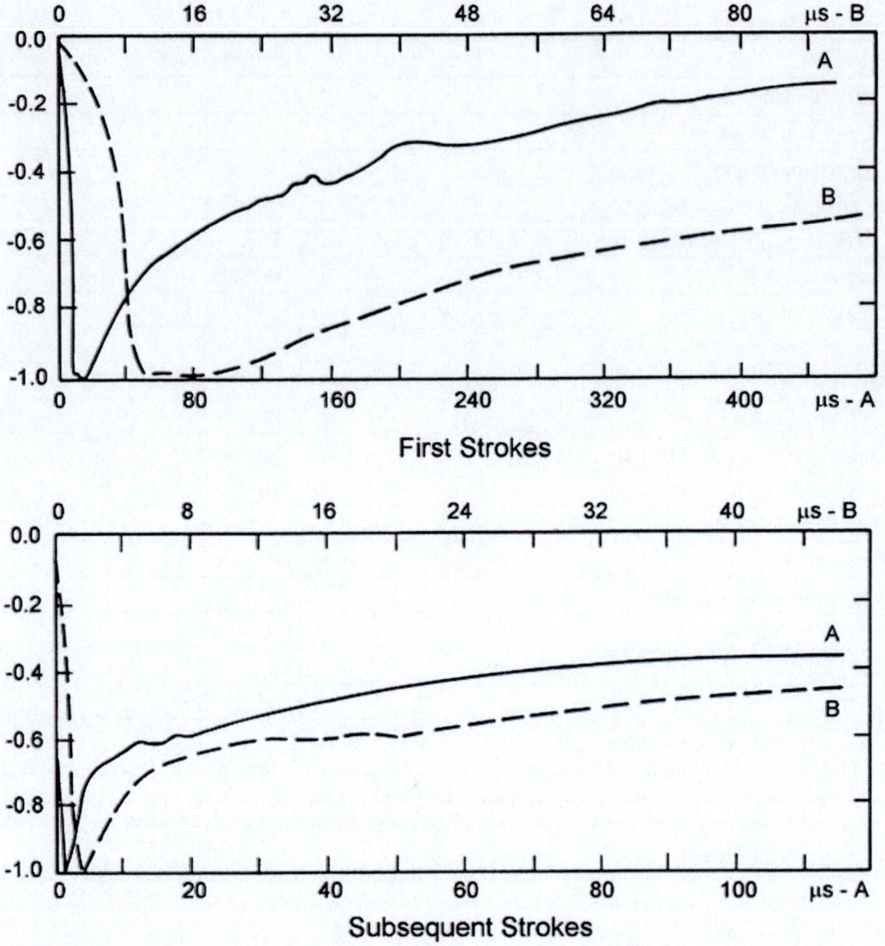

Fig. 11 Average negative first- and subsequent-stroke wave shapes each shown on two time scales, A and B. The lower time scales (A) correspond to solid curves, while the upper time scales (B) correspond to broken curves. The vertical (amplitude) scales are in relative units, the peak values being equal to negative unity. Adapted from Berger et al. [4]

cumulative distribution appears as a slanted straight line, with the horizontal (peak current) scale being logarithmic (base 10). The vertical scale gives the percentage of peak currents exceeding a given value on the horizontal axis. The vertical scale is symmetrical with respect to the 50% value and does not include the 0 and 100% values; it only asymptotically approaches those. For a log normal distribution, the 50% (median) value is equal to the geometric mean value.

The lightning peak current distributions for negative first and subsequent strokes shown in Fig. 12 are also characterized by their 95, 50, and 5% values based on the log normal approximations in Table 2, which contains a number of other parameters

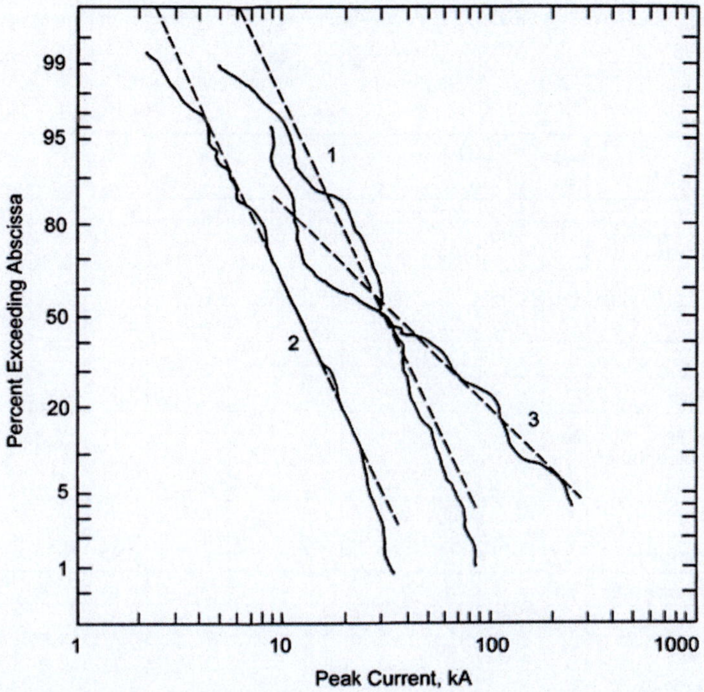

Fig. 12 Cumulative statistical distributions of return-stroke peak current (solid curves) and their log-normal approximations (broken lines) for (1) negative first strokes, (2) negative subsequent strokes, and (3) positive first (and only) strokes, as reported by Berger et al. [4]

derived from the current oscillograms. The minimum peak current value included in the distributions is 2 kA, although no first strokes (of either polarity) with peak currents below 5 kA were observed.

Berger's peak current distributions for first and subsequent negative strokes are generally confirmed by more recent direct current measurements, particularly those with larger sample sizes obtained in Japan (first strokes, N = 120; Takami and Okabe [27]), Austria (subsequent strokes, N = 615, Diendorfer et al. [6]), and Florida (subsequent strokes, N = 165, Schoene et al. [26]). At the same time, direct current measurements in Brazil [28] yielded 50% higher median peak currents for both first (N = 38) and subsequent (N = 71) strokes.

It follows from Fig. 12 and Table 2 that the median return-stroke current peak for first strokes is 2–3 times higher than that for subsequent strokes. Also, negative first strokes transfer about a factor of four larger charge than do negative subsequent strokes. On the other hand, subsequent return strokes are characterized by 3 to 4 times higher current maximum steepness (current maximum rate of rise or maximum dI/dt). It is important to note that the maximum dI/dt reported by Berger et al. [4] and given in Table 2 is an underestimate of the actual value due to the limited time resolution of oscillographic data. [The mean value of maximum dI/dt reported for

Table 2 Parameters of downward negative lightning derived from channel-base current measurements

Parameter	Unit	Sample size	Percent exceeding tabulated value		
			95%	50%	5%
Peak current (minimum 2 kA)	kA				
First strokes		101	14	30	80
Subsequent strokes		135	4.6	12	30
Charge (total charge)	C				
First strokes		93	1.1	5.2	24
Subsequent strokes		122	0.2	1.4	11
Complete flash		94	1.3	7.5	40
Impulse charge (excluding continuing current)	C				
First strokes		90	1.1	4.5	20
Subsequent strokes		117	0.22	0.95	4
Front duration (2 kA to peak)	μs				
First strokes		89	1.8	5.5	18
Subsequent strokes		118	0.22	1.1	4.5
Maximum dI/dt	kA μs^{-1}				
First strokes		92	5.5	12	32
Subsequent strokes		122	12	40	120
Stroke duration (2 kA to half peak value on the tail)	μs				
First strokes		90	30	75	200
Subsequent strokes		115	6.5	32	140
Action integral ($\int I^2 dt$)	A^2s				
First strokes		91	6.0×10^3	5.5×10^4	5.5×10^5
Subsequent strokes		88	5.5×10^2	6.0×10^3	5.2×10^4
Time interval between strokes	ms	133	7	33	150
Flash duration	ms				
All flashes		94	0.15	13	1100
Excluding single-stroke flashes		39	31	180	900

Adapted from Berger et al. [4]

rocket-triggered-lightning strokes (see Sect. 5 of this chapter) by Leteinturier et al. [14] is 110 kA/μs.] As seen in Fig. 12, only a few percent of negative first strokes are expected to exceed 100 kA, while about 20% of positive strokes have been observed to do so. On the other hand, the 50% (median) values of the current distributions for negative first and positive strokes are similar. The action integral (also referred to as specific energy) in Table 2 represents the energy that would be dissipated in a 1 Ω resistor if the lightning current were to flow through it. It is thought that the heating of electrically conducting materials and the explosion of nonconducting materials is, to a first approximation, determined by the value of the action integral. Note that the interstroke interval in Table 2 is likely mislabeled by Berger et al. [4] and is actually the no-current interval, that is, the interstroke interval excluding any continuing current.

Next, we will discuss typical electric and magnetic field waveforms produced by both first and subsequent return strokes at ground level at distances ranging from 1 to 200 km. These waveforms, which are drawings based on many measurements acquired in Florida by Lin et al. [15], are reproduced in Fig. 13.

The electric fields of strokes observed within a few kilometers of the flash are, after the first few tens of microseconds, dominated by the electrostatic component (marked "Ramp" in Fig. 13) of the total electric field, the only field component which is nonzero after the stroke current has ceased to flow. The close magnetic fields at similar times are dominated by the magnetostatic component of the total magnetic field, the component that produces the magnetic field humps marked in Fig. 13. Distant electric and magnetic fields have essentially identical waveshapes and are usually bipolar, as illustrated in Fig. 13. The data of Lin et al. [15] suggest that at a distance of 50 km and beyond, both electric and magnetic field waveshapes are dominated by their respective radiation components.

The initial field peak (marked in some of the waveforms of Fig. 13) is the dominant feature of the electric and magnetic field waveforms beyond about 10 km; this initial peak also is a significant feature of waveforms from strokes between a few and about 10 km and can be identified, with some effort, in waveforms for strokes as close as a kilometer. The initial field peak is due to the radiation component of the total field and, hence, decreases inversely with distance in the absence of significant propagation effects. The field peaks produced by different return strokes at known distances can be range normalized for comparison, for example, to 100 km by multiplying the measured field peaks by $r/10^5$, where r is the stroke distance in meters. The geometric mean of the electric field initial peak value, normalized to 100 km, is typically about 6 V/m for first strokes and 3 V/m for subsequent strokes. Since the initial electric field peak appears to obey a log-normal distribution, the geometric mean value (equal to the median value for a log-normal distribution) is probably a better characteristic of the statistical distribution of this parameter than the mean (arithmetic mean) value. Note that the geometric mean value for a log-normal distribution is lower than the corresponding mean value and higher than the modal (most probable) value.

Lightning peak currents can be estimated from measured electric or magnetic fields, for which a field-to-current conversion procedure (a model-based or empirical

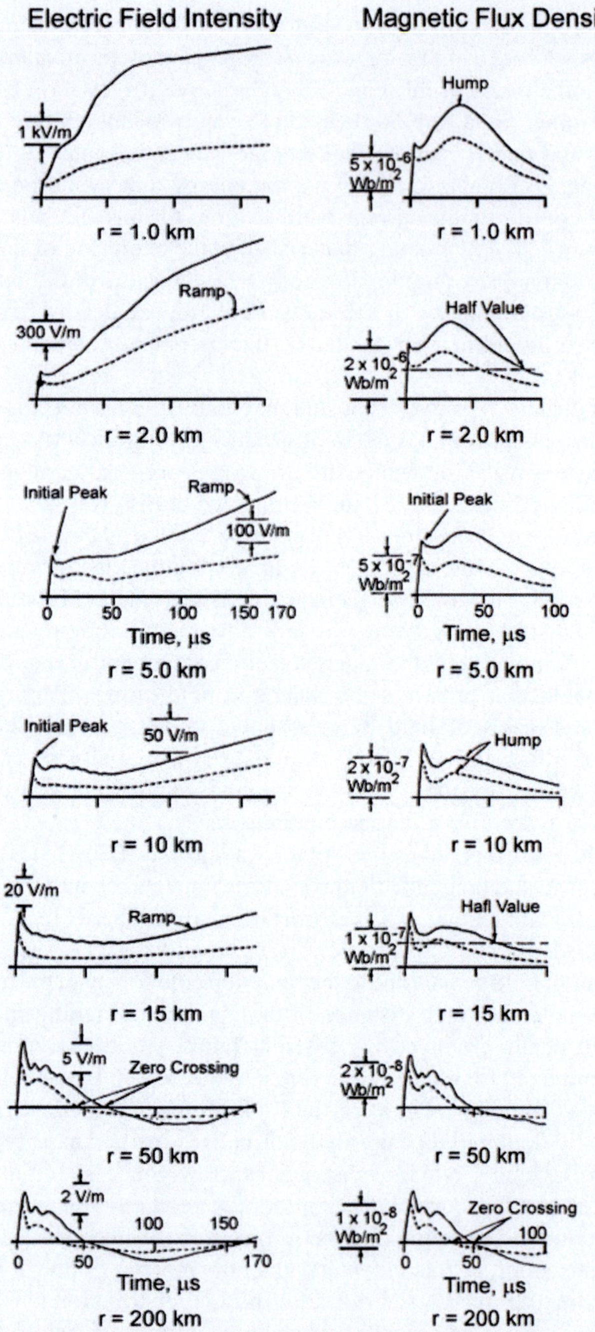

Fig. 13 Typical vertical electric field intensity (left column) and azimuthal magnetic flux density (right column) waveforms for first (solid line) and subsequent (broken line) return strokes at distances of 1, 2, 5, 10, 15, 50, and 200 km. Adapted from Lin et al. [15]

formula) is required. The vertical component of electric field and the azimuthal component of magnetic field are usually employed.

Rakov et al. [21] proposed the following empirical formula (linear regression equation) to estimate the negative return-stroke peak current, I, from the initial electric field peak, E, and distance, r, to the lightning channel:

$$I = 1.5 - 0.037Er \tag{1}$$

where I is in kA and taken as negative, E is positive and in V/m, and r is in km. Equation 1 was derived using data for 28 triggered-lightning strokes acquired by Willett et al. [33] at the Kennedy Space Center (KSC), Florida. The fields were measured at about 5 km and their initial peaks were assumed to be pure radiation. The currents were directly measured at the lightning channel base.

Lightning peak currents can also be estimated using the radiation-field-to-current conversion equation based on the transmission line (TL) model [31], which for the electric field is given by:

$$I = \frac{2\pi\varepsilon_0 c^2 r}{v} E \tag{2}$$

where ε_0 is the permittivity of free space, c is the speed of light, and v is the return-stroke speed (assumed to be constant). The return-stroke speed is generally unknown and its range of variation is from one-third to two-thirds of the speed of light. Both I and E in Eq. 2 are absolute values. The equation is thought to be valid for instantaneous values of E and I at early times (for the initial rising portion of the waveforms, including the peak).

4 Lightning Measurements

In this section, measurements of lightning electric and magnetic fields will be considered. Both the principles and practical aspects will be covered.

A sensor that is commonly used to measure the lightning vertical electric field is a metallic disk placed flush with the ground surface, the so-called flat-plate antenna. Figure 14a schematically shows such an antenna, where it is assumed that the area A of the antenna sensing plate is small enough to consider the electric field E constant over that area and C_a is the capacitance of the antenna. The downward directed electric field induces negative charge Q on the surface of the antenna, which can be found as the product of the surface charge density ρ_s and the area A of the antenna sensing plate. From the boundary condition on the vertical component of electric field on the surface of good conductor, $\rho_s = \varepsilon_0 E$, where ε_0 is the electric permittivity of free space, and hence $Q = \varepsilon_0 EA$. If E is varying with time, there will be current $I = dQ/dt = \varepsilon_0 A dE/dt$ flowing via C_a to ground. This current is proportional to

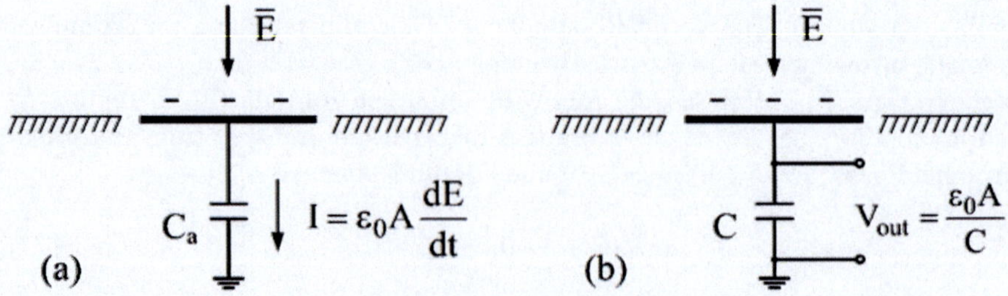

Fig. 14 Illustration of the principle of operation of the flat-plate antenna. **a** Antenna without external circuit. **b** Antenna with external integrating capacitor $C \gg C_a$. Drawing by Potao Sun

dE/dt. In order to measure E, it is necessary to use an integrating capacitor $C \gg C_a$, (see Fig. 14b), since C_a is usually too small for measuring lightning fields, as will be discussed later in this section. Thus, the voltage across the integrating capacitor (capacitive voltage drop) will be

$$V_{out} = \frac{1}{C_a + C_0} \int_0^t I(t')dt' \approx \frac{1}{C} \int_0^t I(t')dt' = \frac{Q}{C} = \frac{\varepsilon_0 AE}{C} \tag{3}$$

Strictly speaking, Eq. 3 applies only to the case of infinitely large input impedance of the recorder. In practice, the input resistance of the recorder (or fiber-optic-link transmitter) plays an important role, limiting the time interval or the lower end of the frequency range over which Eq. 3 is valid. To examine this further, it is convenient to use the Norton equivalent circuit of the antenna, which is the antenna short-circuit current, $I = \varepsilon_0 A j\omega E$ (ideal current source), in parallel with antenna impedance, $1/j\omega C_a$, where $\omega = 2\pi f$ with f being frequency in hertz. The equivalent circuit including the Norton equivalent of the antenna (in the time domain), integrating capacitance, and input resistance R_{in} and capacitance C_{in} of the recorder is shown in Fig. 15.

Since $C \gg C_a$ and usually $C \gg C_{in}$, the current basically splits between C and R_{in}, and Eq. (3) holds when $1/\omega C \ll R_{in}$; that is, $\omega \gg 1/(R_{in}C)$ or $f \gg 1/(2\pi R_{in}C)$. In the time domain, Eq. 3 is valid when the variation time (duration) of the signal of

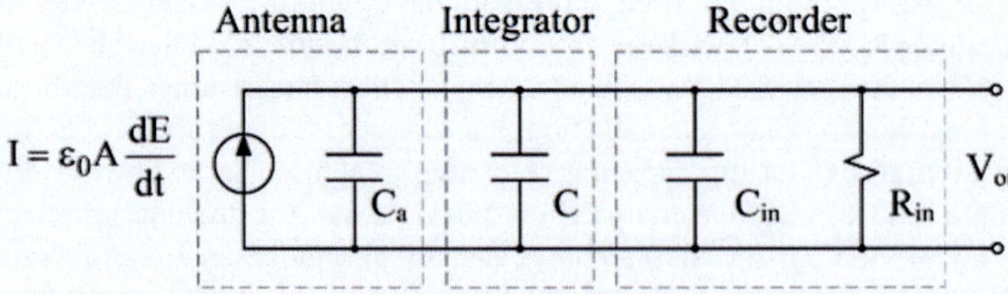

Fig. 15 Norton equivalent circuit of electric field antenna shown along with the integrating capacitance and the input impedance of recorder (usually $C \gg C_a$ and $C \gg C_{in}$). Drawing by Potao Sun

interest $\Delta t \ll \tau$, where $\tau = R_{in}C$ is the decay time constant of the measuring system (when E is a step-function, V_{out} will exponentially decay to 1/e, where e is the base of the natural logarithm, or about 37% of its initial value over the time equal to τ). For example, if $C = 1\ \mu F$ and $R_{in} = 1\ M\Omega$, $\tau = 1$ s, long enough for recording electric fields produced by lightning processes occurring on time scales of the order of tens of milliseconds (for example, stepped leaders or return strokes followed by continuing currents). Typical values of C_a and C_{in} are of the order of tens to hundreds of picofarad (1 pF = 10^{-12} F) or less, clearly much smaller than $C = 1\ \mu F$ (10^{-6} F) in this example. For recording return-stroke pulses, τ of the order of milliseconds is usually sufficient, while for the faithful reproduction of overall flash waveforms it should be of the order of 10 s or so. For recording microsecond- and sub-microsecond-scale pulses, τ shorter than a millisecond or so can be used. Measuring systems with decay time constants of the order of seconds are sometimes referred to as "slow antenna" systems, and those with sub-millisecond time constants as "fast antenna" systems. "Fast-antenna" systems usually have higher gains than "slow-antenna" ones. The terms "slow" and "fast" have nothing to do with the upper frequency response of the system, which is usually determined by the amplifier or fiber-optic link. The measuring system shown in Fig. 15 employs a passive integrator. In the case of active integrator, $\tau = RC$ is determined by R and C connected in parallel in the feedback circuit of the operational amplifier.

We now discuss the situation when the condition of $\Delta t \ll \tau$ is not satisfied. Such situations are not rare. Indeed, since the range of lightning electric field changes is very large (it spans orders of magnitude), it is practically impossible to build a single measuring system that would have a dynamic range suitable for recording all those changes. Smaller field changes require a higher gain that usually leads to system saturation by larger field changes. On the other hand, a lower gain needed to keep the larger field changes on scale would render the smaller field changes unresolved. The larger field changes are usually relatively slow, varying on time scales of the order of milliseconds and longer (e.g., electric field changes produced by long continuing currents), while the smaller field changes are usually microsecond-scale pulses. One way to enable a field measuring system to record relatively small and relatively short pulses is to allow the larger and slower field changes to decay with a relatively short time constant. In order to avoid distortion of the pulses, this time constant should be much longer than the expected duration of the pulses. As discussed above, time constants satisfying the latter requirement are shorter than a millisecond or so. In this case, some associated field changes varying on a millisecond time scale (e.g., overall field changes produced by K- and M-processes) will be distorted. Specifically, ramp-like electric field changes due to lightning K-processes can be converted to pulses, with the falling edge of the pulse being due to instrumental decay, as opposed to occurring in response to source variation. In principle, the instrumental decay can be compensated in post-processing of measured field waveforms, to remove the distortion and reconstruct the undistorted waveform [25].

Placement of flat-plate antenna flush with the ground ensures that the electric field to be measured is not influenced by the antenna. This gives an advantage of theoretical calibration of the measuring system (see Eq. 3). Any antenna elevated above the ground surface will enhance the field that would exist at the same location in the absence of antenna. As a result, experimental calibration is required to determine the field enhancement factor (except for the spherical antenna with isolated cutouts, for which the enhancement factor is known; it is equal to 3) the inverse of which is to be used as a multiplier in Eq. 3. Calibration can be done by placing the antenna in a uniform field of a large parallel-plate capacitor or by comparing the antenna output with that of a flush-mounted reference antenna. When calibration is done experimentally, an antenna of any geometry (e.g., a vertical rod (monopole) with or without capacitive loading at its top or an inverted antenna with a grounded "bowl" above the elevated sensing plate) can be used. However, slender antennas are generally not used for measuring fields at close distances from the lightning channel. Such antennas can enhance the electric field to a degree that corona discharge occurs from the antenna. It is impossible to accurately measure electric fields in the presence of corona from the antenna, since, besides the current charging the antenna, there will be corona current transporting charges into the air surrounding the antenna, both currents flowing through the same integrating capacitor across which the output voltage is measured.

If an essentially flush with the surface flat-plate antenna is installed of the roof of a building or other structure, another field enhancement factor, due to the presence of the building, is to be taken into account. This latter enhancement factor can be calculated numerically. For example, Baba and Rakov [2], who used the 3-D finite-difference time-domain (FDTD) method, estimated that for a building having a plan area of 40×40 m^2 and a height of 20 m the electric field enhancement factor (at the center point of its flat roof) is 1.5 and it is 3.0 if the height of the building is 100 m. For comparison, the enhancement factor on the top of hemispherical structure is independent of its size and equal to 3. The magnitude of vertical electric field at ground level in the immediate vicinity of the building is reduced relative to the case of no building, with this shielding effect becoming negligible at horizontal distances from the building exceeding twice the height of the building. In contrast to the electric field, the magnitude of magnetic field was found to be not much influenced by the presence of building. Note that Baba and Rakov [2] (see their Table VI) showed that the electric field enhancement due to the presence of building is only slightly influenced by building conductivity ranging from 1 mS/m (dry concrete) to infinity and essentially independent of relative electric permittivity ranging from 1 to 10.

The use of long horizontal coaxial cables between the antenna and the associated electronics should be avoided, since the horizontal component of electric field (present due to the finite ground conductivity) can induce unwanted voltages in these cables. The horizontal electric field waveshape is similar to that of the derivative of the vertical field. As a result, the measured field waveform may be a superposition of the vertical field, which is being measured, and the unwanted horizontal field, which causes a distortion of the vertical field waveform by making peaks and valleys sharper than they actually are in the vertical field [29]. The problem can be solved by using a fiber-optic link instead of the coaxial cable. Further, significant reduction of noise can be achieved by digitizing signals at the antenna location and digitally transmitting them to recorder.

One can check if the electric field measuring system is working properly by comparing electric field waveforms produced by individual lightning events (e.g., return strokes) with the corresponding magnetic field waveforms. At large distances (>50 km or so), those waveforms are dominated by their radiation components and, hence, their shapes should be identical. Further, the ratios of electric and magnetic field peaks at large distances for sources near ground (return strokes) should be equal to the speed of light (E/B = c).

If in Figs. 14b and 15 the integrating capacitor C is replaced with the resistor R (such that $R \ll 1/(\omega C_a)$ and $R \ll 1/(\omega C_{in})$), the output voltage is proportional to dE/dt. Measured dE/dt waveforms can be numerically integrated over time to obtain E waveforms, although the integration interval should not be too long in order to avoid accumulation of significant error.

To measure the magnetic field produced by lightning processes a loop of wire can be used as an antenna. According to Faraday's Law, a time varying magnetic field passing through an open-circuited loop of wire will induce a voltage (electromotive force) at the terminals of the loop (see Fig. 16). The induced voltage is proportional to the rate of change of magnetic flux passing through the loop area. Assuming that the loop area, A, is small enough to consider the normal component of magnetic flux density, $B_n = B \cos\alpha$, where α is the angle between the magnetic flux density vector and the normal to the plane of the loop, to be constant over that area, we can express the magnitude of induced voltage as follows:

$$V = A\frac{dB_n}{dt} \quad (4)$$

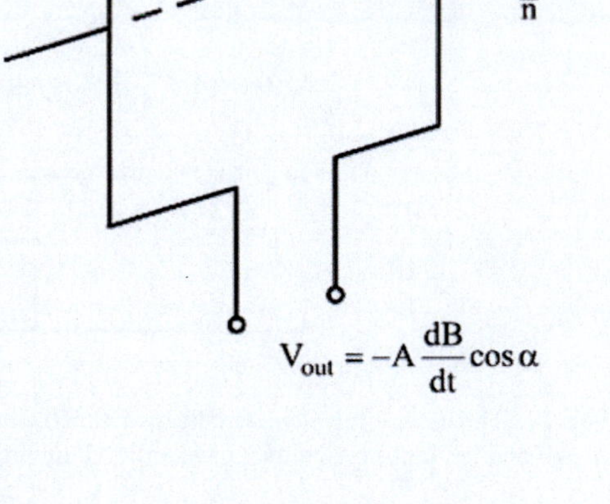

Fig. 16 Illustration of the principle of operation of the loop antenna. Drawing by Potao Sun

When $\cos\alpha = 1$ ($\alpha = 0$), the induced voltage is maximum, and when $\cos\alpha = 0$ ($\alpha = 90°$), the induced voltage is zero. It follows, that a vertical loop antenna in a fixed position is directional in that the magnitude of voltage induced across its terminals is a function of the direction to the source, and two such antennas with orthogonal planes can be used for magnetic direction finding. In order to obtain the horizontal (azimuthal) component of magnetic field, which is the dominant component for essentially vertical lightning channels, two vertical loop antennas are required, unless the direction to the lightning channel is known (for example, in the case of rocket-triggered lightning; see Sect. 5 of this chapter).

Since the signal at the output of a loop antenna is proportional to the magnetic field derivative, the signal must be integrated to obtain the field. This can be accomplished using either an RC or RL circuit, or the measured field derivative signal can be integrated numerically. We will consider below the case of RC integrator. In the following, we will assume that B is normal to the plane of the loop antenna ($\alpha = 0$), so that $B = B_n$. The voltage induced at the terminals of a loop antenna is the open-circuit voltage, AdB/dt or $Aj\omega B$, and, hence, it can be used for building the Thevenin equivalent circuit of the antenna. The source impedance is predominantly inductive, $j\omega L$. The overall equivalent circuit including, besides the antenna, the RC integrator and input impedance (input resistance in parallel with input capacitance) of the recorder is shown in Fig. 17.

In contrast with the electric field antenna (see Fig. 15), the integrating capacitor in Fig. 17 has two discharge paths, one through the input resistance of the recorder (similar to Fig. 15) and the other through resistor R of the integrating circuit and the source (the ideal voltage source has zero impedance). As a result, there are three conditions for undistorted recording of magnetic field with the measuring system shown in Fig. 17. The first one, $R \gg 1/\omega C$ ($\omega \gg 1/(RC)$; C_{in} is neglected), determines the lower frequency limit and is equivalent to the $\Delta t \ll \tau$ ($\tau = RC$) condition. The second one, $R \gg \omega L$ ($\omega \ll R/L$), determines the upper frequency limit. The third one, $R_{in} \gg R$, requires that C discharges through R, not R_{in}. Under those three conditions, the output voltage is independent of frequency and given by

$$V_{out} = \frac{AB}{RC} \quad (5)$$

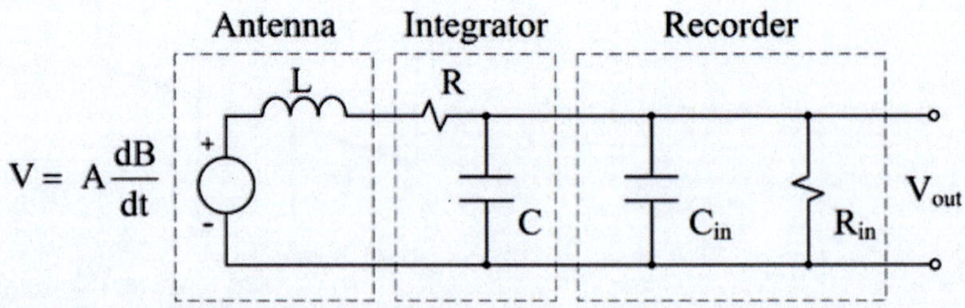

Fig. 17 Thevenin equivalent circuit of magnetic field antenna shown along with the integrating circuit and the input impedance of recorder. Drawing by Potao Sun

Magnetic field measuring circuits are rarely passive; active integrators and amplifiers are usually required. A loop antenna developed by George Schnetzer and used by the University of Florida Lightning Research Group is described in Sect. 7.2 of Rakov [20].

In designing field measuring systems, one needs to know expected magnitudes and durations of signals to be recorded. Different lightning processes produce different electromagnetic signatures, these signatures change with distance, and at the same distance there is large variation in source strength. Both variations in the source and with distance should be considered. Given below is a brief review of characteristics of lightning electric and magnetic fields expected at different distances from the source.

At ground level and at distances greater than a few kilometers, the initial electric field peak is dominated by its radiation component. Typical electric field peak values normalized to 100 km are about 6 and 3 V/m for negative first and subsequent return strokes, respectively. The largest radiation field peaks due to stepped and dart-stepped leaders are typically a factor of 10 smaller than the corresponding return-stroke field peak at the same distance. Radiation fields vary inversely with distance (1/r dependence), if propagation effects due to finite ground conductivity can be neglected. Generally, the typical radiation field peak values normalized to 100 km can be scaled to either smaller or larger distances in the range from about 5 to about 200 km. For example, if the field peak at 100 km is 6 V/m, it is expected to be 60 V/m at 10 km and 3 V/m at 200 km. The corresponding magnetic radiation field peaks can be readily found by dividing the electric field peak by the speed of light (3×10^8 m/s) to find the magnetic flux density (B) and by the intrinsic impedance of free space (377 Ω) to find the magnetic field intensity (H). At a given distance, the field can be at least a factor of 5 greater and a factor of 5 smaller, due to variation in the source.

5 Rocket-Triggered Lightning

An understanding of the physical properties and deleterious effects of lightning is critical to the adequate protection of power and communication lines, aircraft, spacecraft, and other objects and systems. Many aspects of lightning are not yet well understood and are in need of research that often requires the termination of lightning channel on an instrumented object or in the immediate vicinity of various sensors. The probability for a natural lightning to strike a given point on the earth's surface or an object of interest is very low, even in areas of relatively high lightning activity. Simulation of the lightning channel in a high-voltage laboratory has limited application, since it does not allow the reproduction of many lightning features important for lightning protection and it does not allow the testing of large distributed systems such as overhead power lines. One promising tool for studying both the direct and the induced effects of lightning is an artificially initiated (or triggered) lightning discharge from a natural thundercloud to a designated point on ground. The most effective technique for artificial lightning initiation is the so-called rocket-and-wire technique. It allows generation of full-scale lightning discharges with currents up to tens of kiloamperes and potentials of the order of 10 MV. Energy tapped by these discharges is naturally accumulated in the cloud that would otherwise produce natural lightning. In most respects, the rocket-and-wire triggered lightning (often referred to as rocket-triggered or just triggered lightning) is a controllable analog of natural lightning.

The rocket-and-wire technique involves the launching of a small rocket extending a thin wire (either grounded or ungrounded) into the gap between the ground and a charged cloud overhead. In the former case, the triggered lightning is referred to as classical and in the latter case as altitude triggered one. The sequence of processes (except for the transition from leader to return stroke stage that is referred to as the attachment process) in classical triggered lightning is schematically shown in Fig. 18. When the rocket, ascending at about 150–200 m/s, is about 200–300 m high, the field enhancement near the rocket tip launches a positively charged leader that propagates upward toward the cloud. This upward positive leader vaporizes the trailing wire, bridges the gap between the cloud and the ground, and establishes an initial continuous current with a duration of some hundreds of milliseconds that transports negative charge from the cloud charge source region to the triggering facility. After the cessation of the initial continuous current, one or more downward dart-leader/upward return-stroke sequences may traverse the same path to the triggering facility. The dart leaders and the following return strokes in triggered lightning are similar to dart-leader/return-stroke sequences in natural lightning, although the

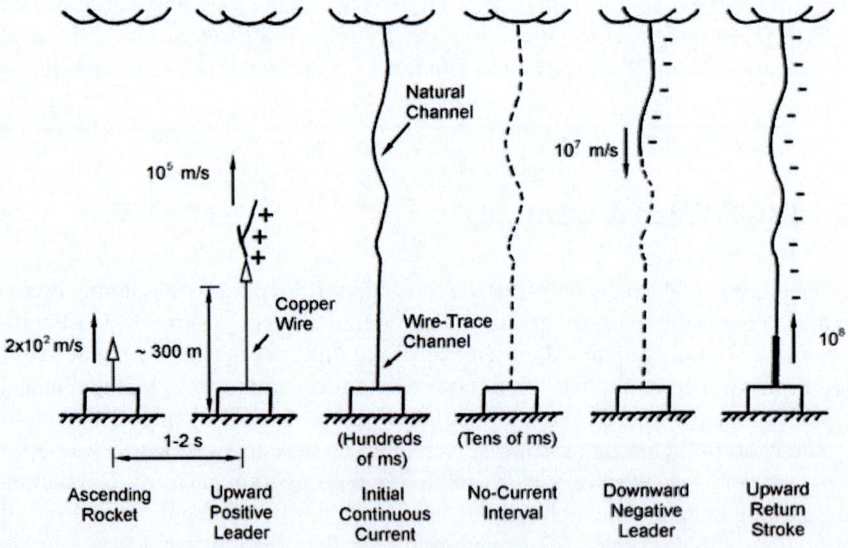

Fig. 18 Sequence of events (except for the attachment process) in classical rocket-triggered lightning. The upward positive leader and initial continuous current constitute the initial stage of a classical rocket-triggered flash. Adapted from Rakov et al. [24]

initial processes in natural downward and rocket-triggered lightning are distinctly different.

First lightning triggering was done in the 1960s over water (inspired by lightning unintentionally initiated by a plume of water resulting from an underwater explosion; see Fig. 1 of Brook et al. [5]) and since the early 1970s has been performed over land. To date, over 1,500 lightning discharges have been triggered by researchers in different countries (United States, France, Japan, China, and Brazil) using the rocket-and-wire technique, with over 450 of them at Camp Blanding, Florida. Presently, there are four facilities, two in China (Binzhou and Conghua) and two in the United States (Florida and New Mexico), where triggered-lightning experiments can be performed. Photographs of two classical rocket-and-wire triggered lightning flashes are shown in Fig. 19. Examples of some results of triggered-lightning experiments are shown in Figs. 20 and 21.

Figure 20 shows a photograph of surface arcing during a triggered-lightning flash from experiments at Fort McClellan, Alabama. The soil was red clay and a 0.3 or 1.3-m steel vertical rod was used for grounding of the rocket launcher. The surface arcing appears to be random in direction and often leaves little if any evidence on the ground. Even within the same flash, individual strokes can produce arcs developing in different directions. In one case, it was possible to estimate the current carried by one arc branch which contacted the instrumentation. That current was approximately 1 kA, or 5% of the total current peak in that stroke. The observed horizontal extent of surface arcs was up to 20 m, which was the limit of the photographic coverage during the Fort McClellan experiments. These results suggest that the uniform ionization of soil, usually postulated in studies of the behavior of grounding electrodes subjected to lightning surges, may be not an adequate assumption.

In 1993, an experiment, sponsored by Electric Power Research Institute (EPRI), was conducted at Camp Blanding by Power Technologies, Inc. to study the effects of lightning on underground power cables. In this experiment three 15 kV coaxial cables with polyethylene insulation between the center conductor and the outer concentric stranded shield (neutral) were buried 5 m apart at a depth of 1 m, and lightning current was injected into the ground at different positions with respect to these cables. The cables differed only in the level of insulation from the surrounding soil. One of the cables (Cable A) had an insulting jacket and was placed in (a) PVC conduit (pipe), another one (Cable B) had an insulating jacket and was directly buried, and the third one (Cable C) had no jacket and was directly buried. About 20 lightning flashes were triggered directly above the cables which were unenergized.

The underground power cables were excavated by the University of Florida researchers in 1994. The damage found ranged from minor punctures of the cable jacket to extensive puncturing of the jacket and melting of nearly all the concentric neutral strands near the lightning attachment point. Some damage to the cable insulation between the center (phase) conductor and the neutral was also observed. In the case of the PVC conduit cable installation, the side wall of the conduit was melted, deformed, and blown open, and the lightning channel had attached to the cable inside and damaged its insulation. Photographs of the damaged parts of the cables are shown in Fig. 21. Note fulgurites (glassy tubes formed when lightning

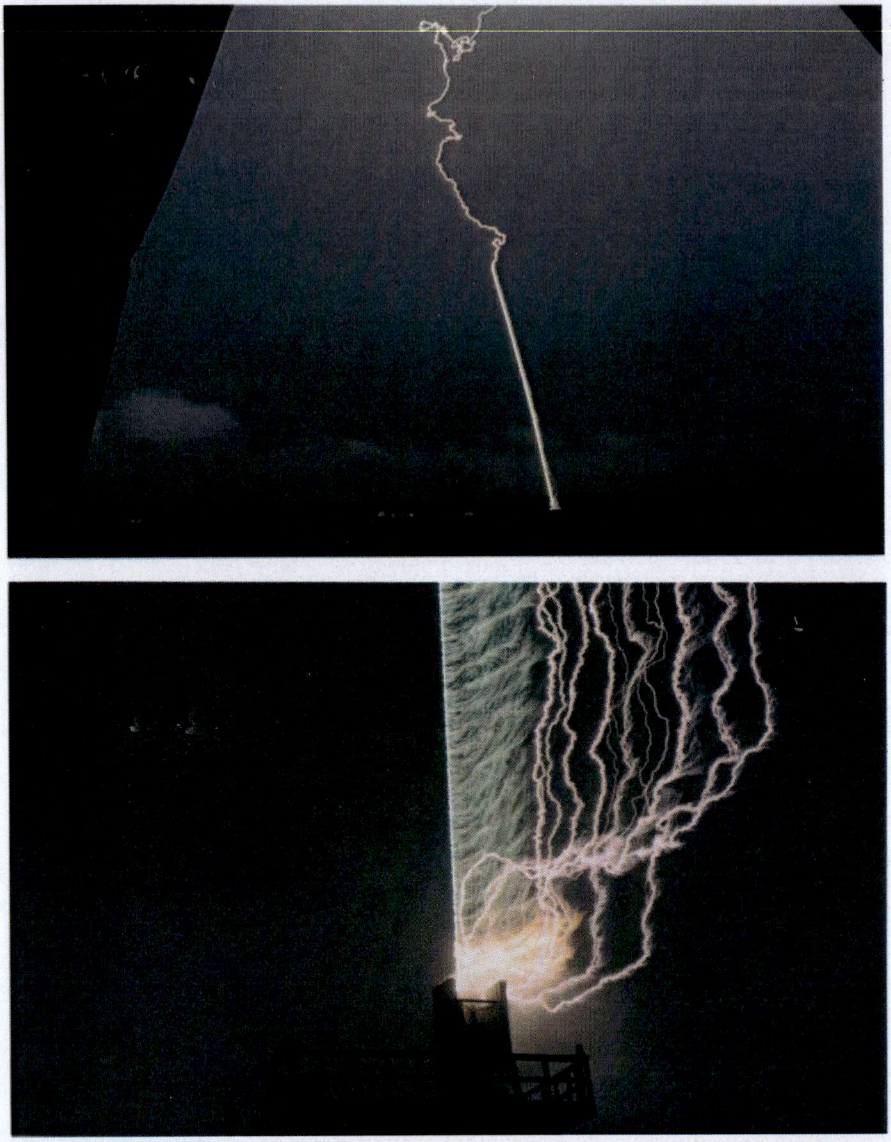

Fig. 19 Photographs of lightning flashes triggered using the rocket-and-wire technique at Camp Blanding, Florida. Top—a distant view of a strike to the test airport runway; bottom—a close-up view of a strike to the test power system

Fig. 20 Photograph of surface arcing associated with the second stroke (current peak of 30 kA) of flash 9312 triggered at Fort McClellan, Alabama. Lightning channel is outside the field of view. One of the surface arcs approached the right edge of the photograph, a distance of 10 m from the rocket launcher. Adapted from Fisher et al. [8]

current flows through sandy soil) in Fig. 21a, b. The presence of fulgurites indicates that the lightning channel continues to extend below the ground surface, in addition to developing along the ground surface in the form of surface arcs (see Fig. 20). Overall, these experiments showed that buried cables attract lightning striking ground within a distance of 10 m or so from the cable and that three layers of insulation (insulating jacket, PVC pipe, and air inside the PVC pipe; see Fig. 21a) plus 1-m layer of soil do not make cables "invisible" to lightning.

Further information on rocket-triggered lightning can be found in works of Horii and Nakano [10], Rakov and Uman [23], Chap. 7), Dwyer and Uman [7], and Qie and Zhang [19].

The results of rocket-triggered-lightning experiments have provided considerable insight into natural lightning processes that would not have been possible from studies of natural lightning due to its random occurrence in space and time. Among such findings are detailed observations of lightning propagation and attachment to ground, discovery that all types of negative lightning leaders produce hard X-rays, identification of the M-component mode of charge transfer to ground, direct measurements of NO_x production by an isolated lightning channel section, estimation of lightning input energy, and many others. The first terrestrial gamma-ray flash (TGF) observed at ground level was associated with triggered lightning. Triggered-lightning experiments have contributed significantly to testing the validity of various lightning models and to providing ground-truth data for testing the performance characteristics of lightning locating systems. Triggered lightning is a very useful tool for studying

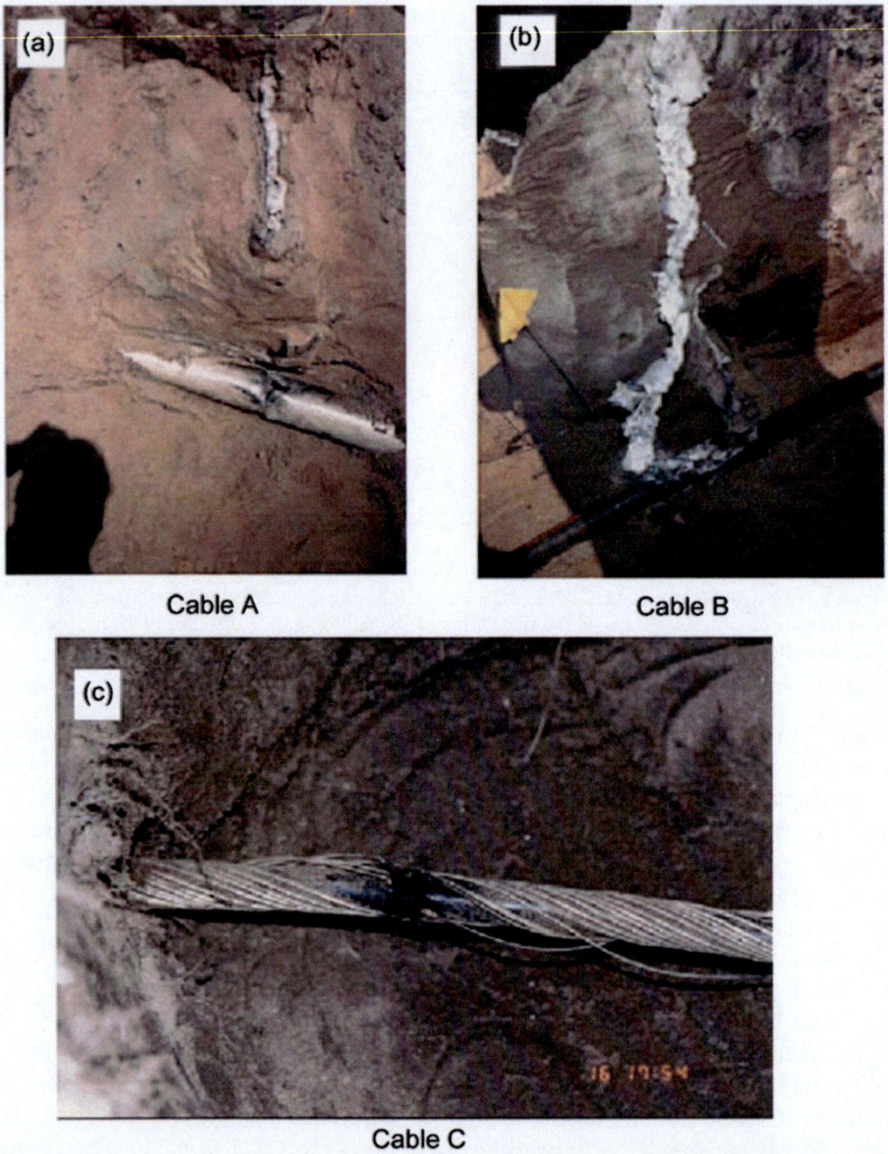

Fig. 21 Lightning damage to underground power cables. **a** coaxial cable in an insulating jacket inside a PVC conduit (pipe); note the section of vertical fulgurite in the upper part of the picture (the lower portion of this fulgurite was destroyed during excavation) and the hole melted through the PVC conduit, **b** coaxial cable in an insulating jacket, directly buried; note the fulgurite attached to the cable, **c** coaxial cable whose stranded neutral was in contact with earth; note that many strands of the neutral are melted through. Photos in **a**, **b** were taken by V. A. Rakov and in **c** by P. P. Barker

the interaction of lightning with various objects and systems and testing lightning protection schemes.

References

1. Anderson RB, Eriksson AJ (1980) Lightning parameters for engineering application. Electra 69:65–102
2. Baba Y, Rakov VA (2007) Electromagnetic fields at the top of a tall building associated with nearby lightning return strokes. IEEE Trans Electromagn Compat 49(3):632–643
3. Berger K (1972) Mesungen und Resultate der Blitzforschung auf dem Monte San Salvatore bei Lugano. der Jahre 1963–1971. Bull SEV 63:1403–1422
4. Berger K, Anderson RB, Kroninger H (1975) Parameters of lightning flashes. Electra 80:223–237
5. Brook M, Armstrong G, Winder RPH, Vonnegut B, Moore CB (1961) Artificial initiation of lightning discharges. J Geophys Res 66:3967–3969
6. Diendorfer G, Pichler H, Mair M (2009) Some parameters of negative upward-initiated lightning to the Gaisberg tower (2000–2007). IEEE Trans Electromagn Compat 51:443–452
7. Dwyer JR, Uman MA (2014) The physics of lightning. Phys Rep 534:147–241. https://doi.org/10.1016/j.physrep.2013.09.004
8. Fisher RJ, Schnetzer GH, Morris ME (1994) Measured fields and earth potentials at 10 and 20 meters from the base of triggered-lightning channels. In: Proceedings of 22nd international conference on on lightning protection, Budapest, Hungary, Paper R 1c-10, 6 p
9. Hendry J (1993) Panning for lightning (including comments on the photos by M.A. Uman). Weatherwise 45(6):19
10. Horii K, Nakano M (1995) Artificially triggered lightning. In: Volland H (ed) Handbook of atmospheric electrodynamics, vol 1. CRC Press, Boca Raton, Florida, pp 151–166
11. Jayaratne ER, Saunders CPR, Hallett J (1983) Laboratory studies of the charging of soft-hail during ice crystal interactions. Q.J.R Meteor Soc 109:609–630
12. Krehbiel PR (1986) The electrical structure of thunderstorms. In: Krider EP, Roble RG (eds) The Earth's electrical environment. National Academy Press, Washington, D.C., pp 90–113
13. Krider EP, Weidman CD, Noggle RC (1977) The electric field produced by lightning stepped leaders. J Geophys Res 82:951–960
14. Leteinturier C, Hamelin JH, Eybert-Berard A (1991) Submicrosecond characteristics of lightning return-stroke currents. IEEE Trans Electromagn Compat 33:351–357
15. Lin YT, Uman MA, Tiller JA, Brantley RD, Beasley WH, Krider EP, Weidman CD (1979) Characterization of lightning return stroke electric and magnetic fields from simultaneous two-station measurements. J Geophys Res 84:6307–6314
16. MacGorman DR, Rust WD (1998) The electrical nature of thunderstorms, 422 p, Oxford Univ. Press, New York
17. Moore CB, Vonnegut B (1977) The thundercloud. In: Golde RH (ed) Lightning, vol 1, Physics of lightning. Academic Press, New York, pp 51–98
18. Nag A, Rakov VA (2009) Some inferences on the role of lower positive charge region in facilitating different types of lightning. Geophys Res Lett 36:L05815. https://doi.org/10.1029/2008GL036783
19. Qie X, Zhang Y (2019) A review of atmospheric electricity research in China from 2011 to 2018. Adv Atmos Sci 36(9):994–1014. https://doi.org/10.1007/s00376-019-8195-x
20. Rakov VA (2016) Fundamentals of lightning. Cambridge University Press, 257 p
21. Rakov VA, Thottappillil R, Uman MA (1992) On the empirical formula of Willett et al. relating lightning return-stroke peak current and peak electric field. J Geophys Res 97:11527–11533
22. Rakov VA, Tran MD (2019) The breakthrough phase of lightning attachment process: From collision of opposite-polarity streamers to hot-channel connection. Electric Power Syst Res 173:122–134. https://doi.org/10.1016/j.epsr.2019.03.018

23. Rakov VA, Uman MA (2003) Lightning: Physics and effects. Cambridge, New York, 687 p
24. Rakov VA, Uman MA, Rambo KJ, Fernandez MI, Fisher RJ, Schnetzer GH, Thottappillil R, Eybert-Berard A, Berlandis JP, Lalande P, Bonamy A, Laroche P, Bondiou-Clergerie A (1998) New insights into lightning processes gained from triggered-lightning experiments in Florida and Alabama. J Geophys Res 103(14):117–130
25. Rubinstein M, Bermúdez J-L, Rakov VA, Rachidi F, Hussein A (2012) Compensation of the instrumental decay in measured lightning electric field waveforms. IEEE Trans EMC 54(3):685–688
26. Schoene J, Uman MA, Rakov VA, Rambo KJ, Jerauld J, Mata CT, Mata AG, Jordan DM, Schnetzer GH (2009) Characterization of return-stroke currents in rocket-triggered lightning. J Geophys Res 114:D03106. https://doi.org/10.1029/2008JD009873
27. Takami J, Okabe S (2007) Observational results of lightning current on transmission towers. IEEE Trans Power Del 22:547–556
28. Visacro S, Mesquita CR, De Conti A, Silveira FH (2012) Updated statistics of lightning currents measured at Morro do Cachimbo station. Atmos Res 117:55–63
29. Uman MA (1987) The Lightning Discharge, 377 p., Orlando, Florida: Academic Press.
30. Uman MA (2001) The lightning discharge. Dover, Mineola, New York, 377 p
31. Uman MA, McLain DK (1969) Magnetic field of the lightning return stroke. J Geophys Res 74:6899–6910
32. Wang D, Rakov VA, Uman MA, Takagi N, Watanabe T, Crawford DE, Rambo KJ, Schnetzer GH, Fisher RJ, Kawasaki Z-I (1999) Attachment process in rocket-triggered lightning strokes. J Geophys Res 104:2143–2150
33. Willett JC, Bailey JC, Idone VP, Eybert-Berard A, Barret L (1989) Submicrosecond intercomparison of radiation fields and currents in triggered lightning return strokes based on the transmission-line model. J Geophys Res 94(13):275–286
34. Williams ER (1995) Meteorological aspects of thunderstorms. In: Volland H (ed) Handbook of atmospheric electrodynamics, vol I. CRC Press, Boca Raton, Florida, pp 27–60

Lightning Detection and Warning

Anirban Guha, Yakun Liu, Earle Williams, Carina Schumann, and Hugh Hunt

Abstract This chapter is concerned with the remote detection and analysis of thunderstorms and lightning flashes by electrostatic, electromagnetic and photographic means, and the use of these methods for public warning of hazardous conditions. Section 1 addresses the measurement of electrostatic fields in fair weather and in response to the stronger fields of electrified shower clouds and thunderstorms. Section 2 reviews various methods in place worldwide for the detection of the electromagnetic radiation from lightning. The observation of the evolution of lightning flashes with video-camera observations is the subject of Sect. 3. The final Sect. 4 addresses the dissemination of the multitude of available observations for purposes of improving lightning safety.

Keywords Electrostatic fields · Video-camera observations · Electromagnetic radiation · Nowcasting · Detection

A. Guha
Department of Physics, Tripura University, Agartala, Tripura, India
e-mail: anirbanguha@tripurauniv.in

Y. Liu · E. Williams (✉)
Department of Civil and Environmental Engineering, Massachusetts Institute of Technology, Cambridge, MA, USA
e-mail: earlew@ll.mit.edu

Y. Liu
e-mail: yakunliu@mit.edu

Y. Liu
Department of Electrical Engineering, Shanghai Jiao Tong University, Shanghai, China

C. Schumann · H. Hunt
Johannesburg Lightning Research Laboratory, School of Electrical and Information Engineering, University of the Witwatersrand, Johannesburg, South Africa
e-mail: carina.schumann@wits.ac.za

H. Hunt
e-mail: hugh.hunt@wits.ac.za

C. Gomes (ed.), *Lightning*, Lecture Notes in Electrical Engineering 780,
https://doi.org/10.1007/978-981-16-3440-6_2

1 Measurement of Electrostatic Fields

One of the earliest quantitative devices for registration of the electrical potential gradient in fair weather conditions was Lord Kelvin's water dropper [1, 2]. Water drops are charged by induction in the electric field to be measured, and emitted from an elevated water container electrically insulated from ground, until the electric potential of the container matches that of the local potential of the atmosphere. The potential of the container is then measured with a high-impedance electrometer (also of Lord Kelvin's design [2]. The value of this potential with respect to local ground is the "potential gradient", and also explains how this terminology entered atmospheric electricity research in the early days.

Other instruments known as electric field mills are most commonly used [3] to record the large departures from fair weather potential gradient (~100 V/m) associated with charge separation in electrified clouds and thunderstorms. Since precipitation is almost invariably present in such circumstances, the use of the water dropper described above is not practical. A field mill consists of a set of fixed stator plates, isolated from ground, which are alternatively exposed and shielded by a motor-driven rotating, grounded conductor. The surface charge density, σ, induced on a stator plate by the applied electric field E (in volt/m) is given simply by in Eq. (1).

$$\sigma = \varepsilon_o E \, \text{Coulombs/m}^2 \tag{1}$$

It is the current flow to and fro from the stator plates that constitutes the field mill "signal", and is directly proportional to the imposed electric field one wishes to measure. The sensitivity of a field mill can be increased by increasing the area of the stator plates, and by increasing the frequency with which they are alternately exposed and shielded. The time resolution of a field mill is set by this latter frequency. All field mills have capability to resolve the abrupt changes in electrostatic field accompanying both intracloud and cloud-to-ground lightning flashes, and so can be used to estimate total lightning flash rates to distances up to 10–20 km from the parent storms.

Field mills can be installed and absolutely calibrated at the Earth's surface by flush-mounting the stator plates with a large grounded metal plate mounted directly on a flat patch of earth's surface. In the typical calibration exercise, a second large metal plate can then be placed on insulator stand-offs over the operating field mill. Well defined voltage differences can then be applied (with ordinary batteries) between the covering plate and ground to replicate specific values of electric field in units of volts per meter.

To protect against short-circuiting of the stator plates in heavy rainfall, so-called inverted field mills can be installed on short grounded conductive masts, with the field mill rotors and stator plates facing downward. The conductive mast serves to distort the local electric field from uniformity, and so additional work is needed in this kind of installation for absolute calibration.

2 Electromagnetic Detection of Lightning Flashes

2.1 Introduction

Lightning discharges produced by thunderclouds can be broadly separated into two main categories, ground flashes and cloud flashes [4, 5]. Cloud flashes are roughly 5 to 10 times more numerous than ground flashes, whereas ground flashes gain more attention as they can cause severe damage to electric power transmission lines, wind turbines, oil storage tanks, etc.. Ground flashes are manifest as a direct highly ionized hot plasma with typical peak temperature reaching 30,000 K, a transient intense current and high voltage injector, and an instantaneous source of electromagnetic radiation [5, 6]. Detection of ground flashes serves to supply basic lightning data for nowcasting and forecasting in specialized applications, protection of human and industrial activity, damage recognition and verification, and for study of climate change and many other lightning-prone areas [7–9].

Ground flashes, also known as cloud-to-ground (CG) flashes, are generally grouped into four categories based on the specific pathway and its current directions within the lightning channel: (1) the downward negative ground flash, (2) upward negative ground flash, (3) downward positive ground flash, and (4) upward positive ground flash [10]. Complex transient electric discharge processes usually occur in ground flashes within hundreds of milliseconds and even a few seconds from cloud to ground. A CG lightning specifically consists of the preliminary breakdown, stepped leader, attachment process, first return stroke, K-process, J-process, dart leader, continuing currents, M-component, and subsequent strokes [10]. These individual physical processes are associated with different characteristic electric and magnetic fields corresponding to significant electromagnetic energy in the radio frequency range from below 1 Hz to ~300 MHz [11]. Although the frequency of electromagnetic radiation emitted by lightning can be up to 10^{20} Hz or more (such as X-rays), this kind of emission is usually undetectable beyond ~1 km from the source at ground level [12]. Large range (a few thousand kilometers) detection technology of ground flashes generally chooses the practically and discernibly observable electromagnetic signals to infer lightning parameters. The acoustic and optical radiation from a lightning source are limited to close detections due to their apparent attenuation with distance. In this Section, the lightning processes and electromagnetic signatures will be introduced in detail in Sect. 2.2. Principles of lightning detection are addressed in Sect. 2.3. Section 2.4 describes the early history of lightning detection and Sect. 2.5 presents some examples of modern lightning detection systems.

2.2 Lightning Processes and Electromagnetic Signatures

Most (accounting for ~90%) of CG discharges are initiated within the cloud charge structure and transfer negative charges to ground in an overall downward developing direction, namely the downward negative lightning discharge. This kind of downward negative lightning discharge is taken as an example to briefly depict the lightning processes and its electromagnetic signature. The basic element in the CG discharge is termed the component stroke, or simply 'stroke'. A ground flash commonly contains several (typically 3–5) strokes with time intervals of tens of milliseconds [13]. These strokes terminate at different ground points spatially separated by up to a few kilometers in the 30–50% of all flashes. For video-camera documentation of the strokes of a flash, see Sect. 3.

Ground flashes are initiated by the preliminary breakdown (PB) process, often referred simply as 'initial breakdown' [14]. Preliminary breakdown occurring within the cloud usually lasts a few tens of milliseconds and tends to produce impulsive narrow radiation bursts on a microsecond scale. The PB process sets the stage for the initiation of a downward-moving stepped leader (SL). This leader is an ionized discharge channel and intermittently develops downward to ground with an average speed of ~2×10^5 m/s over a time interval of a few tens of microseconds between consecutive discrete steps [15]. Each step is associated with a current pulse of 1 kA or greater and has a time duration of typically 1 μs and each develops over tens of meters in length. The entire stepped leader process is usually transpiring within a time scale of tens of milliseconds and with an average leader current of some hundreds of amperes. New lightning channels are forged in the above breakdown stages and in the process emit a dominant electromagnetic energy in the Very High Frequency (VHF) band. During the SL process, a variation in electromagnetic radiation usually occurs from impulsive radiation bursts associated with step development to continuous radiation bursts accompanied by the leader approaching ground with complex branched structure [16].

Generally, when a stepped leader is progressing downward to ground, one or more upward-moving leaders with some tens of meters of length originate from the ground or from the tips of objects and attempt to contact the associated branches of the downward leader and generate single or multiple channel connections [17]. Successful connections represent the attachment process. The last connection stage for the extending plasma channels of downward and upward leaders is known as the break-through phase or 'final jump', indicating that the discharge channel in the air transforms from a relatively low-conductivity streamer into a high-conductivity plasma channel. Meanwhile, one downward moving return-stroke wave and another upward moving return-stroke wave are launched from the junction point and propagate along the lightning channel, thereby initiating the return stroke process. In the case of a subsequent stroke, a second attachment process may occur with an upward connecting leader length of ~10 m shorter to cater to the dart leader [13].

The return stroke is the most recognizable process of a ground flash in terms of both the optical brightness and the electromagnetic signature. The return stroke serves to neutralize the leader charge with a typical average speed of 1/2 to 1/3 of the light speed and is accompanied with a transient brightening of the lightning channel [18]. The peak temperature of the lightning channel rises rapidly to ~30,000 K, creating a high pressure of the order of 10 atmospheres or more and an outward propagating shock wave during the channel expansion. The peak current of the first return stroke is typically some tens or even hundreds of kiloamperes with a risetime (from 10 to 90% of the peak current) of some microseconds and with a decay to half-peak value within several tens of microseconds. By contrast, the current in the subsequent return stroke usually shows a lower peak amplitude but shorter risetime in less than a microsecond and a droptime to its 50% value in a few tens of microseconds. The electric and magnetic field radiated by return strokes shows a conspicuous initial field peak in the recorded waveforms beyond ~10 km. The mean initial electric field ranges typically from 6–8 V m^{-1} for the first return stroke and 3–6 V m^{-1} for subsequent return strokes, with range-normalizing to 100 km [19]. These return strokes emit the most powerful electromagnetic radiation in low frequency (LF) and very low frequency (VLF) ranges. A representative electric field changes radiated from a CG flash (distance = ~60 km) is shown in Fig. 1 [20].

The maximum in radiated electric field for the lightning return stroke in the VLF frequency range (see Fig. 2) can be understood by considering the vertical return stroke channel as resonant antenna with length equal to a half-wavelength. For a typical vertical length of 7 km (a typical height to the main negative charge of a

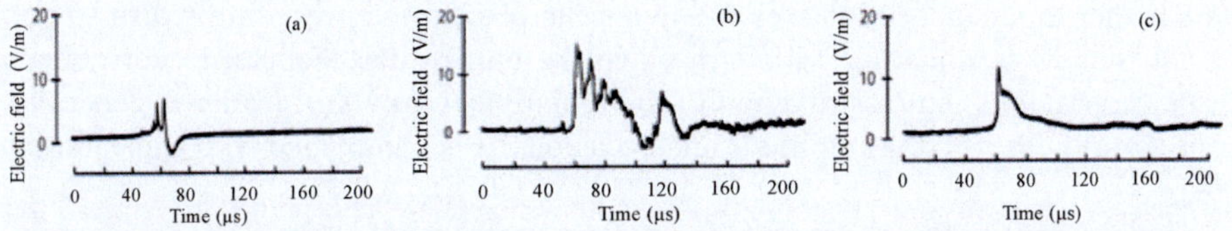

Fig. 1 Representative electric field changes radiated from a CG flash (distance = ~60 km). **a** Preliminary breakdown within the cloud. **b** First return stroke. **c** A subsequent return stroke in the preexisting channel [20]

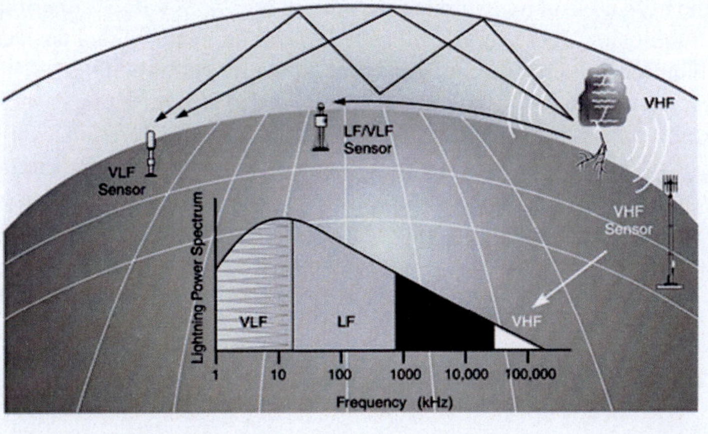

Fig. 2 Illustration of lightning locating techniques and operating frequencies. The spectral maximum lies in the upper VLF frequency range, in agreement with radiation expected from a half-wavelength resonant dipole

thunderstorm), the resonant emission has a dominant wavelength $\lambda = 14$ km and a dominant frequency f given by the dispersion relation in Eq. (2).

$$f\lambda = c \tag{2}$$

where c is the speed of light. Solving for f gives ~20 kHz, in the upper VLF frequency range.

Between the return stroke and the initiation of the dart leader, the J- (for junction) and K- processes occur within the cloud preceding a dart leader and transport additional charge into the existing lightning channel [21]. The J-process is characterized by a relatively steady electric field change lasting tens of milliseconds, which is generally weaker than the field change induced by a continuing current. The K-process generally produces transients and rapidly changing electric fields that are superimposed on the overall electric field associated with the J-process at time intervals of several to tens of milliseconds. The dart leader propagates downward with a speed of ~ 10^7 m/s and typically shows a peak current of ~ 1 kA [22]. The occurrence of the dart-stepped leader or stepped leader as a leader closing with ground depends on whether the leader progresses following the path of the preceding return stroke or not. A new termination will form when the dart or dart-stepped leader deflect from the previous return-stroke path. The radiation from a dart leader is generally continuous with variable amplitude and occasionally accompanied with some pulses [16].

The continuing current process often follows subsequent return strokes and maintains a current magnitude of tens to hundreds of amperes for up to hundreds of milliseconds. 30~50% of the negative ground flashes contain continuing currents with durations in excess of 40 ms, often defined as the long continuing current [23]. The lightning channel during the continuing current process exhibits a relatively faint luminosity and usually exhibits intermittent luminous surges superimposed on that faint luminosity. These surges are called M-components, with a peak current of typically hundreds of amperes and a risetime in the range 300~500 μs. The electric field changes associated with M-components recorded at a close range (mostly 6 km or less) exhibit a characteristic hook-like shape with a typical duration less than a few microseconds [24]. For information on the video-camera detection of lightning M-components, see Sect. 3. A representative E-field of the K-change, J-change, leader and return-stroke, and M-component in a flash adapted from Thottappillil et al. [25] is shown in Fig. 3.

2.3 Principles of Lightning Detection

Lightning information has wide and important applications and is expected to be monitored in all corners of the world. A large number of ground-based and space-based lightning detection sensors or systems have been invented to locate lightning

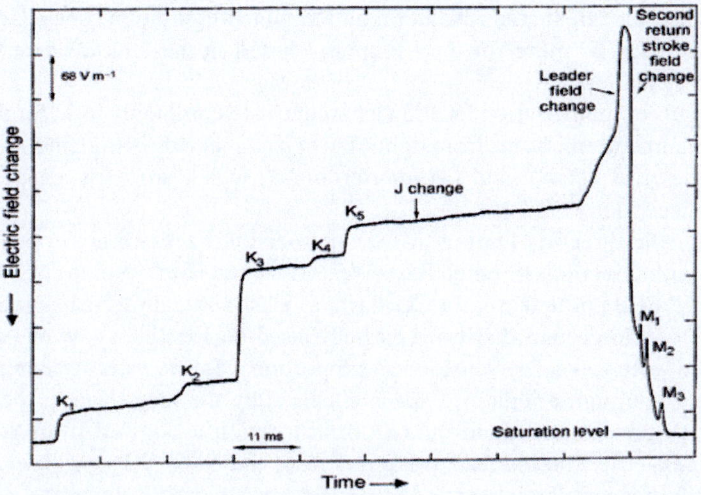

Fig. 3 Partial electric field change of a flash at a distance of 2.5 km (Florida 2228:43 UT 1979). $K_1 - K_5$ are five K-changes. The J-change, leader and return-stroke E-field change are presented. M1−M3 are E-field changes of three M-components [25]

and measure its parameters. Space-based lightning detection sensors mainly monitor total lightning (ICs and CGs) by capturing the optical transient from lightning. The detection sensor is conceptually a high speed event detector onboard an Earth orbiting satellite, which typically consists of a real-time event processor, a high speed Charge Coupled Device (CCD) focal plane, a wide field-of-view lens, and a narrow-band interference filter [26]. It has a wide detection coverage but with a relatively-low spatial resolution (~1 km or less) and is generally incapable of distinguishing ICs and CGs, limiting itself primarily to meteorological service. Space-based sensors can also monitor the electromagnetic emissions (VHF) and detect lightning over large regions [27–29], such as the lightning-associated VHF signal measured by the Fast On-Orbit Rapid Recording of Transient Events (FORTE) satellite [30, 31]. There is no publicly available VHF-based multi-satellite lightning geolocation system.

Ground-based lightning detection sensors monitor electromagnetic radiation in different frequency ranges with selection based on multiple factors. Due to the complexity in the wide-ranging amplitudes and waveforms associated with lightning, it is difficult to accurately locate an individual lightning flash with a single ground-based sensor. Lightning detection geographical coverage and spatial resolution depend on the total number of sensors and the selected frequency. The monitoring of radiation in the VHF range (30~300 MHz) will allow a lightning detection system with a resolution of 1~10 m but will require more sensors more closely deployed because of geometrical attenuation and the absence of over-the-horizon propagation. If a lightning detection system uses the LF (30~300 kHz) or VLF (3~30 kHz) range, the natural global waveguide can be exploited and the lightning detection coverage area can be enlarged efficiently with a more modest numbers of sensors, but with substantial sacrifice of spatial resolution on individual lightning flashes. Uncertainties of order 1 km or more for the LF approach and in the 10 km range for VLF signals can be anticipated.

Three kinds of multi-station locating techniques are commonly used for detecting the electromagnetic emissions from lightning: (1) magnetic direction finding (MDF), (2) time of arrival (TOA), and (3) interferometry, which are each detailed in the following discussion.

The magnetic direction finding (MDF) approach utilizes two horizontal orthogonal loops with directions oriented East–West (EW) and North–South (NS) to detect the magnetic field emitted from a CG flash, a quasi-vertical electromagnetic radiator from cloud to ground discussed earlier. Based on Faraday's law of induction, the acquired voltage of a given loop is proportional to the rate of change of the magnetic flux through a region of space enclosed by the loop. Hence, the tangent of the angle between north and the CG flash location as viewed from the sensor is linearly related to the induced voltage ratio of the NS/EW loops. Two types of crossed-loop magnetic direction finders are commonly used in lightning detection, the narrow band (tuned) and the gated wideband magnetic direction finders. The general frequency employed in narrow band magnetic direction finders ranges from 5–10 kHz to capture the peak frequency spectrum for lightning. The attenuation is relatively low in the Earth-ionosphere waveguide. The narrow band magnetic direction finder has a long application history beginning in the 1920s but it has inherent azimuthal errors (also named polarization errors, or site error corrections) on the order of 10° in close (< ~200 km) lightning detection due to an undesired voltage induced from the non-vertical lightning channel segments and also due to the inhomogeneous conductivity of the ground beneath the station. The gated wideband magnetic direction finder was developed in the 1970s to overcome this shortcoming by adding a gate on the sampling of magnetic field, focusing the analysis on the initial peak from the return stroke. This initial peak contains radiation from the lowest hundreds of meters of the lightning channel, where it tends to be straight and vertical. The gated wideband magnetic direction finder usually operates in the frequency range of a few kHz to 500 kHz and excludes the ionospheric reflections. Magnetic direction finders are susceptible to the unwanted magnetic field from the surroundings. Therefore, it is recommended to select a flat and uniform area, as well as one without conducting structures or buried objects, so as to reduce the site errors.

The time of arrival (TOA) technique locates lightning on the basis of the arrival times of electromagnetic signals at the detection sensors. These sensors usually operate in different frequency ranges of VHF (30~300 MHz), LF (30~300 kHz), and VLF (3~30 kHz), and can be generally divided into three types, (1) very short baseline (tens to hundreds of meters), and (2) short baseline (tens of kilometers) operated at VHF, and (3) long baseline (hundreds to thousands of kilometers) operated at LF and VLF. A very short baseline TOA system consists of two or more TOA receivers and locates lightning from the intersection of hyperboloids deduced from the arrival time differences of every individual VHF pulse. It is capable of resolving air breakdown processes with a time accuracy of tens of nanoseconds. A short baseline TOA system typically uses 5–15 stations as a network to map lightning channels in three-dimension. This system can depict the temporal and spatial development of lightning charges, which has become a major tool for lightning research and operational applications. A long baseline TOA system operates at LF/VLF and generally needs four or more stations to assure a unique lightning location as the hyperbolae from two arrival time differences intersect at two points on the Earth's surface for a remote lightning. The detection system is most sensitive to the return stroke and has a wide coverage of lightning detection of hundreds to thousands of kilometers.

The interferometry technique detects the phase difference between narrow band signals associated with the noise-like bursts produced from lightning by using two or more closely spaced (usually several meters distance) sensors. There is no requirement to identify an individual pulse in this approach. A lightning interferometer is usually composed by two or more identical antennas separated by a few meters and connected by the same narrow-band filter and receiver. The phase difference between two quasi-sinusoidal signals out of the two receivers is then converted into a voltage by a phase detector. Three or more antennas are necessary to form two or more orthogonal baselines in order to obtain the azimuth and elevation of a radiation source. Interferometric systems usually operate in very narrow frequency bands within the VHF band (30~300 MHz)/UHF (ultra-high frequency, 300 MHz–3 GHz) and consist of two or more synchronized interferometers separated by some tens of kilometers or more to locate lightning in three dimensions. These systems have high sensitivity to the signal but also with relatively low signal-to-noise ratio due to the high working frequency and antenna spacing limitation. In the installation of the antenna array, the surrounding obstacles and other noise sources should be cleared to eliminate perturbations and coupling between antennas.

2.4 Early History of Lightning Detection

Lightning has been studied for hundreds of years. In June 1752, Benjamin Franklin performed the famous Kite experiment in Philadelphia and identified lightning as an electrical discharge [32]. In 1895, Popoff used a coherer (a primitive form of radio signal detector) and made the first measurement with purpose to investigate the electromagnetic fields emitted by lightning [33]. The duration of lightning was estimated by De Blois with using an ordinary wireless aerial of feeble damping together with an oscillograph (of natural periodicity 5000–6000) and a rotating mirror [34]. With the advent of the cathode-ray oscilloscope, Watson-Watt and Appleton succeeded in recording the radiation field waveforms visually during 1922–1923 and firstly analyzed the atmospheric radio signals quantitatively [35]. In the development of radio communication, the electromagnetic radiations produced by lightning were generally measured by the narrow-band radio receivers or antennas in the LF and VLF ranges with a motivation to solve the interference problem in long-range radio communication from lightning flashes (the 'sferics') [36].

During World War II, the narrow band VLF lightning detection system was utilized and consisted mainly of two or more magnetic direction finders with a lightning location accuracy of tens of kilometers. The gated wideband magnetic direction finder was developed in the 1970s to overcome this shortcoming of polarization errors in the narrow band magnetic direction finder. The time-of-arrival technique was developed in the 1930s, driven by the demands for marine navigation, and was introduced into lightning location methods in the late 1950s. Oetzel and Pierce first suggested the very short baseline TOA technique to be employed for line-of-sight location of lightning VHF sources in 1969. Three years later, [37] used one pair of antennas with operating frequency in the range 25–35 MHz and successfully verified the TOA direction finding technique. Two short-baseline VHF TOA systems have seen extensive use since the early 1970s, the 253 and 355 MHz systems developed in South Africa and the Lightning Detection and Ranging (LDAR) system (central frequency between 56 and 75 MHz) developed at the NASA Kennedy Space Center [38]. The first long-baseline TOA system, implemented by a pair of receiving stations separated by over 100 km in Massachusetts and operated at VLF and LF (bandwidth of 4–45 kHz), enabled Lewis et al. (1960) to compare time arrival differences for the two stations. Warwick et al. (1979) first designed an interferometer to detect lightning. An improved version of the interferometric system was later developed by the research group at the New Mexico Institute of Mining and Technology [39, 40].

More detailed description on the early history of lightning detection can be found in Uman and Rakov [13] and Cummins and Murphy [36]. The following subsection 2.5 will introduce some modern lightning detection systems together with their respective detection technique and operating frequency.

2.5 *Examples of Modern Lightning Detection Systems*

Substantial demand from the long-standing scientific interest and practical application has motivated the advancement of lightning detection systems. The lightning locating techniques and operating frequencies are illustrated in Fig. 2. The Lightning Mapping Array (LMA) developed by New Mexico Institute of Mining and Technology will be first introduced as it can give the most complete record of the spatial and temporal development of lightning channels. The typical LMA is also equally sensitive to both intracloud and cloud-to-ground lightning detection, and so stands apart from many of the other detection systems that will be described afterward. The globally-oriented Earth Networks Total Lightning Network is also designed for intracloud lightning detection. Many countries have developed their own lightning detection systems, such as the U.S. National Lightning Detection Network (NLDN), the Canadian Lightning Detection Network (CLDN), the Brazilian Lightning Detection Network (BrasilDAT), the LIghtning detection NETwork (LINET) and the European lightning location system EUCLID, the Italian LAMPINET (LAMPI for 'flash' in Italian, NET for network), the Spanish Lightning Detection Network (SLDN) and the Catalan Lightning Detection Network (XDDE), the lightning detection networks in

China, and the South African Lightning Detection Network (SALDN). Three additional lightning detection systems are aimed at the measurement of the global lightning, the World Wide Lightning Location Network (WWLLN), the Global Lightning Dataset (GLD360), and the Earth Networks Total Lightning Network (ENTLN). All these modern lightning detection systems will be briefly reviewed. In addition, single lightning flashes of mesoscale extent (>100 km) can singlehandedly excite the global Schumann resonance intensities 10–100 times the level of the "background" lightning activity, which can be located globally from a single receiving station equipped with sensors for vertical electric field and horizontal magnetic field. The geo-location method and characterization of these ELF transients is also introduced in brief.

(1) Lightning Mapping Array (LMA)

The deployable Lightning Mapping Array (LMA) was developed on the basis of the Lightning Detection and Ranging (LDAR) system used at the NASA Kennedy Space Center [38, 41]. The system uses six or more stations with time synchronized by GPS technology to independently measure the arrival time of impulsive VHF radiation (from both IC and CG flashes) and locate lightning by the VHF TOA technique [42]. Measurement stations are usually deployed over an area typically 60 km in diameter with each station separated by 15–20 km and connected via wireless communication links to a central site for processing. Each station monitors the peak intensity of VHF radiation in a 6 MHz bandwidth centered at 63 MHz (an unused television channel (channel 3)). VHF radiation (time and magnitude) is recorded in every 80–100 μs time interval. The peak signal times are measured with a high-time resolution (~50 ns) by a digitizer accurately phase locked to the 1 pulse-per-second output of a GPS receiver [43, 44].

The LMA has provided unprecedented details on the temporal and spatial evolution of lightning discharges by locating thousands of VHF sources per flash in three-dimensional space. Thomas et al. [44] has experimentally and theoretically investigated the location accuracy of the New Mexico Tech Lightning Mapping Array (LMA) by using balloon sounding measurements, airplane tracks, and observations of distant storms. They found that sources over the network are located with an uncertainty of 6–12 m rms in the horizontal and 20–30 m rms in the vertical. The resultant 3D location errors are less than 100 m for most VHF sources and the location uncertainties for sources outside the network increase with distance [44]. The LMA system is a powerful tool to study lightning and has now been installed in other countries, such as the Brazil, Canada, China, France, Japan, United States, Spain, and others.

(2) U.S. National Lightning Detection Network (NLDN)

The U.S. NLDN has been monitoring lightning in real-time since the early 1980s and has provided continental scale (U.S.) information since 1989 [45, 46]. Its origins lie in the gated wideband Magnetic Direction Finding (MDF, commercialized by Lightning Location and Protection, LLP in the late 1970s) and currently employs the combined TOA and MDF location methods with operating frequency range of 400 Hz–400 kHz. This system is known as the Improved Accuracy through Combined Technology (IMPACT) [36]. The U.S. NLDN now consists of more than 100 stations typically separated by less than 350 km and fully covers the contiguous United States. Both IC and CG lightning discharges can be separately identified and the peak return stroke currents are estimated from the measured fields. The detection efficiencies and location accuracy of the CG stroke and flash have been investigated by using GPS-synchronized video cameras in conjunction with broadband electric field and optical (light pulse) recordings in Southern Arizona, Oklahoma, and Texas [47], in the Central Great Plains [48], and by the ground-truth rocket-triggered lightning at the International Center for Lightning Research and Testing (ICLRT) in Florida [49, 50]. The stroke detection efficiency is estimated to be 76% (N = 3620) in Arizona, 85% (N = 885) in Texas/Oklahoma, 84% (N = 547) in the Central Great Plains, and 76% (N = 139) at ICLRT. And the corresponding flash detection efficiencies are 93% (N = 1097), 92% (N = 367), 91% (N = 342), and 92% (N = 37), respectively. More information about the NLDN, such as the evolution of the NLDN, its enabling methodology, and applications of NLDN data, can be found in Rakov and Uman ([13], Ch. 17), Orville [45], Orville and Huffines [46], Orville et al. [51, 52], Cummins and Murphy [36], Holle [53], and references therein.

(3) Canadian Lightning Detection Network (CLDN)

The Canadian Lightning Detection Network (CLDN), designed in 1997 and managed by Environment Canada since 1998, consists of more than 80 sensors (mainly Vaisala sensors including IMPACT- ES, LPATS-IV, LS7000, LS7001, and LS7002) and detects lightning over most of Canada to approximately 65° N in the far west, 55° N in the far east, and offshore to about 300 km [54, 55]. CLDN sensors utilize both VLF and the LF band to detect cloud-to-ground lightning and a small percentage of cloud-to-cloud lightning. CLDN determines the occurrence time, intensity, and polarity of lightning from electromagnetic (EM) pulses that lightning produces. The EM pulse information measured from each sensor is sent to the network control center in Tucson, Arizona, together with the sensor information to determine the location and other parameters of lightning [56]. The lightning information is transmitted to various clients and Environment Canada's Storm Prediction Centers. The cloud-to-ground flash detection efficiency of the CLDN is better than 90% and less than 500 m in location accuracy for a peak current threshold of 5 kA over its region of coverage [55, 56]. The CLDN and NLDN comprise the North American Lightning Detection Network (NALDN). More detailed information about the CLDN and its evaluation can be found in Burrows et al. [55], Dockendorff and Spring [56], Abreu et al. [54], Shostak et al. [57], Kazazi et al. [58], and references therein.

(4) Brazilian Lightning Detection Network (BrasilDAT)

The first systematic observations of CG lightning in Brazil were enacted in the 1960s based on the number of thunderstorm days at different sites [59]. The first lightning location system in southeast Brazil was a small regional network installed in 1988 and consisted of four Lightning Positioning and Tracking System (LPATS) sensors using the TOA technology [60, 61]. It was later upgraded to include IMPACT sensors in 1996 and named the Brazilian Integrated Lightning Detection Network (RINDAT). In 1999, another lightning location system was installed in northern Brazil to provide ground truth data for the Lightning Imaging Sensor (LIS) in space [59]. BrasilDAT is the integrated result of two main lightning location systems and other regional networks. It has 47 stations at present including LPATS and IMPACT sensors [59]. The CG stroke detection efficiency is investigated by rocket-triggered lightning in the Southeast region of Brazil and with high-speed video camera observations. The detection efficiency is estimated to be about 55% for strokes and 87% for flashes [59, 62]. Additional information about BrasilDAT can be found in Pinto [61, 63] Pinto et al. [64–66] and references therein.

(5) LIghtning Detection NETwork (LINET)

The LINET was developed at the University of Munich beginning in 1994 and operates in the VLF/LF range [67, 68]. It has steadily expanded and now consists of more than 130 sensors (as of March 2014) across 17 countries, covering an area extending from 10°W to 35°E in longitude to 30°N to 65°N in latitude [69]. The LINET receiving station is a simple 4-part modular construction designed for easy-to- handle and economical-to-manufacture or update. Two crossed loops (without any active electronics) make up Module 1 to provide passive sensors for measuring magnetic field components in the frequency range of 1 to 200 kHz. Module 2 supplies the timing signal by using a GPS clock with an accuracy of <100 ns. Module 3 is a single plug-in device consisting of the signal amplifier, filter, and A-to-D-converter. Module 4 is a separately positioned processing unit. The incoming signal is sampled at a rate of 1 MHz and recorded with 14-bit resolution in a continuous mode. The pre-trigger time is 100 μs and standard time window length is 512 μs to enable inspection for occurrence of one pulse. A fast Fourier analysis and time coincidence considerations among sensors are performed to discriminate signals and eliminate noise. IC-CG classification is achieved by a specially adopted 3D-algorithm based on height (instead of the waveform differences) in the central processing unit, but the waveforms and other data are stored locally [70]. Each sensor collects data and transmits packets of condensed information to the central station at Munich. Real-time lightning location can be performed. The statistical average lightning location accuracy is ~150 m verified by strikes to towers. More details about LINET can be found in Betz et al. [6, 67, 68].

(6) European Lightning Location System (EUCLID)

The European lightning location system EUCLID was initiated by several countries (Austria, France, Germany, Italy, Norway, and Slovenia) in 2001 and then expanded to the European-wide region (Schulz et al. 2016a). The EUCLID network is now a consortium of 19 national lightning detection networks and in 2014 has 149 sensors manufactured by Vaisala Inc., 7 LPATS, 10 IMPACT, 31 IM- PACT ES/ESP and 101 LS700x sensors, and operated in the same frequency range with individually calibrated sensor gains and sensitivities. The data are processed locally by each national lightning location system (LLS) [71, 72]. Additionally, the total data from all 149 sensors are handled by a central processor in Austria at ALDIS (Austrian Lightning Detection and Information System) in real-time. A full backup EUCLID processing center is set in Germany with independent and direct data connections to all sensors. Based on the ground truth data from direct lightning current measurements at the Gaisberg Tower (GBT), the detection efficiency is 96% (93% for validation by Säntis Tower, [73] and 70% for negative flashes and strokes. The median location accuracy is 89 m for the 100 strokes recorded at the GBT. The detection efficiency is 98% for negative flashes and 84% for strokes based on video and E-field recordings [74]. More details about EUCLID can be found in Pohjola and Mäkelä [75], Schulz et al. [72], Azadifar et al. [71], and Poleman et al. [76].

(7) Lightning Detection Network in Italy (LAMPINET)

The Italian Air Force Meteorological Service built the Italian lightning detection network-LAMPINET and started operation during 2004 [79] for the comprehensive detection of atmospheric discharges. LAMPINET consists of 15 IMPACT ESP[3] sensors quasi-uniformly distributed over Italy and utilizes both MDF and TOA techniques [77]. The sensors monitor lightning electromagnetic field signatures and waveforms with a frequency bandwidth 1–350 kHz by means of a parallel plate capacitor for the electric field and crossed loops for the magnetic field. A GPS clock, signal analyzer, and electronics for telecommunications link each sensor to the central processing server located in the national weather center of the Italian Air Force Meteorological Service (Centro Nazionale di Meteorologia e Climatologia, C.N.M.C.A.). CG and IC classification are based on waveform criteria. C.N.M.C.A. periodically conducts data comparisons with other systems that share the same operational area with LAMPINET [78]. The LAMPINET has a detection efficiency of 90% for normalized currents higher than 50 kA and a location accuracy of 500 m in the central areas of Italy (claimed by the sensor manufacturer [78]. Larger errors are evident in the border areas of Italy and beyond the network. More detailed information about LAMPINET can be found in Biron [77], Biron et al. [78, 79].

(8) Spanish Lightning Detection Network (SLDN) and Catalan Lightning Detection Network (XDDE)

The first lightning measurement in Spain appeared in about 1904 and a keraunograph was installed at the Observatori de l'Ebre [80]. In 1992, the modern Spanish Lightning Detection Network (SLDN) was built, consisting of 14 sensors and making use of joint MDF and TOA techniques in the LF radio frequency range (similar to NLDN sensors) [81]. Now the SLND has more than 30 sensors with several expansions and covers the Iberian Peninsula, Canary Islands, and Balearic Islands [82]. The Catalan Lightning Detection Network (XDDE) was set up to detect both IC and CG discharges with 2-D location of VHF sources and initially consisted of three VHF interferometers (108–116 MHz) and covered the area of Catalonia in Northeast Spain. The experimental evaluation of the XDDE was carried out with two field measurement campaigns by means of electrostatic and electromagnetic field measurements, and digital video recordings. Its CG flash detection efficiency was found to lie between 86 and 92%. The XDDE can supply valuable data to evaluate the detection efficiency, stroke discrimination and location accuracy of the SLDN [83]. More detailed information about the SLDN and XDDE can be found in Pineda et al. [81, 84].

(9) Lighting Detection Network in China (BLNet and GHMLLS)

There are several 3D positioning systems for lightning radiation pulses (based on the TOA or interferometry techniques) and long baseline lightning location networks operating in China. The first 3D VHF lightning radiation source mapping technique installed in China operates at 270 MHz with a 3-dB bandwidth of 6 MHz and processes peak events in a consecutive time window of 50 μs [85, 86]. A broadband electric field location system is synchronously running with an operational bandwidth from 1.5 kHz to 10 MHz. This system has a horizontal error of 12–48 m and a vertical uncertainty of 20–78 m for radiation sources validated by a balloon-borne VHF transmitter [87]. A multiband 3D lightning location network installed in Beijing, the Beijing Lighting Network (BLNet), is deployed with one data center and sixteen substations utilizing the fast antenna, slow antenna, magnetic antenna, and VHF antenna to cover a wide bandwidth from VLF to VHF [88]. The average detection efficiency of the BLNet is 93.2% for total flashes. The location error in the horizontal direction is 52–250 m based on lightning flashes to tall towers [89]. Another lightning location network in Chongqing Province is composed of 14 lightning sensors to detect VLF/LF and VHF sources radiated by lightning, and was introduced in Liu et al. [90]. The performance of lightning location systems in Guangdong Province can be found in Chen et al. [91]. The information of the Guangdong-Hong Kong-Macao Lightning Location System (GHMLLS) is detailed in Zhang et al. [92]. Additionally, two lightning observatories of the Guangdong Comprehensive Observing Experiment on Lightning Discharge (GCOELD) and the Tall-Object Lightning Observatory in Guangzhou (TOLOG) contribute comprehensive observational data to understand lightning physical processes [92–94]. More detailed information about the lightning location system and lightning observatory in China can be found in Qie et al. [95], Shi et al. [96] and Zhang et al. [92].

(10) South African Lightning Detection Network (SALDN)

The South African Lightning Detection Network (SALDN) consists of 24 Vaisala LS7000 sensors [97]. The network was originally installed in 2005 and consisted of 19 sensors. In late 2009-early 2010, three more sensors were added. A second upgrade was performed in mid-year 2011 in which two more sensors were added to the network and three of them relocated for better coverage [98]. A new flash density map was created utilizing data from the SALDN from 2006–2011 [97]. Here, ground-flash densities of 15–20 flashes/km^2/year were found, indicating that the previous map was underestimating the lightning incidence in the country. The network has been evaluated against photographs of tall tower lightning events and high-speed video lightning observations. These studies show that the detection efficiency of the network has greatly improved with the addition of sensors and detects 92% of the observed downward flashes in Johannesburg, South Africa with a median location accuracy of less than 100 m [99–101].

(11) World Wide Lightning Location Network (WWLLN)

R. Dowden and is now operated by the University of Washington. This network now consists of more than 70 sensors around the globe (maintained by different participating institutions). This system utilizes a time-of-group-arrival (TOGA) method based on the fact that lightning VLF signals propagating in the Earth-ionosphere waveguide experience dispersion, in that the higher-frequency components arrive earlier than the lower-frequency components. The sensors operate in the VLF range (3–30 kHz) and measure the waveforms from sferics to calculate the TOGA [102]. Each lightning stroke location requires the TOGA from at least five WWLLN sensors. The timing accuracy is 100 ns, maintained by GPS receivers, and the uncertainty in the stroke timing is less than or equal to 30 us. Lightning location are characterized by an accuracy of about 5 km [103, 104]. WWLLN can detect the majority of all lightning-producing storms, even in regions with inter-station distances larger than 2000 km [105, 106]. The NLDN data have been used as the ground truth to investigate the CG flash detection efficiency of the WWLLN. The DE has increased from ~3.88% in 2006–2007 to 10.3% in 2008–2009, as the number of sensors increased from 28 in 2006 to 38 in 2009 [103]. For events with NLDN-reported peak currents ≥130 kA, the detection efficiency is 35% [107, 108]. Recent research indicates a detection efficiency for strokes with peak current greater than 30 kA is approximately 30% globally. More detailed information about the WWLLN can be found in http://wwlln.net and in Lay et al. [109].

(12) Global Lightning Dataset (GLD360)

The Global Lightning Dataset (GLD360) utilizes both TOA and MDF methods in conjunction with a lightning waveform recognition algorithm [110]. The sensors are strategically placed around the world but the exact total number of sensors is proprietary information and is unspecified. Each sensor stores a local empirical waveform bank (derived by using a VLF receiver and known lightning location data from NLDN), which catalogs the expected sferic waveform shape, each indexed by distance and ionospheric profile. The most reliable repetitive features (either the rising portion of the ground wave or the zero-crossing of the first or second ionospheric reflection) are used to establish the precise arrival time of the sferic at the receiver. Measured information is sent back to a central processor. All arrival time data are aggregated to make a determination of the event's time and location using an optimization routine that minimizes the root mean squared error from all time and azimuth measurements. The polarity of each stroke is inferred via the cross correlation with the waveform bank. The stroke peak current is deduced from the measured magnetic field, which is first corrected by the source-receiver distance using a propagation model [110]. A lightning event must be simultaneously detected by at least three sensors to be geolocated, though most are detected by more than three. Thunderstorm detection efficiency for the GLD360 is better than 99%, and event timing precision is 1 microsecond RMS. The median location accuracy is 1.5–2.5 km. The present flash detection efficiency is greater than 80% in most areas of the Northern Hemisphere and between 10 and 80% in the Southern Hemisphere, with the lower efficiencies at latitudes south of 42 S. More detailed information about GLD360 can be found in Said et al. [110, 111].

(13) Earth Networks Total Lightning Network (ENTLN)

The Earth Networks Total Lightning Network (ENTLN) is a unique total lightning detection system specifically targeting the signals emitted from both IC and CG flashes based on over 1700 wideband sensors with frequency reception ranging from 1 Hz to ~12 MHz and now deployed in more than 100 countries worldwide [112]. This system utilizes the TOA technique and extends the frequency range of the sensors into the MF and HF frequency domains, aiming to detect weaker pulses at longer distances than other VLF/LF systems with similar baselines. The sensors record whole waveforms for each flash. The central server employs sophisticated digital signal processing technologies to extract the location, polarity, amplitude, and other stroke parameters from the rich signal measurement. The precise arrival times are calculated by correlating the waveforms from all sensors that detect the strokes of a flash. The type of discharge and polarity are determined by the polarity in the initial half cycle for bipolar pulses and by the waveform in the measured electric field pulse [113]. Strokes (or individual K-change cloud events) are clustered into a flash if they are within 700 ms in time and within 10 km in space of the first detected stroke (or cloud event). The ENTLN processors show a detection efficiency (DE) higher than 99% and 96% for flash and stroke detection, respectively, evaluated using ground truth from natural and rocket-triggered lightning experiments in Florida [114]. More detailed information about the ENTLN can be found in Heckman [115], Liu and Heckman [112], Stock et al. [116], Marchand et al. [117].

(14) Geo-Location and Characterization of Q-burst Transients at ELF Frequencies

At ELF frequencies, the attenuation of the electromagnetic wave propagation in the Earth-ionosphere cavity is sufficiently small (~0.2 dB/Mm) that global resonances are possible, and are now identified as "Schumann resonances", after their predictor W.O. Schumann [118]. Single lightning flashes of mesoscale extent (>100 km), and often in the form of positive ground flashes, can singlehandedly excite the global Schumann resonance intensities 10–100 times the level of the "background" lightning activity, for periods of one hundred milliseconds or more. Ogawa et al. [119] named these ELF events Q-bursts, because they are relatively Quiet at higher frequencies. These extraordinary flashes can be located globally from a single receiving station equipped with sensors for vertical electric field and horizontal magnetic field.

Theoretical frequency domain representations of the electric E and magnetic H fields throughout a uniform Earth-ionosphere cavity, and produced by a single lightning flash with vertical current moment IdS (coul-km/sec) are given by Eq. (3)

$$
\begin{aligned}
\mathrm{E}(\omega, \theta) &= \frac{\mathrm{Ids\,v(v+1)}}{4\mathrm{R}^2\varepsilon\omega\mathrm{h}} \frac{p_\nu^0(-\cos\theta)}{\sin(\pi\nu)} \\
\mathrm{H}(\omega, \theta) &= \frac{-\mathrm{Ids}}{4\mathrm{Rh}} \frac{p_\nu^1(-\cos\theta)}{\sin(\pi\nu)}
\end{aligned}
\tag{3}
$$

where ω is the angular frequency, R is the radius of the Earth, h is the height of the waveguide, ε is the permittivity of free space, *p* is the Legendre function, ν is the complex eigenvalue defining the ionosphere, and θ is the source-receiver angular separation.

So long as the measurements are confined to the lower ELF frequency band from 3 Hz to the waveguide cutoff (1600 Hz at nighttime), only a single TEM waveguide mode is present.

One can eliminate the (unknown) current moment source term IdS by dividing the expression for the E field by the like expression for the H field, to form the wave impedance Z [120], as shown in Eq. (4)

$$Z(\omega, \theta) = E(\omega, \theta)/H(\omega, \theta) = \frac{(\nu + 1)\, P_{\nu}^{0}(-\cos\theta)}{R\varepsilon o\omega P_{\nu}^{1}(-\cos\theta)} \tag{4}$$

The wave impedance then becomes a meaningful indicator of the source-receiver distance θ, the only other unknown quantity in Eq. (3). Once θ is known, one can recover the current moment IdS from either of Eq. (3). The direction to the source from the receiving station is determined by comparison of two perpendicular calibrated magnetic field antennas, aligned with geographical EW and NS axes. A Poynting vector measurement ($S = E \times H$) can be used to resolve the directional ambiguity along the great circle path inferred from the magnetic direction finding. Given the distance estimate and this great circle bearing, the geographical location of the Q-burst source is uniquely obtained. In this fashion, global lightning maps can be constructed from single station observations [121, 122]. No great locational accuracy can be claimed for this procedure since the relevant EM wavelengths are comparable to the size of the Earth, and the day-night asymmetry of the waveguide (ignored in this application) causes degradation of the linear polarization of the magnetic field that is used for direction finding. Uncertainties in location are of the order of a few hundred km.

The majority of lightning flashes worldwide, and especially the delicate intracloud flashes, do not stand out strongly against the background Schumann resonances, and so cannot be located by the method described above. However, geophysical inversion methods working on the background Schumann resonance spectra recorded at multiple locations [123], [124] can be used to assign an integrated lightning "activity" (in units of $coul^2\ km^2/sec$) for regional "chimney" zones like South America, Africa and the Maritime Continent. This lightning source is expected to have contributions from all lightning flashes, because they all are expected to show some vertical component of charge transfer, given that charge separation in thunderstorms is gravity-driven.

3 Lightning Photography

3.1 Introduction

While detection of lightning events through the propagation of radiated electromagnetic fields is of great value, photographic and video observations also provides us with valuable knowledge about physical lightning processes. In fact, with the advent of modern high-speed video technology, such video studies have become integral to analyzing the structural characteristics and development in time of the lightning process, with thousands of frames per second *(fps)* allowing for the lightning process to be viewed in microseconds. Such observations allow us to clearly view cloud-to-ground lightning events, whether they initiate in the cloud or begin as upward propagating leaders and it is even possible to infer the polarity from such videos. Additionally, when coupled with GPS timestamps, high-speed video studies provide invaluable ground-truth data for comparison with lightning detection and location systems, allowing for detection efficiency and location accuracy to be evaluated.

Using photography to characterize lightning has in fact been attempted since 1926 [125]. Sir Basil Schonland captured images of a lightning leader progression using a two lens streak camera known as a Boys camera (Scholand 1934, [126]). Since then, other studies involving photographing or filming lightning have been conducted involving some initial high-speed technology as well as lower frame rate video recordings [127–130]. Some of the first modern high-speed studies were done by Moreau et al. [131] observing lightning at 200 frames per second. Since then, Ballarotti, Saba and Warner have pioneered high-speed studies of lightning, characterizing both downward and upward lightning flashes, as well as both positive and negative polarities [132–140]. This has led to high-speed studies being conducted around the world [141–145].

3.2 Downward Flashes

Downward flashes are lightning flashes where the leader originates in the cloud and propagates towards the ground. Propagation characteristics are based on the polarity of the flash. Figure 4 shows a frame sequence of a negative downward flash. Negative flashes will appear as branched (or forked) leaders, that continuously branches until one of these branches connects to the ground (Fig. 4a–c). Negative leaders are brighter on the tips than the channel formed during their propagation. The propagation of positive flashes as captured by the cameras will depend on the frame rate used to record. Positive flashes recorded at lower frame rates appear as continuous leaders with no branching; at higher frame rates, repetitive pulses ahead of the leader tip can be seen. Some of these pulses connect and extend the leader propagation towards the ground, and are referred to as recoil leaders. The propagation leader for both positive and negative flashes increases as the leader approaches the ground. From

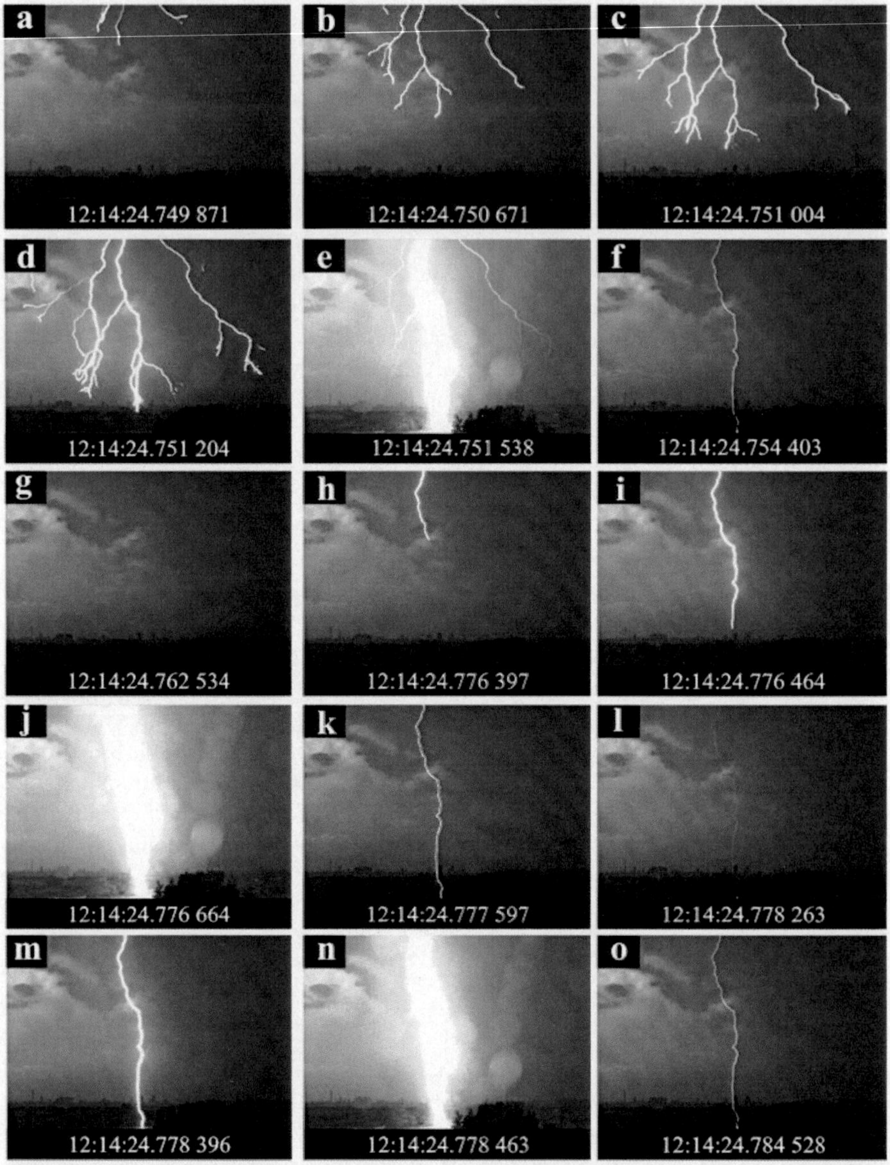

Fig. 4 Photo sequence of a downward flash. Images (**a**)–(**d**) show the downward leader approaching the ground; Image (**e**) shows the return stroke; Image (**f**) shows the continuing current after the first stroke; Image (**g**) shows a period of no current prior to the downward leader (**h**)–(**i**) that uses the same path to reach the ground and is called the subsequent return stroke (**j**); Image (**k**) shows the presence of continuing current in the subsequent return stroke and images (**l**) and (**m**) show the current variation and (**n**) the sudden intense current variation known as the M-component

2D analysis of videos, negative flashes appear to have an average leader speed of 3.30×10^5 m s^{-1} and positive flashes have an average leader speed of 2.76×10^5 m s^{-1}.

When the leader arrives close to the ground, upward connecting leaders may appear from the potential contact points (Fig. 4d). Only one of these upward connecting leaders will in fact connect and the others will collapse back to ground. These upward connecting leaders are seen to have an average 2D propagation speed from 0.27×10^5 to 2.9×10^5 m/s before connecting to the downward leader. This connection is known as the return stroke, when thousands of Coulombs are transferred to the ground in a few microseconds. From the camera view, a few very bright frames appear and a fully formed channel is shown in the posterior frames (Fig. 4e and f).

Kitagawa et al. [23], determined that some flashes were able to keep connected to the ground for a longer period than the return stroke. This is seen as a permanence of luminosity in the channel for longer periods and is called continuing current. The continuous luminosity in the photography was correlated with the current through the channel. When there is luminosity, there is charge being transferred between the cloud and the ground. The duration of continuing current can be separated into long (longer than 40 ms), short (between 10 and 40 ms) and very short (between 3 to 10 ms) [23, 60, 146]. The continuing current can be as long as hundreds of milliseconds.

For negative flashes registered in Brazil, 55% of the return strokes were followed by some continuing current [133]. For positive flashes, 97% of the strokes were followed by continuing current (long, short or very short continuing current). To be more specific, 68% of these cases had continuing currents longer than 40 ms [137]. On this continuing current, current pulses may be superimposed and are called M-components (Fig. 4m–o). These superimposed pulses may be detected by lightning location systems.

In addition to the first connection, followed by a continuous current or not, a flash may present a time with no-current followed by a second downward leader using the channel (fully or partially, or even creating a new path) to transfer more charge (called as subsequent return stroke—Fig. 4h–j). Negative flashes commonly have multiple return strokes—on average 3–4 times, and a maximum of 26 subsequent return strokes has been recorded. The majority of positive flashes (84%) only have one and only the return stroke.

The total duration of a flash is considered from the beginning of the first return stroke until the end of the last return stroke (followed by continuous current or not). For negative flashes the median total duration was 163 ms [133]. For positive flashes, the median duration was 125 ms [137]. Negative flashes have longer total duration due to the multiple return strokes, while positive strokes mostly have only one return stroke with long continuous current (Fig. 5).

As the luminosity in the channel is associated with the current flowing between cloud and ground, the luminosity schematics for downward flashes (Fig. 5) represent the different lightning processes and how the channel current varies.

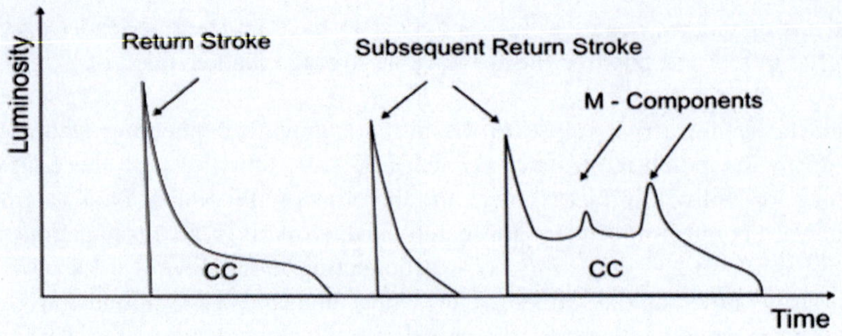

Fig. 5 Luminosity schematics for downward flashes

3.3 Upward Flashes

Upward flashes are flashes that initiate from tall structures and propagate toward the cloud. These flashes may be self-initiated or triggered by nearby lightning activity [145]. The specific component of nearby activities have been determined by recent studies. An intracloud flash, return stroke from a downward flash nearby or a triggering leader over the tower (T-leader) may trigger an upward leader from the tower if it causes enough electric field vriation [136, 147]. Figure 6 shows the conditions on the electric field variation to the occurrence of an upward leader from tower tip: self-initiated or triggered by a nearby activity.

When the electric field was reached, an upward leader will initiate from the tower tip towards the cloud base if the electric field variation is enough to pass the critical value of the electric field (E_c). This critical electric field value will be a function of the meteorological conditions (wind, temperature, etc.) and tall structure characteristics (effective height, material, etc.).

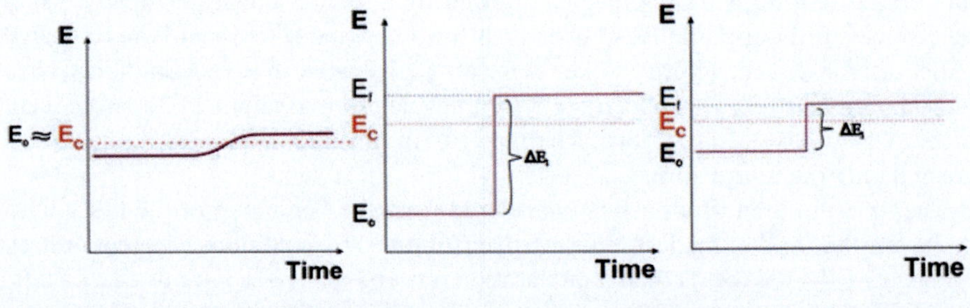

Fig. 6 **a** ***Upward leader self-initiated***: conditions of the initial electric field during thunderstorm are close enough to the electric field critical (Ec) that a slow variation will cause an upward leader from the tall structure. (b1 and b2) ***Upward leader triggered by nearby activity***: initial electric field is not enough to cause an upward leader from the tower tip, but due to a fast variation in the electric field nearby (eg. intracloud, return stroke, T-leader) the electric field level surpasses the critical value (Ec). (based on Schumann [136])

For upward flashes, polarities are labelled based on the direction of charge transfer to the ground. Therefore, positive upward flashes will have a negative upward leader propagating towards the cloud base (Fig. 9). Similarly, negative upward flashes will have a positive upward leader propagating toward the cloud base. The propagation leader characteristics will depend on the polarity. Negative upward leaders (Fig. 8) will propagate up with branching as the negative downward flashes do and the positive upward leaders will propagate continuously and may exhibit recoil leaders (repetitive pulses ahead of the leader tip as described in the positive downward flashes leader propagation). These leaders will complete a path, also called a channel (multiple channels in some cases—Fig. 8c), from the tower tip to the cloud, transferring charge. When this process starts with an upward leader and transfers charge between the cloud and the tower, it is called Initial Continuous Current (ICC)—Fig. 8a–k. This process is not fast enough to be detected by most of the lightning detection systems discussed in Sect. 2, although during this transfer some impulsive (ICC Pulses) events may occur and these events can be detected [148]. As shown in the Fig. 8, ICC may have multiple ICC pulses, the first one in Fig. 8 is the sequence (d–f) and the second sequence of ICC pulses (g–i). Even though the connection used different channels, the tower was receiving charge from at least one channel. The duration of the initial continuous current is hundreds of milliseconds and ICC pulses are present in 50% of upward flashes.

Figure 7 shows the luminosity schematic for an upward flash. Sudden events appear brighter in the videos and are more likely to be detected by lightning location network [100]

The majority of the upward flashes consist only of initial continuous current. But when the ICC process finish, a period with no current in the channel can be observed. After a while, a downward leader may use the path to transfer charge. This process is similar to that of subsequent return strokes in the downward flashes (Fig. 8l–o). An upward flash may have multiple return strokes using the same channel of the initial continuous current. Subsequent return strokes are present in 25% of upward flashes [135]. These subsequent return strokes may have M-components (Fig. 8q).

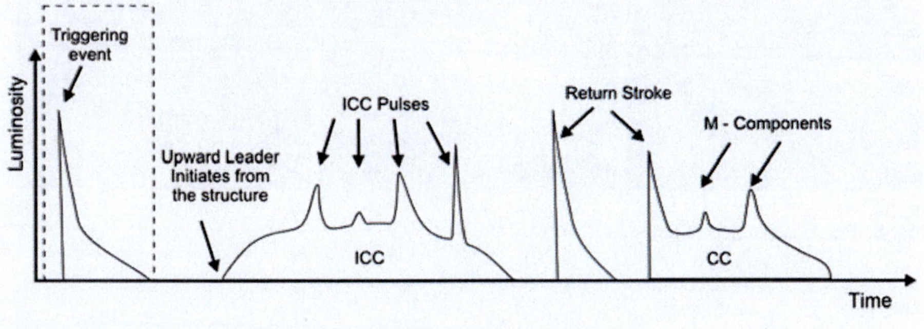

Fig. 7 Luminosity schematics for upward flashes

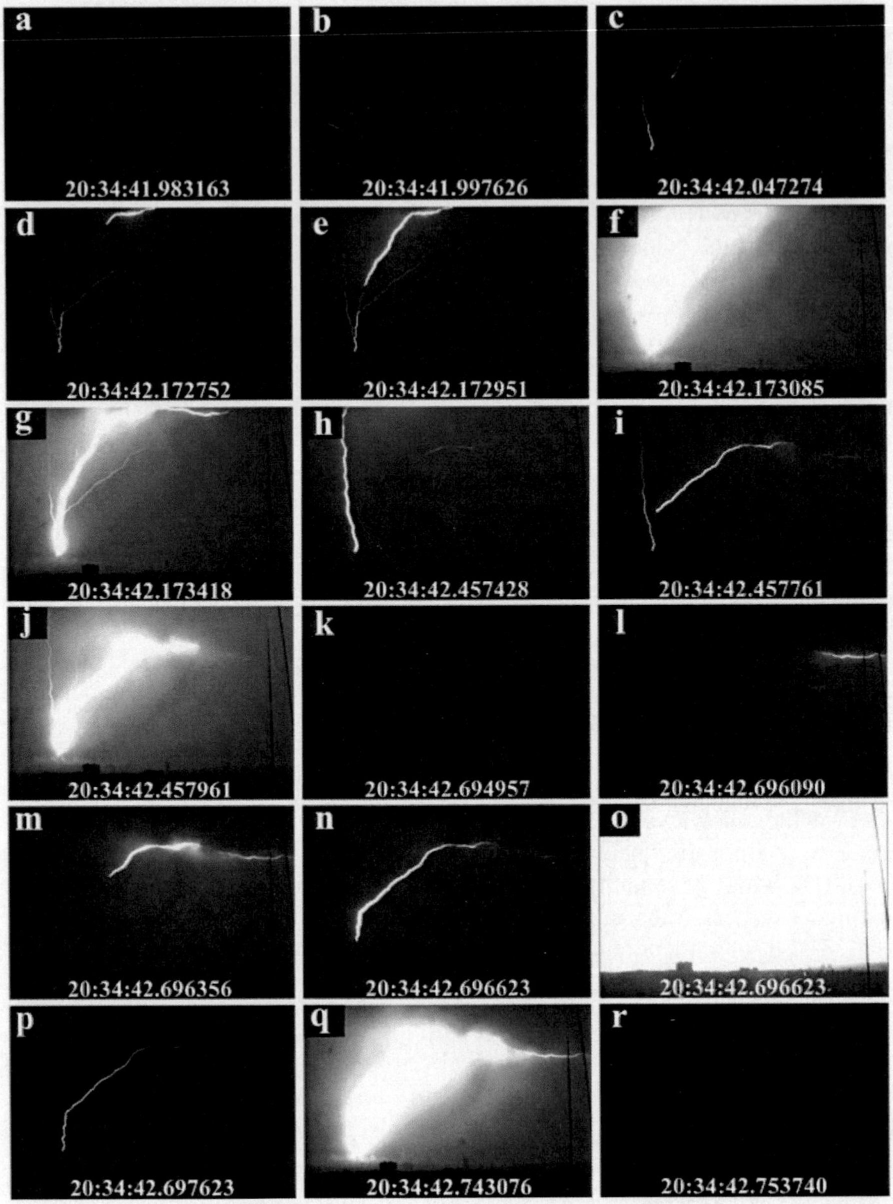

Fig. 8 Photo sequence of a negative upward flash

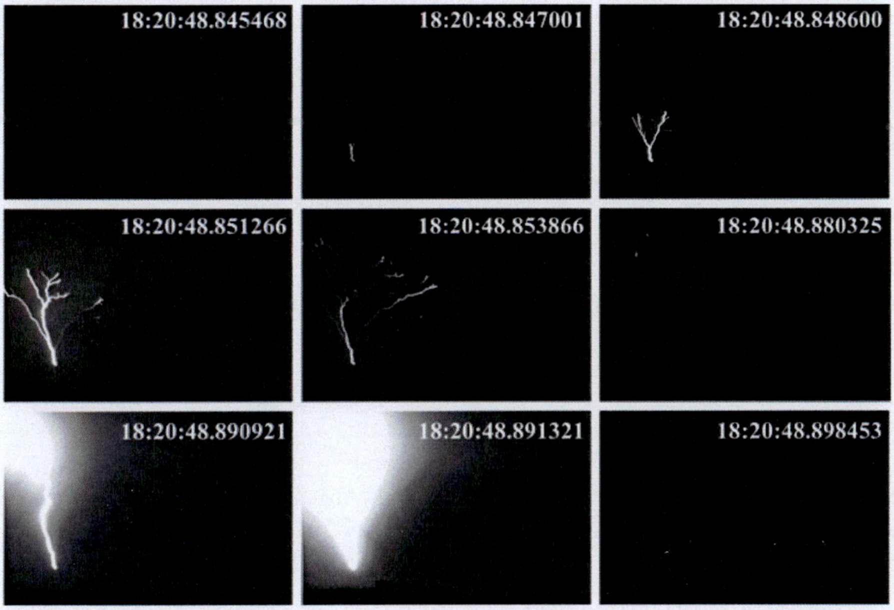

Fig. 9 Negative upward leader

The total average duration of upward flashes is 427 ms (with a maximum of 1143 ms) and is considered from the beginning of the initial continuous current to the end of the continuous current of the last subsequent return stroke [135, 147].

3.4 *Comparison with Lightning Detection Networks*

As discussed in Sect. 2, the detection of ground lightning flashes through radiated electromagnetic fields relies on multiple sensors detecting the event and then TOA and MDF methods to geolocate the event. As such, it is possible that events may be inaccurately located or not detected at all. Understanding how often this occurs given a detection network is important to understanding the quality of data coming from the network and is referred to as the performance of the network. The performance of a network is usually determined from two parameters: detection efficiency—the number of true lightning events versus the number of detected lightning events and location accuracy—the median error in meters from where true lightning events attached to where the events were reported to have attached.

In order to determine values for such performance criteria, a detection network needs to be compared with ground-truth lightning events. Most notably, rocket-triggered lightning studies in Florida have provided ground-truth cases with a known location to evaluate both detection efficiency and location accuracy [50]. These two studies showed how the U.S. National Lightning Detection Network improved the flash detection efficiency from 84 to 92% the median location accuracy from 600 to 308 m from 2003 to 2009. Other evaluations have been done with instrumented tall towers—the European Cooperation for Lightning Detection (EUCLID) as regularly evaluated against current measurements made at the Gaisberg tower [74, 149]. High-speed videos with GPS time-stamping provide invaluable ground-truth data for such evaluations as well. Studies in Rapid City, USA by Warner et al. in Brazil by Saba et al., in Austria by Schwaltz et al. and by Fensham et al. in South Africa compare lightning detection network reports with high-speed observations of the same lightning events [99, 100, 134, 135, 138, 139, 148, 150].

3.5 *Detection Efficiency*

A high-speed video study in Johannesburg, South Africa yielded 206 filmed flashes in 24 thunderstorms from February 2017 to February 2018 [100, 147]. This included downward and upward flashes of both polarities. From the high-speed footage, 667 strokes could be seen in the 206 flashes. Each of these flashes was GPS time-stamped meaning the exact timing of every event (stroke, M-component etc.) in each flash is known. The South African Lightning Detection Network (SALDN)—a lightning detection network covering the entirety of South Africa, consisting of 26 Vaisala LS7000 sensors [97]—was then queried for the times of the filmed flashes to correlate the SALDN stroke reports with the ground-truth high-speed video records.

This is shown in Table 1, where can see that 175 of the 206 filmed flashes could time-correlated with SALDN reports (or 457 stroke reports could be time-correlated with the 667 filmed strokes). In other words, the SALDN detected 85% of the filmed flashes and 69% of the filmed strokes. This is to be expected, as a flash only needs

Table 1 Number of SALDN reports correlated with high-speed lightning videos filmed in Johannesburg, South Africa from February 2017 to February 2018 [100, 147]

Lightning events	High speed videos	SALDN detections	Detection efficiency %
Flashes	206	175	85
Strokes	667	457	69
Downward flashes	163	151	93
• Strokes	604	417	69
• M-components	101	12	12
Upward flashes	43	24	56
• Upward leaders	55	0	0
• ICC pulses	387	57	15
• SRSs	63	40	63
• M-components	10	3	30

one stroke to be detected and can be seen in the results presented by Nag et al., where the U.S.NLDN had a flash detection efficiency of 92% but a stroke detection efficiency of 76% compared with rocket-triggered lightning [50]. However, given the detail that can be ascertained about lightning events from high-speed footage, such performance values can be interrogated further. Firstly, the number of downward and upward events can be distinguished—it is often observed that lightning detection networks do not detect upward events as well as downward events due to the slow rise of the initial continuing current. Furthermore, the high-speed video allows us to distinguish M-components from return strokes and subsequent return strokes as well as identify upward leaders and ICC pulses [139, 140, 148].

Table 1 further breaks done the 206 flashes into 163 downward flashes and 43 upward flashes. This distinction already yields interesting results when compared with the SALDN reports with 93 and 56% of the upward flashes being detected. As expected, the detection of upward events is where most of the missed events occurred. The upward events are further described and fractionated as in Fig. 7. Here we once again confirm what was expected, that none of the 55 upward leaders (or initial continuing current) resulted in a detection by the SALDN. However, some of these upward events were detected due to ICC pulses. In other cases, subsequent return strokes occurring after the initial continuing current were detected by the SALDN meaning the upward flash was detected.

Table 2 shows a breakdown of the 43 upward flashes that were filmed. Of these flashes, 16 were followed by subsequent return strokes and 27 were subsequent return strokes (only initial continuing current). 94% of the upward flashes followed by subsequent return strokes were detected whereas only 33% of the upward flashes without subsequent return strokes were detected (due to ICC pulses instead). This confirms the observation that lightning detection networks are able to detect most lightning events, but the slow rise of the initial continuing current leads to non-detected upward flashes.

It is apparent that the detection of subsequent return strokes is key to the ability of a lightning detection network to detect lightning flashes—particularly upward flashes. However, in the case of downward flashes, is it more likely that the first return stroke will be detected than subsequent strokes? And will the multiplicity of a flash mean a higher chance of detection? High-speed video can once again assist in investigating these questions as all first and subsequent return strokes are observable and time-stamped. Figure 10 shows the 163 downward flashes (or 604 strokes) from the study conducted in Johannesburg, South Africa. These are further categorized

Table 2 Upward flash detection efficiency [100, 147]

Lightning event	High speed videos	SALDN detections	Detection efficiency %
Upward flashes	43	24	56
• With SRS	16	15	94
• Without SRS (ICC Only)	27	9	33

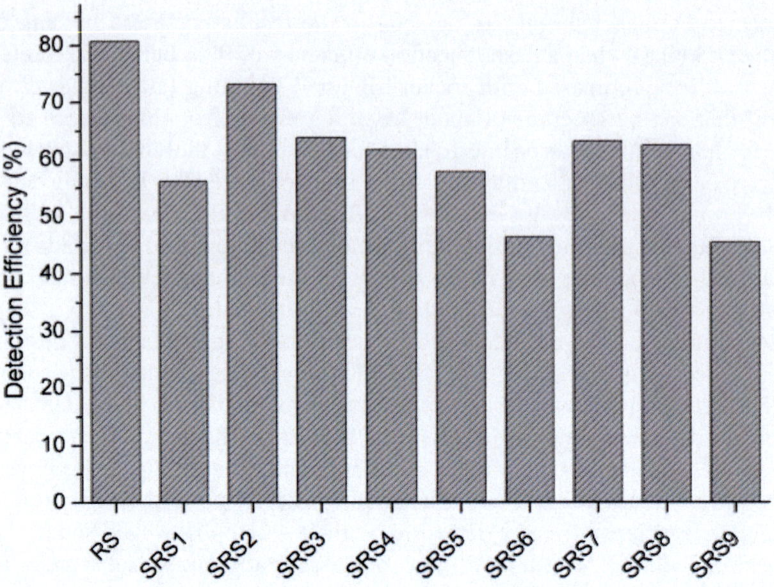

Fig. 10 Detection efficiency of subsequent return strokes. (Adapted from [100])

by subsequent return stroke number—first return stroke (RS), first subsequent return stroke (SRS1), second subsequent return stroke (SRS2) etc. The number of each SRS detected by the SALDN is determined and the stroke detection efficiency is then plotted against the subsequent return stroke order.

It is clear that the first return stroke is detected more often than any of the subsequent return strokes with a detection efficiency of 80%. However, this is not the 93% detection efficiency seen in Table 1 for downward flashes meaning that a number of the downward flash detections are due to subsequent strokes. Interestingly, it appears the first subsequent return stroke (SRS1) is detected less often than the second subsequent return stroke (SRS2).

3.6 Location Accuracy

Ground-truth evidence of lightning attaching to a known location is needed in order to evaluate the location accuracy of a lightning detection network. While high-speed footage of lightning provides excellent ground-truth cases for detection comparisons, it is not always clear where filmed lightning events attached. However, in some cases, particularly events to tall towers, it is clear to see where the lightning channel attached.

The upward lightning flashes filmed in the study by Schumann et al. all initiated from two tall towers in the Johannesburg city center—the Sentech tower and the Hillbrow tower [99, 100, 147]. These are two tall communications towers both approximately 250 m in height and about 5 km apart. Figure 11 shows a geographical plot of the Johannesburg region. The Sentech and Hillbrow towers are clearly indicated in this figure at 26.1925° South, 28.0068° East and 26.1869° South, 28.0494° East respectively. The location at which the high-speed cameras were placed during the study is also indicated in the plot at 26.1628° South, 27.9598° East, and the field of view is indicated looking South-East through the city with both towers clearly in view.

The figure also shows the reported locations of the SALDN strokes time-correlated with the upward lightning events filmed on the Sentech and Hillbrow towers. As can be seen, a number of reported strokes are clustered around the location of the Sentech tower (N = 45) and a number are around the Hillbrow tower (N = 53). The cluster around the Sentech tower has a median location error of 75 m with a maximum location error of 1.7 km. Similarly, the cluster around the Hillbrow tower as a median location error of 51 m and a maximum location error of 3.8 km.

In both cases, the maximum error is due to a single stroke outlier, which asks the question—why would one such stroke suddenly be so inaccurately located? We can examine the high-speed footage. Figure 12 shows one frame from the high-speed footage which correlates with the reported time of the stroke reported 1.7 km northwest of the Sentech tower. We can see an intra-cloud (IC) flash as well as an

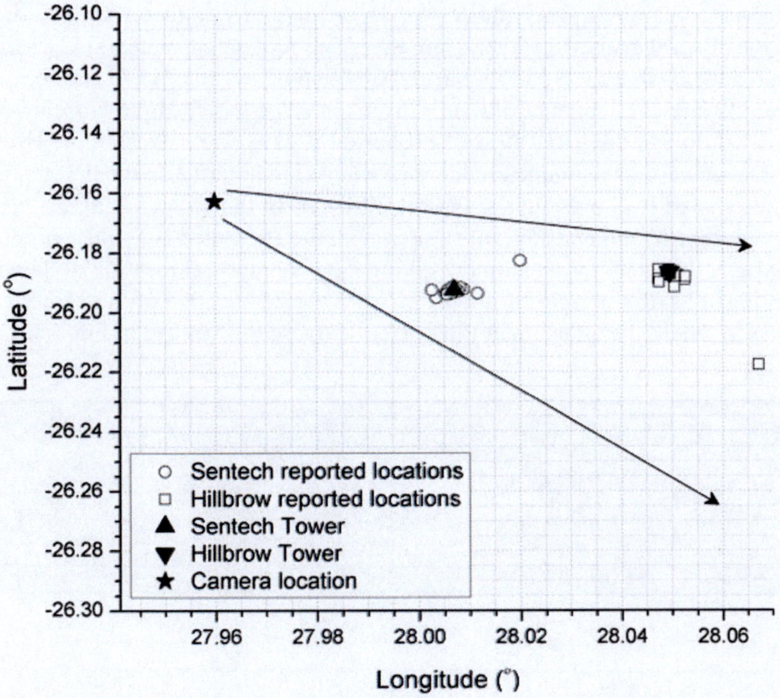

Fig. 11 SALDN reported stroke locations for detections of lightning flashes to the Sentech and Hillbrow towers, Johannesburg, South Africa. (Adapted from [100])

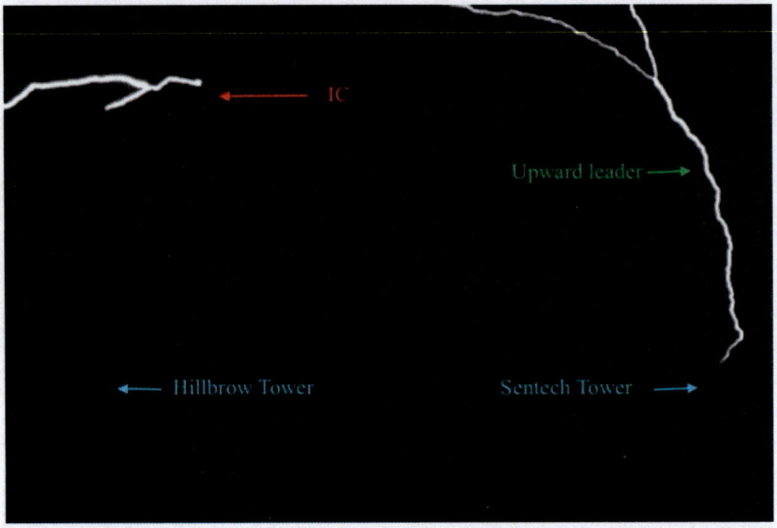

Fig. 12 Image explaining the outlier in the Sentech Tower data [100]

upward leader from the Sentech tower. It appears that this IC flash occurred within the same time that the upward flash was occurring and was therefore classified as part of the upward event. However, from the video footage, it is clear it is a separate flash and appears that the SALDN misclassified an IC event as a CG stroke.

Figure 13 shows a sequence of frames from the video footage time-correlated with the reported time of the SALDN stroke reported 3.8 km South-West of the Hillbrow tower. The upward flash in question was of a negative polarity. In the first image we see an attempted ICC pulse to the Hillbrow Tower. In the second image, a positive subsequent return stroke makes attachment with the attempted ICC pulse in mid-air. The SALDN detected this attachment giving the geographical location of the outlier seen in Fig. 11—to the right of the tower from the perspective of the camera. This agrees with where the attachment took place in the second image in Fig. 13.

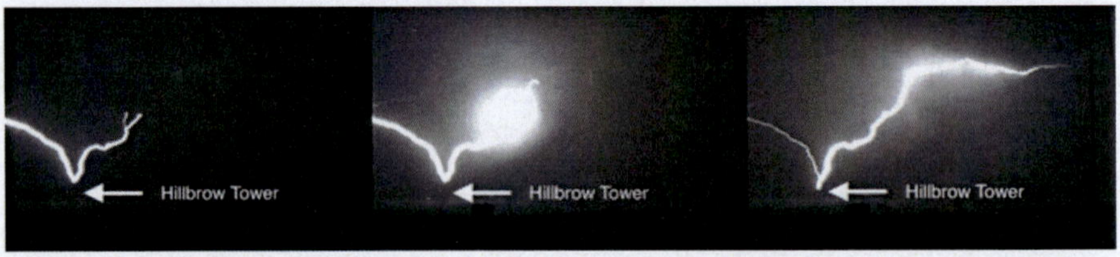

Fig. 13 Sequence of images explaining the Hillbrow Tower outlier [100]

4 Thunderstorm Warning by Media and Mobile Apps

4.1 Introduction

Lightning is recognized as one of the most powerful natural hazards. Worldwide, lightning is a more potent killer than any other natural disaster, causing the deaths of at least 6000 people annually, in addition to causing huge damage to livestock and property. In recent years, many laboratory and field experiments have confirmed that a robust relationship exists between lightning flash characteristics and thunderstorm dynamics and other microphysical parameters. Therefore, information on lightning derived from detection networks described in Sect. 2 can be used for meteorological applications such as warning of severe weather and for improving numerical weather prediction. Many agencies have launched many apps aimed at prediction and have successfully provided safety measures to the benefit of society. The main purpose of this section is to decrease deaths and injuries caused by lightning around the world by encouraging gathering and disseminating data to public safety planners, NGOs, policy makers, and others who are in positions to improve safety from lightning for all of their citizens. This section of the chapter will cover methods for reaching the public, particularly the most vulnerable populations, through media and mobile apps.

4.2 Thunderstorm Warning by Media and Mobile Apps

A huge explosion of mobile phone and social media usage has occurred in recent years throughout the developing world. Mobile phones are used for accessing health messages on HIV, child care, and other areas, for information on crops and animal husbandry, for checking market and fuel prices, and for transferring money [151–154]. This mobile phone and social platform may be used effectively for the purpose of thunderstorm warning. The Arab Spring and daily political protests in Venezuela were driven by text messages, Snapchat, and other social media mainly because all other media sources have been shut down [155–157]. In partnership with Airtel, free 3–2-1 service on Human Networks International's started in Madagascar in 2010 and was generating over 250,000 calls per month by 2015 (HNI.org). These are all examples of person-accessed messages, not broadcast warnings.

4.3 Thunderstorm Warning by Media and Mobile Apps in India

The Indian Institute of Tropical Meteorology (IITM) launched 'Damini', a free mobile-based application that can warn people about lightning 30–45 min lead times Whenever a person is within 20-km radius of a lightning event, the app will send

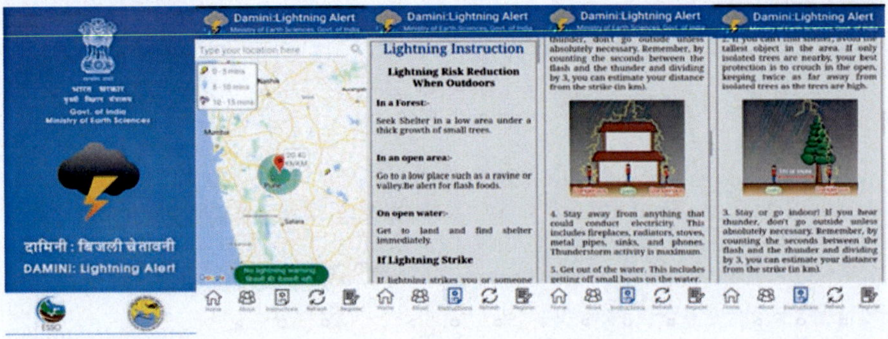

Fig. 14 Sample snapshot from the Damini app showing safety measures to society. With help from this app one may get a warning about lightning hazard within 20 km radius from one's current location

warnings. The alert can be sent 30 min to 45 min before the event and can help people get to safer locations. The warnings can be given in Hindi and English.

IITM has installed 85-sensor Lightning Location Network over India to investigate the damages caused by lightning over the country. Lightning detection networks use the 'time of arrival (TOA)' technique to estimate flash location, as discussed previously in Sect. 2. The network gives some vital information about lightning characteristic such as the location and peak power of lightning. The output from this network is being used to study the relation between different lightning parameters and number of lightning deaths and to generate Lightning alerts which are shared with different authorities dealing with disaster management [158] (Fig. 14).

https://play.google.com/store/apps/details?id=com.lightening.live.damini&hl=en_IN

4.4 Thunderstorm Warning by Other Mobile Apps

In addition to mobile apps developed by Governmental agencies, other commercial and community based warming tools also extend their services to the common people. One such app named 'WeatherBug' integrates lightning data from Earth Networks and provides advisories to common people for an impending thunderstorm. The app may be downloaded from the following link.

https://play.google.com/store/apps/details?id=com.aws.android&hl=en_IN

Another community based project www.blitzortung.org serves the same purpose. The mobile app may be downloaded from the link below.

https://community.windy.com/topic/6605/real-time-lightning-strikes-on-windy-com

4.5 Television, Radio, and Print Media

Television and radio are probably the most effective means of delivering safety messages and other information about lightning and thunderstorms. They require no literacy; the broadcasts are usually free and often in the prevailing language in the area where the safety message is needed. The disadvantage is the frequent unavailability of electricity to power them and the cost of television sets. Radios are much less expensive and more easily battery powered. Everyone recognizes that print media are dying, at least paper versions. Although, the print media may be important for educating people in advance where access to electricity is poor. Effective dissemination is dependent on the popularity of the print venue, literacy, and where the message is located in the publication. If it is buried in the middle of the publication, it is unlikely to be seen.

4.6 The Challenges

Trengove and Jandrell [153] posited that the use of mobile phone texting to issue lightning warnings and education would have the following impacts:

1. Reach a large number of people
2. Reach rural people
3. Bridge the digital divide by providing the same service to rich and poor
4. Could use existing mobile telephone infrastructure
5. Could geographically target lightning warning messages.

Some of these hypotheses have been verified, while others are still questionable. Unfortunately, despite decreasing costs in many countries, smartphones remain prohibitively expensive for the poor and rural populations of most countries. Tushemereirwe and Cooper [159] noted that while 92% of Ugandans surveyed had mobile phones, only 4% had smartphones. Floods, drought, and severe storms tend to disproportionately affect women since they are more commonly responsible for farm labor, food security, and household management in developing countries. The majority of farmers in Uganda are women, yet gender disparities limit their access to information on which to base decisions and adjust to climate shocks [160]. This circumstance is caused by women's restricted access to technology and communication channels, by lower education levels, and by culturally defined roles in household chores such as raising children and cooking. In Africa, women are 23% less likely to own a mobile phone than are men [161]. Financial barriers such as the inability to pay fees or even to own a mobile phone or radio can leave them uninformed of weather-related impacts. Airtime can be expensive and unreliable. Electricity may be available only erratically, leading to mismatches between a person or home's allotted electricity window and the internet provider's window. As in most countries, there are multiple airtime providers so that not all of the population would likely be covered, and it would require funding by each company. For the HNI-Airtel partnership mentioned earlier, HNI, through private funding, provides the translated messages and Airtel provides a monthly allowance of free calls to promote customer loyalty. Not all telephone services are this benevolent [162].

Of course, targeted warnings in either of these settings would depend on GPS coordinates, reliable and consistent internet or cell phone availability, how long thunderstorms last in this region, and the speed of movement of a thunderstorm. Additional factors include the quality and timeliness of the forecast and the willingness and funding of the meteorological service to implement such a system.

References

1. Chalmers JA (1967) Atmospheric electricity, 2nd Edition, Elsevier
2. Kelvin L (1884) Reprints on papers in electrostatics and magnetism. MacMillan and Company
3. Rust WD, MacGorman DR (1987) Techniques for measuring electrical parameters of thunderstorms, Chapter 8. In: Kessler E (Editor) Instruments and techniques for thunderstorm observation and analysis, vol 3, University of Oklahoma Press
4. Mazur V (2016) Principles of lightning physics. IOP Publishing
5. Uman MA (2001) The lightning discharge. Courier Corporation
6. Berger K (1967) Novel observations on lightning discharges: Results of research on Mount San Salvatore. J Franklin Inst 283(6):478–525
7. Cooray GV (2009) Lightning protection. Inst Eng Technol
8. Uman MA, Krider EP (1989) Natural and artificially initiated lightning. Science 246:457–464
9. Williams E (2020) Lightning and climate change, Chapter 1 of lightning interaction with power systems. In: Piantini A (ed) Fundamentals and modelling, vol 1
10. Uman MA (2012) Lightning. Courier Corporation
11. Cummins KL, Murphy MJ, Bardo EA, Hiscox WL, Pyle RB, Pifer AE (1998) A combined TOA/MDF technology upgrade of the US National Lightning Detection Network. J Geophys Res Atmos 103(D8):9035–9044
12. Dwyer JR, Rassoul HK, Al-Dayeh M, Caraway L, Chrest A, Wright B, Kozak E, Jerauld J, Uman MA, Rakov VA, Jordan DM, Rambo KJ (2005) X-ray bursts associated with leader steps in cloud-to-ground lightning. Geophys Res Lett 32(1):1–4
13. Rakov VA, Uman MA (2003) Lightning: physics and effects. Cambridge University Press
14. Clarence ND, Malan DJ (1957) Preliminary discharge processes in lightning flashes to ground. Q J R Meteorol Soc 83(356):161–172
15. Campos LZ, Saba MM, Warner TA, Pinto O Jr, Krider EP, Orville RE (2014) High-speed video observations of natural cloud-to-ground lightning leaders-A statistical analysis. Atmos Res 135:285–305
16. Krider EP, Weidman CD, Noggle RC (1977) The electric fields produced by lightning stepped leaders. J Geophys Res 82(6):951–960
17. Warner TA (2012) Observations of simultaneous upward lightning leaders from multiple tall structures. Atmos Res 117:45–54
18. Idone VP, Orville RE (1982) Lightning return stroke velocities in the thunderstorm research international program (TRIP). J Geophys Res Oceans 87(C7):4903–4916
19. Cooray GV (2013) The lightning flash (No. 34). IET Digital Libr
20. Krider EP, Noggle RC, Pifer AE, Vance DL (1980) Lightning direction-finding systems for forest fire detection. Bull Am Meteor Soc 61(9):980–986
21. Kitagawa N (1957) On the mechanism of cloud flash and junction or final process in flash to ground. Pap Meteorol Geophys 7(4):415–424
22. Jordan DM, Idone VP, Rakov VA, Uman MA, Beasley WH, Jurenka H (1992) Observed dart leader speed in natural and triggered lightning. J Geophys Res Atmos 97(D9):9951–9957

23. Kitagawa N, Brook M, Workman EJ (1962) Continuing currents in cloud-to-ground lightning discharges. J Geophys Res 67(2):637–647
24. Malan DJ, Schonland BFJ (1947) Progressive lightning. VII Directly-correlated photographic and electrical studies of lightning from near thunderstorms. Proc R Soc London. Ser Math Phys Sci 191(1027):485–503
25. Thottappillil R, Rakov VA, Uman MA (1990) K and M changes in close lightning ground flashes in Florida. J Geophys Res Atmos 95(D11):18631–18640
26. Goodman SJ, Blakeslee RJ, Koshak WJ, Mach D, Bailey J, Buechler D, Carey L, Schultz C, Bateman M, McCaul E Jr, Stano G (2013) The GOES-R geostationary lightning mapper (GLM). Atmos Res 125:34–49
27. Jacobson AR, Shao XM, Holzworth RH (2011) Satellite triangulation of thunderstorms, from fading radio fields synchronously recorded on two orthogonal antennas. Radio Sci 46(06):1–17
28. Kotaki M, Katoh C (1983) Global distribution of atmospheric radio noise derived from distribution of lightning activity. Radio Res Lab J 30:35–57
29. Morimoto T, Kikuchi H, Sato M, Suzuki M, Yamazaki A, Ushio T (2011) VHF lightning observations on JEM-GLIMS mission. IEEJ Trans Fundam Mater 131(12):977–982
30. Boeck WL, Suszcynsky DM, Light TE, Jacobson AR, Christian HJ, Goodman SJ, Buechler DE, Guillen JLL (2004) A demonstration of the capabilities of multisatellite observations of oceanic lightning. J Geophys Res Atmos 109(D17):1–8
31. Jacobson AR, Knox SO, Franz R, Enemark DC (1999) FORTE observations of lightning radio-frequency signatures: Capabilities and basic results. Radio Sci 34:337–354
32. Jernegan MW (1928) Benjamin Franklin's "electrical kite" and lightning rod. N Engl Q 1(2):180–196
33. Norinder H (1953) Long distance location of thunderstorms. Thunderstorm Electricity, 276–327
34. Blois De (1914) Amer InstElec Eng 33:563–579
35. Watt RW, Appleton EV (1923) On the nature of atmospherics. I. Proc R Soc London. Ser Containing Pap Math Phys Charact 84–102
36. Cummins KL, Murphy MJ (2009) An overview of lightning locating systems: History, techniques, and data uses, with an in-depth look at the US NLDN. IEEE Trans Electromagn Compat 51(3):499–518
37. Cianos N, Oetzel GN, Pierce ET (1972) A technique for accurately locating lightning at close ranges. J Appl Meteorol 11(7):1120–1127
38. Maier L, Lennon C, Britt T, Schaefer S (1995) LDAR system performance and analysis. In: Proceedings of the international conference on cloud physics. Am Meteorol Soc. Dallas, Tex
39. Rhodes CT, Shao X-M, Krehbiel PR, Thomas RJ, Hayenga CO (1994) Observations of lightning phenomena using radio interferometry. J Geophys Res 99:13059–13082
40. Shao XM, Krehbiel PR, Thomas RJ, Rison W (1995) Radio interferometric observations of cloud-to-ground lightning phenomena in Florida. J Geophys Res 100:2749–2783.
41. Rison W, Thomas RJ, Krehbiel PR, Hamlin T, Harlin J (1999) A GPS-based three-dimensional lightning mapping system: initial observations in central New Mexico. Geophys Res Lett 26(23):3573–3576
42. Krehbiel PR, Thomas R, Rison W, Hamlin T, Harlin J, Davis M (1998) Lightning mapping observations during MEaPRS in central Oklahoma. EOS Trans AGU 79:F127
43. Thomas RJ, Krehbiel PR, Rison W, Hamlin T, Boccippio DJ, Goodman SJ, Christian HJ (2000) Comparison of ground-based 3-dimensional lightning mapping observations with satellite-based LIS observations in Oklahoma. Geophys Res Lett 27(12):1703–1706
44. Thomas RJ, Krehbiel PR, Rison W, Hunyady SJ, Winn WP, Hamlin T, Harlin J (2004) Accuracy of the lightning mapping array. J Geophys Res Atmos 109(D14)
45. Orville RE (1994) Cloud-to-ground lightning flash characteristics in the contiguous United States: 1989–1991. J Geophys Res Atmos 99(D5):10833–10841
46. Orville RE, Huffines GR (2001) Cloud-to-ground lightning in the United States: NLDN results in the first decade, 1989–98. Mon Weather Rev 129(5):1179–1193

47. Biagi CJ, Cummins KL, Kehoe KE, Krider EP (2007) NLDN performance in Southern Arizona, Texas and Oklahoma in 2003–2004. J Geophys Res 112:D05208
48. Fleenor SA, Biagi CJ, Cummins KL, Krider EP, Shao XM (2009) Characteristics of cloud-to-ground lightning in warm-season thunderstorms in the Central Great Plains. Atmos Res 91:333–352
49. Jerauld J, Rakov VA, Uman MA, Rambo KJ, Jordan DM, Cummins KL, Cramer JA (2005) An evaluation of the performance characteristics of the U.S. National Lightning Detection Network in Florida using rocket-triggered lightning. J Geophys Res Atmos 110(D19)
50. Nag A, Mallick S, Rakov VA, Howard JS, Biagi CJ, Hill JD, Uman MA, Jordan DM, Rambo K, Jerauld JJE, Decarlo BA, Cummins KL, Cramer JA (2014) Evaluation of U.S. national lightning detection network performance characteristics using rocket-triggered lightning data acquired in 2004–2009. J Geophys Res Atmos 116(D2)
51. Orville RE, Huffines GR, Burrows WR, Holle RL, Cummins KL (2002) The North American lightning detection network (NALDN)—First results: 1998–2000. Mon Weather Rev 130(8):2098–2109
52. Orville RE, Huffines GR, Burrows WR, Cummins KL (2011) The North American lightning detection network (NALDN)-Analysis of flash data: 2001–09. Mon Weather Rev 139(5):1305–1322
53. Holle RL (2014) Diurnal variations of NLDN-reported cloud-to-ground lightning in the United States. Mon Weather Rev 142(3):1037–1052
54. Abreu D, Chandan D, Holzworth RH, Strong K (2010) A performance assessment of the World Wide Lightning Location Network (WWLLN) via comparison with the Canadian Lightning Detection Network (CLDN). Atmos Meas Tech 3(4):1143–1153
55. Burrows WR, King P, Lewis PJ, Kochtubajda B, Snyder B, Turcotte V (2002) Lightning occurrence patterns over Canada and adjacent United States from lightning detection network observations. Atmos Ocean 40(1):59–80
56. Dockendorff D, Spring K (2005) The Canadian lightning detection network-novel approaches for performance measurement and network management. In: Proceedings, WMO technical conference on instruments and methods of observation (TECO). Bucharest, Romania
57. Shostak V, Bormotov O, Pavanello D, Janischewskyj W, Rachidi F (2012) Analysis of lightning detection network data for selected areas in Canada. In: 2012 international conference on lightning protection (ICLP). IEEE, pp 1–12
58. Kazazi S, Hussein AM, Liatos P (2015) Evaluation of the performance characteristics of the North American Lightning Detection Network based on recent CN Tower lightning data. In: 2015 international symposium on lightning protection (XIII SIPDA). IEEE, pp 327–333
59. Ballarotti MG, Saba MMF, Pinto Jr O (2006) A new performance evaluation of the Brazilian Lightning Location System (RINDAT) based on high-speed camera observations of natural negative ground flashes. In: 19th international lightning detection conference, Vaisala, Tucson, Arizona
60. Ballarotti MG, Saba MMF, Pinto Jr O (2005) High-speed camera observations of negative ground flashes on a millisecond-scale. Geophys Res Lett 32(23)
61. Pinto Jr O (2005) The art of war against lightning (in Portuguese), Oficina de Texto Press
62. Solorzano NN (2003) Triggered lightning study in Brazil. Ph.D. Thesis, INPE, 178 p (in Portuguese)
63. Pinto Jr O (2003) The Brazilian lightning detection network: a historical background and future perspectives, paper presented at VII International Symposium on Lightning Protection (SIPDA), Inst.de Eletr. e Energ., Curitiba, Brazil
64. Pinto Jr O, Pinto IRCA, Gomes MASS, Vitorello I, Padilha AL, Diniz JH, Carvalho AM, Filho AC (1999) Cloud-to-ground lightning in southeastern Brazil in 1993: 1. Geographical distribution. J Geophys Res Atmos 104 (D24):31369–31379
65. Pinto O Jr, Naccarato KP, Pinto IRCA, Fernandes WA, Neto OP (2006a) Monthly distribution of cloud-to ground lightning flashes as observed by lightning location systems. Geophys Res Lett 33(9):1–4

66. Pinto Jr O, Naccarato KP, Saba MMF, Pinto IRCA, Abdo RF, Garcia SDM, Cazetta Filho A (2006b) Recent upgrade to the Brazilian integrated lightning detection network. In: Proceedings of the 19th international lightning detection conference (ILDC). Tucson, AZ
67. Betz HD, Schmidt K, Oettinger WP (2009a) LINET-An international VLF/LF lightning detection network in Europe. In: Lightning: principles, instruments and applications. Springer, Dordrecht, pp 115–140
68. Betz HD, Schmidt K, Laroche P, Blanchet P, Oettinger WP, Defer E, Dziewit Z, Konarski J (2009b) LINET-An international lightning detection network in Europe. Atmos Res 91(2–4):564–573
69. Betz HD, Meneux B (2014) LINET systems-10 years experience. In: 2014 international conference on lightning protection (ICLP), pp 1553–1557
70. Betz HD, Schumann U, Laroche P (eds) (2008) Lightning: principles, instruments and applications: Review of modern lightning research. Springer Science & Business Media, Berlin
71. Azadifar M, Rachidi F, Rubinstein M, Paolone M, Diendorfer G, Pichler H, Schulz W, Pavanello D, Romero C (2016) Evaluation of the performance characteristics of the European Lightning Detection Network EUCLID in the Alps region for upward negative flashes using direct measurements at the instrumented Säntis Tower. J Geophys Res Atmos 121(2):595–606
72. Schulz W, Cummins K, Diendorfer G, Dorninger M (2005) Cloud-to-ground lightning in Austria: A 10-year study using data from a lightning location system. J Geophys Rese Atmos 110(D9)
73. Romero C, Paolone M, Rachidi F, Rubinstein M, Rubinstein A, Diendorfer G, Schulz W, Bernardi M, Nucci CA (2011) Preliminary comparison of data from the säntis tower and the EUCLID lightning location system. In: 2011 international symposium on lightning protection. IEEE, pp 140–145
74. Schulz W, Diendorfer G, Pedeboy S, Poelman DR (2016) The European lightning location system EUCLID part 1: Performance analysis and validation. Nat Hazard 16(2):595–605
75. Pohjola H, Mäkelä A (2013) The comparison of GLD360 and EUCLID lightning location systems in Europe. Atmos Res 123:117–128
76. Poelman DR, Schulz W, Kaltenboeck R, Delobbe L (2017) Analysis of lightning outliers in the EUCLID network. Atmos Meas Tech 10(11)
77. Biron D (2009) LAMPINET-lightning detection in Italy. In: Lightning: principles, instruments and applications, Springer, Dordrecht, pp 141–159
78. Biron D, De Leonibus L, Betz HD (2007) Campaign, with locations produced by two different detection network in Central Europe: LAMPINET and LINET. European Geosciences Union, 2007. In: Geophysical research abstracts, 9, 02500
79. Biron CD, De Leonibus LCL, Zauli LCF (2006) The lightning network LAMPINET of the Italian Air Force Meteorological Service. In: Proceeding of 19th international lightning detection conference. Tucson, USA
80. García Mollá PJ (1910) La section électrique, Mèmories de l'Observatoire de l'Èbre, Ed. Gustavo Gili, Barcelona
81. Pineda N, Montanyà J (2009) Lightning detection in Spain: the particular case of Catalonia. In: Lightning: principles, instruments and applications. Springer, Dordrecht, pp 161–185
82. Pérez PF, Rodríguez CZ (AEMET) (2008) La frecuencia de las tormentas eléctricas en España. Bol de la quinta etapa de la Asoc Meteorológica Esp 21:37–44
83. Montanyà J, Pineda N, March V, Illa A, Romero D, Solà G (2006) Experimental evaluation of the Catalan lightning detection network. In: 19th international lightning detection conference, Tucson, Apr 2006
84. Pineda N, Rigo T, Bech J, Soler X (2007) Lightning and precipitation relationship in summer thunderstorms: case studies in the North Western Mediterranean region. Atmos Res 85(2):159–170
85. Li YJ, Zhang GS, Wang YH, Wu B, Li J (2017) Observation and analysis of electrical structure change and diversity in thunderstorms on the Qinghai-Tibet Plateau. Atmos Res 194:130–141

86. Zhang GS, Wang YH, Qie XS, Zhang T, Zhao YX, Li YJ, Cao DJ (2010) Using lightning locating system based on time-of-arrival technique to study three-dimensional lightning discharge processes. Sci China Earth Sci 53(4):591–602
87. Li YJ, Zhang GS, Wen J, Wang DH, Wang YH, Zhang T, Fan XP, Wu B (2013) Electrical structure of a Qinghai-Tibet Plateau thunderstorm based on three-dimensional lightning mapping. Atmos Res 134:137–149
88. Wang Y, Qie X, Wang D, Liu M, Su D, Wang Z, Liu D, Wu Z, Sun Z, Tian Y (2015) Beijing Lightning NETwork (BLNET): Configuration and preliminary results of lightning location. Chin J Atmos Sci 39(3):571–582
89. Srivastava A, Tian Y, Qie X, Wang D, Sun Z, Yuan S, Wang Y, Chen Z, Xu W, Zhang H, Jiang R (2017) Performance assessment of Beijing Lightning Network (BLNET) and comparison with other lightning location networks across Beijing. Atmos Res 197:76–83
90. Liu H, Qiu S, Dong W (2018) The three-dimensional locating of VHF broadband lightning interferometers. Atmosphere 9(8):317–331
91. Chen LW, Zhang YJ, Lu WT, Zheng D, Zhang Y, Chen SD, Huang ZH (2012) Performance evaluation for a lightning location system based on observations of artificially triggered lightning and natural lightning flashes. J Atmos Oceanic Tech 29:1835–1844
92. Zhang H, Lu G, Qie X, Jiang R, Fan Y, Tian Y, Sun Z, Liu M, Wang Z, Liu D, Feng G (2016) Locating narrow bipolar events with single-station measurement of low-frequency magnetic fields. J Atmos Solar Terr Phys 143:88–101
93. Lyu W, Ma Y, Qi Q, Chen L, Wu B, Su Z, Wu S (2018) Lightning optical observation at TOLOG. In: 34th international conference on lightning protection (ICLP). IEEE, pp 1–4
94. Wu B, Lyu W, Qi Q, Ma Y, Chen L, Zhang Y, Zhu Y, Rakov VA (2019) Synchronized two-station optical and electric field observations of multiple upward lightning flashes triggered by a 310-kA+ CG flash. J Geophys Res Atmos 124(2):1050–1063
95. Qie X, Zhang Y, Yuan T, Zhang Q, Zhang T, Zhu B, Lu W, Ma M, Yang J, Zhou Y, Feng G (2015) A review of atmospheric electricity research in China. Adv Atmos Sci 32(2):169–191
96. Shi D, Zheng D, Zhang Y, Zhang Y, Huang Z, Lu W, Chen S, Yan X (2017) Low-frequency E-field detection array (LFEDA)-construction and preliminary results. Sci China Earth Sci 60(10):1896–1908
97. Gijben M (2012) The lightning climatology of South Africa. S Afr J Sci 108(3–4):44–53
98. Gill T (2008) A lightning climatology of South Africa for the first two years of operation of the South African Weather Service Lightning Detection Network: 2006–2007. In: 20th international lightning detection conference, pp 1–12
99. Fensham HG, Schumann C, Hunt HGP, Nixon KJ, Warner TA, Gijben M (2018) Performance evaluation of the SALDN using high- speed camera footage of ground truth lightning events over Johannesburg, South Africa. In: 34th international conference on lightning protection (ICLP). Rzeszow, Poland
100. Fensham HG (2019) Performance evaluation of the SALDN using high-speed camera footage. Master Dissertation, University of the Witwatersrand
101. Hunt H, Liu YC, Nixon K (2014) Evaluation of the South African Lightning Detection Network using photographed tall tower lightning events from 2009–2013. In: 32rd international conference on lightning protection (ICLP). Shanghai, China
102. Dowden RL, Brundell JB, Rodger CJ (2002) VLF lightning location by time of group arrival (TOGA) at multiple sites. J Atmos Solar Terr Phys 64(7):817–830
103. Abarca SF, Corbosiero KL, Galarneau Jr TJ (2010) An evaluation of the worldwide lightning location network (WWLLN) using the national lightning detection network (NLDN) as ground truth. J Geophys Res Atmos 115(D18):2010
104. Hutchins ML, Holzworth RH, Brundell JB, Rodger CJ (2012) Relative detection efficiency of the World Wide Lightning Location Network. Radio Sci 47(6):1–9
105. Jacobson AR, Holzworth R, Harlin J, Dowden R, Lay E (2006) Performance assessment of the world wide lightning location network (WWLLN), using the Los Alamos sferic array (LASA) as ground truth. J Atmos Oceanic Tech 23(8):1082–1092

106. Rodger CJ, Werner S, Brundell JB, Lay EH, Thomson NR, Holzworth RH, Dowden RL (2006) Detection efficiency of the VLF World-Wide Lightning Location Network (WWLLN): initial case study. Ann Geophys 24(12):3197–3214
107. Mezuman K, Price C, Galanti E (2014) On the spatial and temporal distribution of global thunderstorm cells. Environ Res Lett 9(12):124023
108. Virts KS, Wallace JM, Hutchins ML, Holzworth RH (2013) Highlights of a new ground-based, hourly global lightning climatology. Bull Am Meteor Soc 94(9):1381–1391
109. Lay EH, Holzworth RH, Rodger CJ, Thomas JN, Pinto O Jr, Dowden RL (2004) WWLLN global lightning detection system: regional validation study in Brazil. Geophys Res Lett 31(3):1–5
110. Said RK, Inan US, Cummins KL (2010) Long-range lightning geolocation using a VLF radio atmospheric waveform bank. J Geophys Res Atmos 115(D23)
111. Said RK, Cohen MB, Inan US (2013) Highly intense lightning over the oceans: Estimated peak currents from global GLD360 observations. J Geophys Res Atmos 118(13):6905–6915
112. Liu CL, Heckman S (2011) The application of total lightning detection and cell tracking for severe weather prediction. In: 91st annual meeting of the American meteorological society. pp 1–10
113. Mallick S, Rakov VA, Hill JD, Ngin T, Gamerota WR, Pilkey JT, Jordan DM, Uman MA, Heckman S, Liu C (2015) Performance characteristics of the ENTLN evaluated using rocket-triggered lightning data. Electr Power Syst Res 118:15–28
114. Zhu Y, Rakov VA, Tran MD, Stock MG, Heckman S, Liu C, Sloop CD, Jordan DM, Uman MA, Caicedo JA, Kotovsky DA (2017) Evaluation of ENTLN performance characteristics based on the ground truth natural and rocket-triggered lightning data acquired in Florida. J Geophys Res Atmos 122(18):9858–9866
115. Heckman S (2014) ENTLN status update. In: XV international conference on atmospheric electricity, 15–20 June 2014
116. Stock M, Lapierre J, Heckman S, Borges M, Anderson J (2017) Estimating detection efficiency in the absence of satellites: ENTLN detection efficiency in 2015 and 2016. In: IEEE international symposium on lightning protection (XIV SIPDA), pp 333–335
117. Marchand M, Hilburn K, Miller SD (2019) Geostationary lightning mapper and earth networks lightning detection over the contiguous United States and dependence on flash characteristics. J Geophys Res Atmos 124(11):552–567
118. Schumann WO (1952) Über die strahlungslosen Eigenschwingungen einer leitenden Kugel, die von einer Luftschicht und einer Ionosphärenhülle umgeben ist, Z. Naturforsch., 7a, 149
119. Ogawa T, Tanaka Y, Miura T, Yasuhara M (1966) Observations of natural ELF and VLF electromagnetic noises by using ball antennas. J Geomagn Geoelectr 18(4):443–454
120. Kemp DT, Jones DL (1971) A new technique for the analysis of transient ELF electromagnetic disturbances within the Earth-ionosphere cavity. J Atmos Terr Phys 33(4):567–572
121. Hobara Y, Hayakawa M, Williams E, Boldi R, Downes E (2006) Location and electrical properties of sprite-producing lightning from a single ELF site. In: Sprites, elves and intense lightning discharges. Springer, Dordrecht, pp 211–235
122. Huang E, Williams E, Boldi R, Heckman S, Lyons W, Taylor M, Nelson T, Wong C (1999) Criteria for sprites and elves based on Schumann resonance observations. J Geophys Res Atmos 104(D14):16943–16964
123. Williams E, Mareev E (2014) Recent progress on the global electrical circuit. Atmos Res 135:208–227
124. Prácser E, Bozóki T, Sátori G, Williams E, Guha A, Yu H (2019) Reconstruction of global lightning activity based on Schumann Resonance measurements: Model description and synthetic tests. Radio Sci 54(3):254–267
125. Boys CV (1926) Progressive lightning. Nature 118:749–750
126. Schonland BFJ (1956) The lightning discharge. Hand Phys 22:576–628
127. Brantley RD, Tiller JA, Uman MA (1975) Lightning properties in Florida thunderstorms from video tape records. J Geophys Res 80(24):3402–3406

128. Jordan DM, Rakov VA, Beasley WH, Uman MA (1997) Luminosity characteristics of dart leaders and return strokes in natural lightning. J Geophys Res Atmos 102(D18):22025–22032
129. Waldteufl P, Metzger P, Boulay JL, Laroche P, Hubert P (1980) Triggered lightning strokes originating in clear air. Journal of Geophysical Research: Oceans 85(C5):2861–2868
130. Winn WP, Aldridge TV, Moore CB (1973) Video tape recordings of lightning flashes. J Geophys Res 78:4515–4519
131. Moreau JP, Alliot JC, Mazur V (1992) Aircraft lightning initiation and interception from in situ electric measurements and fast video observations. J Geophys Res Atmosp 97(D14):15903–15912
132. Ballarotti MG, Medeiros C, Saba MM, Schulz W, Pinto O Jr (2012) Frequency distributions of some parameters of negative downward lightning flashes based on accurate-stroke-count studies. J Geophys Res Atmos 117(D6):1–8
133. Saba MMF, Ballarotti MG, Pinto Jr O (2006) Negative cloud-to-ground lightning properties from high-speed video observations. J Geophys Res Atmos 111(D3)
134. Saba MM, Schulz W, Warner TA, Campos LZ, Schumann C, Krider EP, Cummins KL, Orville RE (2010) High-speed video observations of positive lightning flashes to ground. J Geophys Res Atmos 115(D24)
135. Saba MM, Schumann C, Warner TA, Ferro MAS, de Paiva AR, Helsdon J Jr, Orville RE (2016) Upward lightning flashes characteristics from high-speed videos. J Geophys Res Atmos 121(14):8493–8505
136. Schumann C (2016) Estudo dos raios ascendentes a partir de observações de câmeras de alta resolução temporal e de medidas de campo elétrico—Ph.D. thesis 2016
137. Schumann C (2012) Caracterização dos raios positivos através de câmeras de alta velocidade e sensores de campo elétrico, Master's dissertation, Department of Space Geophysic, Atmospheric Eletricity Group, National Institute for Space Research, São José dos Campos, Brazil
138. Warner TA, Saba MMF, Rudge S, Bunkers M, Lyons WA, Orville RE (2012a) Lightning-triggered upward lightning from towers in Rapid City, South Dakota. In: 2012 International lightning detection conference. Boulder, Colorado
139. Warner TA, Cummins KL, Orville RE (2012b) Upward lightning observations from towers in Rapid City, South Dakota and comparison with National Lightning Detection Network data, 2004–2010. J Geophys Res Atmos 117(D19)
140. Warner TA, Saba MM, Schumann C, Helsdon JH Jr, Orville RE (2016) Observations of bidirectional lightning leader initiation and development near positive leader channels. J Geophys Res Atmos 121(15):9251–9260
141. Flache D, Rakov VA, Heidler F, Zischank W, Thottappillil R (2008) Initial-stage pulses in upward lightning: leader/return stroke versus M-component mode of charge transfer to ground. Geophys Res Lett 35(13)
142. Jiang R, Qie X, Wu Z, Wang D, Liu M, Lu G, Liu D (2014) Characteristics of upward lightning from a 325-m-tall meteorology tower. Atmos Res 149:111–119
143. Lu W, Wang D, Zhang Y, Takagi N (2009) Two associated upward lightning flashes that produced opposite polarity electric field changes. Geophys Res Lett 36(5)
144. Mazur V, Ruhnke LH (2011) Physical processes during development of upward leaders from tall structures. J Electrostat 69(2):97–110
145. Wang D, Takagi N, Watanabe T, Sakurano H, Hashimoto M (2008) Observed characteristics of upward leaders that are initiated from a windmill and its lightning protection tower. Geophys Res Lett 35(2)
146. Brook M, Kitagawa N, Workman EJ (1962) Quantitative study of strokes and continuing currents in lightning discharges to ground. J Geophys Res 67(2):649–659
147. Schumann C, Hunt HG, Tasman J, Fensham H, Nixon KJ, Warner TA, Saba MM (2018a) High-speed video observation of lightning flashes over Johannesburg, South Africa 2017–2018. In: 2018 34th international conference on lightning protection (ICLP), pp 1–7
148. Paiva AR, Saba MMF, Naccarato KP, Schumann C, Jaques R, Ferro MAS, Warner TA (2014) Detection of upward lightning by lightning location systems, lightning protection (ICLP). In: International conference. Shanghai, pp 1824–1826

149. Diendorfer G, Pichler H, Schulz W (2014) EUCLID located strokes to the Gaisberg tower-accuracy of location and its assigned confidence ellipse. In: 2014 international lightning detection conference (ILDC)
150. Schwalt L, Pack S, Schulz W (2020) Ground truth data of atmospheric discharges in correlation with LLS detections. Electr Power Syst Res 180:106065
151. Chiumbu S (2012) Exploring mobile phone practices in social movements in South Africa–the Western Cape Anti-Eviction Campaign. Afr Identities 10(2):193–206
152. Southwood R (2009) Less walk more talk: how Celtel and the mobile phone changed Africa. Wiley, New York
153. Trengove E, Jandrell IR (2012) Leveraging a mobile culture for lightning awareness: the African context. In: Preprints of the 31st international conference on lightning protection. Vienna, pp 2–7
154. Wasserman H (2011) Mobile phones, popular media, and everyday African democracy: transmissions and transgressions. Int J Media Cult 9:146–158
155. Gire S (2017) The role of social media in the Arab Spring. https://sites.stedwards.edu/pangaea/ the-role-of-social-media-in-the-arab-spring/. Accessed 8 Sept 2017
156. Lopez V (2017) On the frontline of Venezuela's punishing protests. https://www.theguardian.com/world/2017/may/25/venezuela-protests-riots-frontline-caracas-nicolas-maduro. Accessed 30 Oct 2017
157. Wilson P (2014) Social media key for Venezuelan protesters. https://www.usatoday.com/story/news/world/2014/02/19/venezuela-uprisingprotests/5606899/. Accessed 8 Sept 2017
158. Gopalakrishnan V (2019) Early warning of lightning: lightning location network over India, RMLT(2019). Tripura University
159. Tushemereirwe R, Cooper MA, Tuhebwe D (2017) The most effective methods for delivering severe weather early warnings to fishermen on Lake Victoria. PLOS Current Disasters, 16, 22 Feb 2017
160. Kyazze F, Kristjanson P (2011) Summary of baseline household survey results: Rakai district, south central Uganda, CGIAR research program on climate change, and food security (CCAFS)
161. GSMA Development Fund (2010) Women & mobile: a global opportunity—a study on the mobile phone gender gap in low-and middle-income countries
162. Cooper MA, Holle RL (2019) Reducing lightning injuries worldwide. Springer Natural Hazards, Berlin

Risk Assessment for Lightning Protection

Alain Rousseau

Abstract It is frequently needed to analyze the possible damages caused by lightning to a structure and its connected lines and evaluate their frequency of occurrence as well as the extent of the associated damages. The method to perform this is named Lightning Risk Assessment and is based on an IEC Standard 62305-2 that is based on a methodology with 25 years of experience. This analysis will aim to determine what is the most efficient protection strategy, where to install the selected lightning protection components and with which rating. Lightning may cause many types of damages to a structure. They are usually named "losses" and there are four types of losses: loss due to injury to human beings, loss due to physical damage of the structure and its contents, loss due to failure of internal systems and loss of economic value including cost to repair and production loss. To each of this loss is associated a risk and there are then a maximum of four risks that can be calculated. Four sources of damage are considered in the calculation of the risk: lightning flash to the structure, lightning flash near the structure, lightning flash to the connected lines and lightning flash near the connected lines. Risk component are related to the sources of damage as well as to the type of losses. There is a maximum of eight risk components to calculate, the risk being the sum of the considered components. Each risk component is calculated in the same way as a multiplication of the number of events by the probability this event creates a damage and by the amount of damage generated. The number of events depends on the structure or lines dimensions and parameters and also on the severity of the area in terms of lightning flashes per year and per km^2. The probability this event creates a damage is related to the resilience of the structure and its contents and connected lines, to a lightning event as well as lightning protection measures such as lightning rods or Surge Protective Devices. Then the amount of damage generated, can be calculated based on the existing mitigating measures (for example fire detection) and to the time of presence of people in a dangerous zone inside the structure. When the appropriate risk components are calculated, the risk

A. Rousseau (✉)
SEFTIM, Vincennes, France
e-mail: alain.rousseau@seftim.fr

Chair Committee 81X Lightning Protection, CENELEC (Europe), Brussels, Belgium

C. Gomes (ed.), *Lightning*, Lecture Notes in Electrical Engineering 780,
https://doi.org/10.1007/978-981-16-3440-6_3

can be evaluated as a sum of these components and compared to a tolerable risk level. If the calculated risk is lower than this tolerable level, no additional lightning protection measures are needed and the level of risk can be accepted. If this is not the case, lightning protection measures should be added until the risk decreases below the tolerable level. The efficiency of the protection measures is related to a level of protection ranging from IV to I, I being the best that defines the percentage of cases for which the structure will be protected. Of course, a Lightning Protection System at level I will cost more than at level IV and a specific economic risk method is proposed for evaluating the cost/benefit of these measures. To collect the input parameters that are required for a proper evaluation of the risk level, needs time and shouldn't be under estimated, especially for old structures for which documents may not exist anymore or complex cases such as a factory where many lines are connected to a structure. For a few cases, simplified methods have been developed to avoid the burden of collecting all these parameters.

Keywords Ground flash density · IEC 62305-2 · Damage source · Structural damage · Human risk

1 Need of Risk Assessment

Very often, when people think about lightning protection for buildings, they first think about lightning rods on the roof. Lightning rods were first created by Benjamin Franklin but at this time electricity and furthermore electronics were of course not a concern. Since this time, it is clear that surge protection should be provided at the same time than lightning protection. Furthermore, it is frequent that there are more possible damages coming from the connected lines (power or telecom/data) that from direct lightning on the structure. Thus, it is needed to analyse the possible sources of damages and evaluate their frequency of occurrence as well as the extent of the associated damages. This applies to buildings and more generally to structures but not to open spaces. It is possible to determine by calculation how many lightning impacts are expected on a given open surface but generally this surface is quite large (field, golf course, football stadium …) and protection is not provided by lightning rods or Surge Protective Devices but by Thunderstorm Warning devices and risk prevention rules. Specific aspects of lightning risk related to TWS will be presented below.

Due to multiple factors involved, the risk analysis for structures, is not straight through, except for simple structures and use, and the conclusions cannot be obtained without performing a few calculations. The old and simple philosophy to install lightning protection components everywhere, expecting that they will adequately protect people and goods, is over. It is generally needed to perform calculations that are simple or complex depending on the complexity of the case as well as associated risks. This analysis will aim to determine:

- What is the most efficient protection strategy (Lightning Protection System, Surge Protective Devices, shielding, equipotentiality, Thunderstorm Warning Systems etc.);
- Where to install the selected lightning protection components;
- With which rating.

Lightning may cause many types of damages to a structure. They are usually named "losses" and we may define up to four types of losses:

- Loss due to injury to human beings. This includes injuries and fatalities for people inside the building, on its roof and terraces and also in close vicinity of the building.
- Loss due to physical damage of the structure and its contents. This includes mechanical damages, fire and in the worst-case explosion.
- Loss due to failure of internal systems. This includes damages to electronic and electrical equipment or installations including telecommunication and data systems.
- Loss of economic value: this is the economic loss associated to physical damage and failure of internal systems including cost to repair and production loss.

The mechanism that creates such damages may be different form one loss to another.

A direct strike to people will of course cause an electric shock to human beings. But even if not stuck directly by lightning, human beings can be exposed to an electric shock due to resistive and inductive coupling (step voltage, touch voltage or sparkover between the lightning circuit and a nearby human being).

When a lightning strikes a building, it may create a mechanical damage (hole on the roof, destruction of concrete walls or roofs ...) or trigger a fire due to sparking on flammable materials (thatched roof, insulating material, oil or chemical product, electrical cables ...) and trigger an explosion if there is an explosive area (gas, dust ...).

When a lightning strikes a building or its connected systems (power, telecom, data, metal pipes, covered by the generic name of "services") the surges due to the lightning or partial lightning current can also trigger a fire or an explosion and they will of course damage electrical and electronic devices. But even when lightning occurs near the building or near the services, the associated induced surges can damage equipment and especially sensitive equipment. If this piece of equipment is related to people's health (hospitals, medical centre ...) such a surge can also create a hazard for human beings. When such a surge damages a safety equipment (fire alarm, smoke detectors, pollution detectors ...) the damage can also create a hazard for human beings (being not warned that a fire is occurring for example) or for the environment (pollution of rivers by releasing chemical product or stored polluted water, due to electromagnetic disturbances on control circuit for example).

Lightning can also, at the same time, be the origin of fire and cause the destruction of fire detectors that can create cumulative hazards.

There is then a need to:

- Determine the possible sources of damages;

- Determine the associated amount of damages;
- Establish the damage occurrence;
- Determine is this is acceptable and if not propose solutions;
- Determine the cost efficiency of proposed solutions especially when there is more than one protection solution.

The methodology to achieve this is named “lightning risk assessment” or sometimes “lightning risk management”. This is covered by an international standard: IEC 62305-2 that is widely recognized and used.

IEC 62305-2 [1] is the reference standard for lightning risk calculation. Its formulas are used by many other standards including electrical installation (IEC 60364 part 443[2]), photovoltaic systems (IEC 60364 Part 7-712 [3]) and wind turbines (IEC 61400-24[4]) or risk for simple structures at national level. The first 62305-2 standard has been published in 2006 but is based on IEC report dated 1995, so it has a long experience. Present edition is edition 2 (published in 2010) and edition 3 is in preparation (expected by the end of 2021). However, this standard may appear too complex for simple cases and too simple for complex cases. The complexity can be addressed by software that are available on the market and it is not the purpose of the present publication to address very complex cases. It is important to note that IEC has made available in the past a software for edition 1 of the 62305-2 standard that is no more applicable. It was, at this time, just addressing a simple version of the standard for allowing users to get familiar with the method.

It is important that the risk is well estimated because it results in a level of protection and the price and efficiency of the lightning protection measures is directly related to this level of protection. Overprotecting is causing an economical issue when under protecting may create a safety hazard.

As it is said previously, there are also simpler methods derived from this main method:

- Risk for electrical installations of buildings;
- Risk for photovoltaic (PV) systems;

For simplicity’s sake, these simple methods will also be presented.

2 The Overall Concept & Levels of Protection

The lightning risk is evaluated for a one-year period. The method is applicable to a single structure. When there are many structures in a site that need to be studied (for example in industrial sites) the calculations should be performed for each structure, taking into account the other structures that may reduce the risk (concept of location factor Cd—see below) or that are connected to the studied structure by lines (named adjacent structures).

The risk due to lightning is the sum of different risk components, differing in their source of damage (S1, S2, S3, S4) and type of damage (D1, D2, D3) defined as follows:

- S1: lightning flash to the structure;
- S2: lightning flash near the structure (flash at a distance up to a few hundred metres but too far to be captured by the structure itself);
- S3: lightning flash to the lines connected to the structure (including lightning flash to the adjacent structures, a part of this flash will flow, thanks to the line, to the studied structure). This also includes underground cables even if generally one falsely believes that being underground a service is protected;
- S4: lightning flash near the lines connected to the structure (flash at a distance up to a few hundred metres but too far to be captured by the line itself).

and:

- D1: injury to living beings by electric shock (mainly human being but in a few occasions, livestock may be concerned as well especially for economic calculations);
- D2: physical damage (fire, explosion, mechanical destruction ...) due to lightning current effects, including sparking;
- D3: failure of internal systems due to lightning electromagnetic pulse (LEMP) i.e. the magnetic field generated by lightning and associated conducted or induced surges.

Using these sources of damage and types of damage, the user can calculate up to eight risk components R_A, R_B, R_C, R_M, R_U, R_V, R_W and R_Z. The fact that the components are ranging from A to Z in spite of being only eight altogether is unfortunately misleading. Training and experience have shown, that the awkward denomination of these risk components, is one of the sources of the complexity of the method if not the main one. Only an experienced user will remember what is the concept behind R_W for example. There is no way to guess what R_W means and if the user doesn't practice risk calculation frequently, he will need a summary of the concepts close to him when performing risk calculation or a dedicated software.

The total risk R is defined as the sum of a few risk components (number of risk component to calculate is depending on the type of risk the user wants to cover and the structure parameters).

If the structure is partitioned in individual zones having specific characteristics (for example a building having an explosive area when the remaining part of the building is having a low risk of fire), each risk component shall be evaluated for each zone (assuming the whole building is an explosive area would overestimate the risk, ignoring the small explosive area, would under estimate the risk). The total risk R of the structure is the sum of all the considered risk components over all the zones which constitute the structure. Zoning may be interesting to estimate more precisely what is the real risk level but implies more calculations and more work to get the raw data.

Each of the risk components R_X (X = A, B, C, M, U, V, W or Z) is expressed by the following equation:

$$R_X = N_X \times P_X \times L_X \quad (1)$$

where:

N_x is the number of dangerous events per year;
P_x is the probability of damage to a structure;
L_x is the consequent loss.

The number N_x of dangerous events is influenced by the lightning ground flash density (N_g), by the dimensions of the structure to be protected, its surroundings, the connected lines and adjacent buildings connected to the structure to be protected by these lines. N_x is based on physical parameters (length, width, height …), that are generally easy to obtain, as well as parameters related to lightning physics and lightning statistics.

You cannot do much generally to reduce the risk with N_x because nobody will accept to move its structure to a safer place or bury its connected line when there are aerial, only to reduce lightning risk.

The probability of damage P_x is influenced by the characteristics of the structure to be protected, the connected lines and the protection measures provided.

This is the key place to reduce the risk, when needed. You generally start the process assuming that there are no lightning protection measures. If the calculated risk is below what is considered as acceptable (see below) the structure is considered as self-protected and no additional protection measures are needed. This doesn't mean that no lighting damage will occur but the statistical risk is low enough to allow to ignore it. If the calculated risk is exceeding what is considered as acceptable, you need to add protection means from the easiest to the more complex one until the risk becomes below the tolerable level. Protection means are related to a protection level ranging generally from IV (lowest level, weaker protection) to I (highest level, stronger protection).

The consequent loss L_x is influenced by the use to which the structure is assigned, the attendance of persons, the type of service provided to public, the value of goods involved in the damage and the measures provided to limit the amount of loss.

You cannot play so much either with L_x to reduce the risk but you can reduce time of presence in dangerous areas thanks to a Thunderstorm Warning System or you can also install firefighting means such as fire detection or automatic fire extinguisher such as sprinklers, to help reducing the consequent loss.

The values of the acceptable amount of loss L_X should be evaluated and fixed by the lightning protection designer or the owner of the structure, unless otherwise fixed by local authority having jurisdiction. Typical mean values of loss L_X in a structure are proposed in the IEC 62305-2 standard. Different values of acceptable loss may be assigned by each country or other authority having jurisdiction or after a detailed investigation.

In the same way, tolerable risk values are proposed by the standard but local authority having jurisdiction can fix other values even if generally most of the users apply the standard proposed values.

Each part of the risk equations, the number of dangerous events per annum N, the probability of damage to a structure P and the consequent loss L, will be explained successively.

But before discussing the risk procedure in more details, it is needed to introduce the level of protection concept. In lightning engineering, you rarely describe a Lightning Protection System in terms of kA or kV. The normal way to characterize an LPS is a figure ranging from 4 to 1, 1 being the more efficient, that is written in roman characters i.e. from IV to I. This figure is named the Lightning Protection Level. As soon as you say to a lightning technician that the LPS should be for example LPL II, he knows exactly how to define and build the LPS.

LPL is then a very important parameter and we need to explain it in details. A LPL is related to two parameters:

- Capture lightning efficiently: as it is easier to capture a high lightning current than a small one due to the electro-geometric model, the critical parameter will be the lowest lighting current the LPS can capture.
- Once captured by the LPS, the LPS should be able to handle this current. As direct lightning current are supposed to have all the same wave shape (10/350 μs) the most severe case is for the highest current magnitude the LPS can handle.

Related probabilities are given based on CIGRE internationally recognized lightning current distribution (Table 1).

This is to be related to lightning current values. Table 2 presents the maximum levels of lightning current associated to each LPL when Table 3 presents the minimum lightning current that will be captured by a LPS for each LPL.

Then the complete Table 4 can be derived from these tables:

Then for LPL I you can expect that up to 2% of lightning event will create a damage to the structure because the current is greater than 200 kA (and LPS could be damaged and thus structure badly protected) or because the current is lower than

Table 1 Probabilities for the lightning current parameters

Probability that lightning current parameters	LPL			
	I	II	III	IV
Are smaller than the maximum values	0.99	0.98	0.95	0.95
Are greater than the minimum values	0.99	0.97	0.91	0.84

Table 2 Maximum lightning current associated to LPL

Maximum lightning current associated to LPL	LPL		
	I	II	III–IV
Peak current/(kA)	200	150	100

Table 3 Minimum lightning current associated to LPL

Minimum lightning current associated to LPL	LPL			
	I	II	III	IV
Peak current/(kA)	3	5	10	16

Table 4 Combined probabilities associated to LPL

Probabilities associated to LPL	LPL			
	I	II	III	IV
Lowest current %	0.99	0.98	0.95	0.95
Largest current %	0.99	0.97	0.91	0.84
Combination %	98	95	90	98
Lightning currents not considered by the LPL (i.e. accepted damage rate due to either a current lower than lowest current considered or a current greater than largest current considered) (%)	2	5	10	20

3 kA and thus not captured by the LPS and the building damaged at the impact location.

It should be noted than low current will generally create small damages and thus the experienced overall efficiency of the LPS may be higher.

For LPL IV, the failure rate is up to 20% so this means that 20% of lighting event could create a damage to the building in spite of the installed LPS.

Lowest level considered by standard is 80% of efficiency but of course a protection providing 50% of efficiency only would be better than nothing if budget is restricted.

3 The Procedure in Brief

Each type of damage may produce a different consequential loss in the structure to be protected. The type of loss that may appear, depends on the characteristics of the structure itself and its content. The following types of loss are considered:

- L1: loss of human life: this include permanent injury and more generally any injury due to lightning, on or near a structure;
- L2: loss of service to the public: this includes power and telecommunication stations as well as water and gas distribution. Basically, this covers any application for which, a damage to a structure (for example a power station) will create a loss to many people located outside the structure. Radio and TV could be included in that list as well as for example train stations or airports or traffic controls where a structure damage would create a trouble to the public. What is considered as service to the public is often different from one country to another and IEC offers only a very restricted list of what it covers.

- L3: loss of cultural heritage: this applies to museum, old buildings and any national heritage for which partial destruction would be a cultural loss. At the moment, this applies only to the structure itself and not to the content that may be as important as the structure itself or even more in case of a museum for example. In that case, the whole structure should be considered as national heritage.
- L4: loss of economic value: this applies to structure, content, and loss of activity. In general, L4 is applied to determine which protection means are most cost efficient. But it is also possible to use this method to determine if a structure that is not considered in the list above (none a service to the public nor a national heritage) deserve a lightning protection. Obviously, most of the structures are occupied, even if not permanently, by human being (employees, customers, safety or maintenance team) and the key risk to calculate is related to L1. In such a case, a few structures may deserve, an additional analysis, based on L4, but with the target to provide more protection that issued from the analysis based only on L1. It may be the case for example for a shopping centre.

Risks are evaluated based on the type of loss that may be concerned. Risk R1 is associated to type of loss L1 and so on. So, a maximum of four risks may be calculated for a structure: R1, R2, R3 and R4.

As explained previously, each risk, R, is the sum of its risk components. When calculating a risk, the risk components may be grouped according to the source of damage and the type of damage.

i. Risk components associated to lightning flashes to the structure

R_A: component related to injury to living beings caused by electric shock due to touch and step voltages inside the structure and outside in the zones up to 3 m around down-conductors. It applies also to people that may be located on the roof and terraces but obviously it is safer to stay inside when thunder roars. Early warning may be given by a TWS to avoid staying in an exposed place.

R_A is associated to loss L1. But in the case of structures with livestock such as farm, cowshed, factory farming loss L4 may also be considered.

R_B: component related to physical damage caused by dangerous sparking inside the structure triggering fire or explosion which may also endanger the environment. This also applies to physical damage to the structure (hole in the roof, wall head damages, concrete spalling and so on).

R_B is associated to all losses: L1, L2, L3 and L4.

R_C: component related to failure of internal systems caused by LEMP.

R_C is normally associated to loss L2 and L4. This can also apply for L1 when there is a risk of explosion or for hospitals or other structures where failure of internal systems immediately endangers human life (retirement home, medical assistance etc.).

ii. Risk component associated to lightning flashes near the structure

R_M: component related to failure of internal systems caused by LEMP.

R_M is normally associated to loss L2 and L4. This can also apply for L1 when there is a risk of explosion or for hospitals or other structures where failure of internal systems immediately endangers human life (retirement home, medical assistance etc.).

iii. Risk components associated to lightning flashes to a line connected to the structure

The lines taken into account are only the lines entering the structure. A line that would connect two parts of the same structure would be ignored.

Lightning flashes to metal pipes are not considered due to the grounding of pipes at structure entrance that is requested by IEC standards. If direct grounding is not accepted or possible (for example in case of cathodic protection) an insulating spark gap will provide grounding when a surge occurs.

R_U: component related to injury to living beings caused by electric shock due to touch voltage inside the structure. This can happen when a spark is created near people between an electrical appliance or power socket and a metal grounded part. This can also happen, when maintenance is performed on electrical or telecommunication system inside the structure in a stormy weather. People inside structure may not be aware that lightning will occur soon, if they work far from a window.

R_U is associated to loss L1. But in the case of structures with livestock such as farm, cowshed, factory farming loss L4 may also be considered.

R_V: component related to physical damage (fire or explosion triggered by dangerous sparking between external lines and metallic parts generally at the entrance point of the line into the structure) due to lightning current flowing in the lines.

R_V is associated to all losses: L1, L2, L3 and L4.

R_W: component related to failure of internal systems caused by overvoltages induced on incoming lines and transmitted to the structure.

R_W is normally associated to loss L2 and L4. This can also apply for L1 when there is a risk of explosion or for hospitals or other structures where failure of internal systems immediately endangers human life (retirement home, medical assistance …).

iv. Risk component associated to lightning flashes to the structure for a structure due to flashes near a line connected to the structure

The lines taken into account are only the lines entering the structure. A line that would connect two part of the same structure would be ignored.

Lightning flashes to metal pipes are not considered due to the grounding of pipes at structure entrance that is requested by standards. If direct grounding is not accepted or possible (for example in case of cathodic protection) an insulating spark gap will provide grounding when a surge occurs.

R_Z: Component related to failure of internal systems caused by overvoltages induced on incoming lines and transmitted to the structure.

R_Z is normally associated to loss L2 and L4. This can also apply for L1 when there is a risk of explosion or for hospitals or other structures where failure of internal systems immediately endangers human life (retirement home, medical assistance …).

v. Summary of risk components associated to each type of loss

R1: Risk of loss of human life:

$$R1 = R_A + R_B + R_C^x + R_M^x + R_U + R_V + R_W^x + R_Z^x \quad (2)$$

[x]For structures with risk of explosion and for hospitals and similar structures with life-saving electrical equipment or other structures when failure of internal systems may endanger human life.

R2: Risk of loss of service to the public:

$$R2 = R_B + R_C + R_M + R_V + R_W + R_Z \quad (3)$$

R3: Risk of loss of cultural heritage:

$$R3 = R_B + R_V \quad (4)$$

R4: Risk of loss of economic value:

$$R4 = R_A^y + R_B + R_C + R_M + R_U^y + R_V + R_W + R_Z \quad (5)$$

[y]For properties where animals may be lost (farms, etc.)

The risk components corresponding to each type of loss are also combined in Table 5.

vi. Tolerable risks RT

Representative values of tolerable risk RT per year are given in Table 6. As explained previously this value could be changed by site owner, authority having jurisdiction and so on. They are just indicative but used extensively worldwide.

Table 5 Risk components associated to each type of loss

Source of damage	Flash to a structure S1			Flash near a structure S2	Flash to a line connected to a structure S3			Flash near a line connected to a structure S4
Risk component	R_A	R_B	R_C	R_M	R_U	R_V	R_W	R_Z
R1	X	X	X[x]	X[x]	X	X	X[x]	X[x]
R2		X	X	X		X	X	X
R3		X				X		
R4	X[y]	X	X	X	X[y]	X	X	X

[x]For structures with risk of explosion and for hospitals and similar structures with life-saving electrical equipment or other structures when failure of internal systems may endanger human life
[y]For properties where animals may be lost (farms, etc.)

Table 6 Typical values of tolerable risk RT

Risk considered		RT (per year)
R1	Risk of loss of human life	10^{-5}
R2	Risk of loss of service to the public	10^{-3}
R3	Risk of loss of cultural heritage	10^{-4}
R4	Risk of loss of economic value[a]	10^{-3}

[a]In principle, for risk of loss of economic value, a cost/benefit comparison should be performed but very often the data for this analysis are not available and thus a representative value of tolerable risk has also be defined

vii. Procedure to determine the protection means to be used for Risk R1–R3

For each risk R1, R2 and/or R3 to be considered the following steps shall be taken:

- calculation of these risk components R_X as defined above;
- calculation of the risk R as sum of these risk components;
- comparison of the risk R with the tolerable value RT.

If $R \leq RT$, lightning protection means are not necessary.

If $R > RT$, lightning protection means shall be adopted in order to reduce the risk until $R \leq RT$.

In cases where the risk cannot be reduced below the tolerable level in spite of using the highest levels of protection (I), better protection means (more efficient than LPL

I, this will be explained below) or alternate solutions such as Thunderstorm Warning Systems should be used. Reduction of the risk may also be achieved by using zones inside the structure. Dividing a structure into zones allows the designer to take into account the characteristics of each zone of the structure in the evaluation of risk components and to select the most suitable protection measures tailored zone by zone. A more accurate evaluation of input parameters may also be useful. If all these methods fail to reduce enough the risk, the site owner should be informed and the highest level of protection (i.e. I) should be used.

Generally, for structures with a risk of explosion, a minimum LPL II is needed.

When the damage to a structurse due to lightning may also involve surrounding structures or the environment (e.g. chemical or radioactive emissions, toxic fumes, pollution of soil or water …) a specific risk evaluation is needed. It is based on an additional evaluation of the loss L1.

viii. Procedure to determine the protection means to be used for risk R4 when the simplified method of tolerable risk is not used

Whether or not there is need to determine protection to reduce risks R1, R2, and R3, it is useful to evaluate the economic justification of protection measures by calculating risk R4.

The cost of loss may be calculated by the following equation:

$$CL = R4 \times Ct \tag{6}$$

where Ct is the total value of the structure (animals, building, content and internal systems including their activities) and R4 is the risk related to loss of value in the structure, without protection measures.

The cost CRL of residual loss in spite of protection means may be calculated by the equation:

$$CRL = R'4 \times Ct \tag{7}$$

where R'4 is the risk related to loss of value in the structure, with protection measures.

The annual cost CPM of protection measures may be calculated by means of the equation:

$$CPM = CP \times (i + a + m) \tag{8}$$

where

CP the cost of protection measures including raw protection material and installation cost;

i the interest rate (to be obtained from financial authorities);

Table 7 Typical values of i, a and m

Rate	Symbol	Typical values
Interest	i	0.04
Amortization	a	0.05
Maintenance	m	0.01

a the amortization rate (to be based on how many years are considered until lightning protection is amortized. With a linear amortization rate over 20 years the rate is 0.05);

m the maintenance rate (to be obtained from lightning protection installer or designer).

When i, a and/or m are not known, the following values can be used instead (Table 7).

The annual saving SM in currency is then:

$$SM = CL - (CPM + CRL) \tag{9}$$

Protection is justified if the annual saving SM is greater than 0.

Note that the economic risk can also be evaluated by zones as explained above.

It should also be noted that when all parameters are not known to perform such a complete economic risk calculation, a simpler calculation of risk based on assessment of R4 only and comparing it to the tolerable risk define above (see Table 6) can be performed instead.

ix. Formulas used to calculate risk components

Table 8 summarize the various equations used to calculate the risk components.

Where:

N_D Average annual number of dangerous events due to lightning flashes to the structure;

N_M Average annual number of dangerous events due to lightning flashes near the structure;

N_L Average annual number of dangerous events due to lightning flashes to a line entering the structure;

N_i Average annual number of dangerous events due to lightning flashes near a line entering the structure;

N_{DJ} Average annual number of dangerous events due to lightning flashes to the adjacent structure;

P_A Probability that a flash to the structure will cause injury to living beings by electric shock;

P_B Probability that a flash to the structure will cause physical damage;

Table 8 Risk components equations

Damage	Source of damage			
	S1 Lightning flash to a structure	S2 Lightning flash near a structure	S3 Lightning flash to an incoming line	S4 Lightning flash near a line
D1 Injury to living beings by electric shock	$R_A = N_D \times P_A \times L_A$		$\mathbf{R_U = (N_L + N_{DJ}) \times P_U \times L_U}$	
D2 Physical damage	$\mathbf{R_B = N_D \times P_B \times L_B}$		$\mathbf{R_V = (N_L + N_{DJ}) \times P_V \times L_V}$	
D3 Failure of electrical and electronic systems	$\mathbf{R_C = N_D \times P_C \times L_C}$	$\mathbf{R_M = N_M \times P_M \times L_M}$	$\mathbf{R_W = (N_L + N_{DJ}) \times P_W \times L_W}$	$\mathbf{R_Z = N_i \times P_Z \times L_Z}$

P_C	Probability that a flash to the structure will cause failure of internal systems;
P_M	Probability that a flash near the structure will cause failure of internal systems;
P_U	Probability that a flash to a line will cause injury to living beings by electric shock;
P_V	Probability that a flash to a line will cause physical damage;
P_W	Probability that a flash to a line will cause failure of internal systems;
P_Z	Probability that a flash near a line will cause failure of internal systems;
$L_A = L_U$	Loss due to injury to living beings by electric shock;
$L_B = L_V$	Loss due to physical damage;
$L_C = L_M = L_W = L_Z$	Loss due to failure of internal systems.

Equations to evaluate all these parameters are given below based on input parameters.

4 Input Parameters

To obtain accurate input parameters may be tricky and a long process. When calculating risk for a new structure the drawing and specification are generally available but for older structures it may be difficult to obtain a few parameters such as for example the cable shield resistance. In that case, the more conservative value should be used. If too many parameters are unknown the validity of risk calculation could be discussed. In a risk calculation, the collect of input parameters is sometimes a challenge and time to spend collecting these parameters should not be underestimated. For a few cases, it is known that input parameters will be difficult to obtain and a few simplified methods have been developed to concentrate only on the key parameters. They will be presented in V.

The list of input parameters is given below and then how they can be obtained and how they are used, is also presented.

N_G	number of lightning flashes per km^2 per year;
L	structure length in m (studied or adjacent);
W	structure width in m (studied or adjacent);
H	structure height in m (this height may vary along the roof and should be considered at each point around the structure, this applies to the studied or adjacent structure);
C_D	location factor of the studied or adjacent structure (see Table 9);
L_L	length of lines connected to the studied structure in m;
C_L	installation factor of these lines (see Table 10);
C_T	lines type factor (see Table 11);
C_E	environmental factor of these lines (see Tables 12, 13, 14 and 15);

Table 9 Structure location factor C_D

Relative location	C_D
Structure surrounded by higher objects	0.25
Structure surrounded by objects of the same height or smaller	0.5
Isolated structure: no other objects in the vicinity	1
Isolated structure on a hilltop or a knoll	2

Table 10 Line installation factor C_L

Routing	C_L
Aerial	1
Buried	0.5

Table 11 Line type factor C_T

Installation	C_T
LV power, telecommunication or data line	1
HV power (with HV/LV transformer)	0.2

Table 12 Line environmental factor C_E

Environment	C_E
Rural	1
Suburban	0.5
Urban	0.1
Urban with buildings higher than 20 m	0.01

C_{LD} factor depending on shielding, grounding and isolation conditions of these lines related to direct flashes (see Table 16);

C_{Li} factor depending on shielding, grounding and isolation conditions of these lines related to induced surges (see Tables 16, 17 and 18);

P_{TA} protection measures against step and touch voltages outside structure (see Table 13);

Table 13 Values of probability P_{TA} that a flash to a structure will cause shock to living beings due to dangerous touch and step voltages

Additional protection measure	P_{TA}
No protection measures	1
Warning notices	10^{-1}
Electrical insulation (e.g. at least 3 mm cross-linked polyethylene) of down-conductors	10^{-2}
Effective soil equipotentialization	10^{-2}
Physical restrictions or building framework used as a down-conductor system	0

Table 14 Values of probability P_B depending on the LPS used to reduce physical damage

Characteristics of structure	LPL of LPS	P_B
Structure not protected by LPS	–	1
Structure protected by LPS	IV	0.2
	III	0.1
	II	0.05
	I	0.02
Structure with an air-termination system conforming to LPS I and a continuous metal or reinforced concrete framework acting as a natural down-conductor system		0.01
Structure with an air-termination system conforming to LPS I, with complete protection of any roof installations against direct lightning strikes and a continuous metal or reinforced concrete framework acting as a natural down-conductor system		0.001

Table 15 Values of probability P_{SPD} for coordinated SPDs

LPL	P_{SPD}
No coordinated SPD system	1
III–IV	0.05
II	0.02
I	0.01
SPD having better having better protection characteristics (lower protective level UP …) compared with the requirements defined for LPL I	0.005–0.001

Table 16 Values of factors C_{LD} and C_{Li}

External line type	Connection at entrance	C_{LD}	C_{Li}
Any type	Unshielded internal systems	1	0
Aerial or buried line unshielded		1	1
Multi grounded neutral power line		1	0.2
Shielded buried line (power or telecom)	Shield not bonded to the same bonding bar as equipment	1	0.3
Shielded aerial line (power or telecom)	Shield not bonded to the same bonding bar as equipment	1	0.1
Shielded aerial or buried line (power or telecom)	Shield bonded to the same bonding bar as equipment	1	0
Wiring in lightning protective cable ducts, metallic conduit, or metallic tubes	Shield bonded to the same bonding bar as equipment	0	0
Any type	Isolating interface according to IEC 62305-4 (this is generally called SIT in IEC 61643 series of standards)	0	0

Table 17 Values of factors K_{S3}

Type of internal wiring	K_{S3}
Unshielded cables—no routing precaution in order to avoid loops (loop area in the order of 50 m2 typical of a device connected to two different services such as power and telecom, in a large building)	1
Unshielded cables—routing precaution in order to avoid large loops (loop area in the order of 10 m2 typical of conductors routed in the same conduit or conductors with different routing in a small building)	0.2
Unshielded cable—routing precaution in order to avoid loops (loop area in the order of 0.5 m^2 typical of conductors routed in the same cable)	0.01
Shielded cables or cables running in metal conduits when shield and metal conduits are bonded to an equipotential bonding bar at both ends and equipment is connected to the same bonding bar	0.0001

Table 18 Typical values of Uw for various power systems

Nominal voltage of installation (V)	Rated impulse withstand voltage (kV)			
	Voltage category IV (origin of installation)	Voltage category III (distribution equipment and circuits)	Voltage category II (equipment)	Voltage category I (sensitive equipment)
127/220	4	2.5	1.5	0.8
230/400 277/480	6	4	2.5	1.5
400/690	8	6	4	2.5
1000	12	8	6	4

P_{TU} protection measures against touch voltages inside structure (see Table 19);
P_B probability associated to the LPL of the LPS (see Table 14);
P_{SPD} probability associated to coordinated SPD (see Table 15);
P_{EB} probability associated to equipotential bonding SPD (see Table 20);
K_{S1} screening effectiveness of the structure, LPS or other shields at boundary LPZ 0/1;
K_{S2} screening effectiveness of shields internal to the structure at boundary LPZ 1/2;

Table 19 Values of probability P_{TU}

Protection measure	P_{TU}
No protection measures	1
Warning notices	10^{-1}
Electrical insulation	10^{-2}
Physical restrictions	0

Table 20 Values of probability P_{EB} for SPD at entrance of installation (equipotential bonding)

LPL	P_{EB}
No SPD	1
III–IV	0.05
II	0.02
I	0.01
SPD having better having better protection characteristics (lower protective level UP …) compared with the requirements defined for LPL I	0.005–0.001

Table 21 Values of probability P_{LD}

Power line or telecom line characteristics		Withstand voltage U_W in kV				
		1	1.5	2.5	4	6
Aerial or buried line, unshielded or shielded whose shield is not bonded to the same bonding bar as equipment		1	1	1	1	1
	Shield resistance					
Shielded aerial or buried whose shield bonded to the same bonding bar as equipment	5 Ω/km < Rs ≤ 20 Ω/km	1	1	0.95	0.9	0.8
	1 Ω/km < Rs ≤ 5 Ω/km	0.9	0.8	0.6	0.3	0.1
	Rs ≤ 1 Ω/km	0.6	0.4	0.2	0.04	0.02

K_{S3} characteristics of internal wiring (see Table 17);
K_{S4} impulse withstand voltage of the system to be protected;
U_w rated impulse withstand voltage of system to be protected in kV;
R_S shield resistance of connected shielded lines in Ω/km (see Table 21).

i. Assessment of annual number N of dangerous events

The key parameter for risk calculation is the lightning ground flash density N_G (number of lightning flashes per km^2 per year) that depends on the thunderstorm activity of the region where the structure or lines are located.

In many areas of the world this value can be derived from data provided by Lightning Location Systems (LLS) complying with IEC 62858.

In areas where a N_G map is not available, the recommended estimate of ground flash density is

$$N_G = 0.25 \times N_T \tag{10}$$

where N_T is the total (cloud to ground and intra clouds flashes) density of optical recorded flashes per km^2 per year, obtained through the website http://lightning.nsstc.nasa.gov/data/data_lis-otd-climatology.html (NASA web site).

(a) Calculation of N_D

Once, Ng is obtained, it is needed to assess the equivalent collection area of the studied structure (it applies also to the adjacent structure).

For simple structures on flat ground, the collection area of the structure A_D (in m^2) is the area defined at ground level by a curve located at a distance 3 × H from the structure border.

For a rectangular (L, W, H) structure the collection area can then be easily calculated:

$$A_D = L \times W + 2 \times (3 \times H) \times (L + H) + \pi \times (3 \times H)^2 \tag{11}$$

If the structure has a complex shape such as roof protrusions a graphical method should be used to evaluate A_D always taking care of 3 times the height of the structure.

An example, is given below for a structure with a tower or a chimney.

By considering only the highest part of the structure, the collection area is under estimated. It is represented in orange on figure below. It is a circle of radius 3 × HT (the diameter of the chimney itself is negligible in that example, HT is measured from ground level). When considering only the main part of the structure, the collection area is also under estimated. It is represented in blue in figure below and is defined based on L, W and H as in Figs. 1 and 2. The correct collection area is the combined surface between the blue and orange one and is represented in green on Fig. 3.

An acceptable approximate value of the collection area is the greater between the collection area in blue and the collection area attributed in orange. Very often, when the chimney is much higher than the main part of the structure, considering only the orange area will give a pretty good approximation of the correct collection area.

When the structure is not isolated on a flat ground, it is needed to take care of surrounding objects within a distance of 3 × H from the structure, taking care of the location factor C_D defined in Table 9.

Then N_D may be evaluated as follows:

$$N_D = N_G \times A_D \times C_D \times 10^{-6} \tag{12}$$

10^{-6} is used to allow the use of structure dimensions in m when Ng is given in flashes/year/km^2.

(b) Calculation of N_M

N_M may be evaluated as follows:

$$N_M = N_G \times A_M \times 10^{-6} \tag{13}$$

where A_M is the collection area of flashes striking near the structure (in m^2), represented by a line located at a distance of 500 m from the perimeter of the structure.

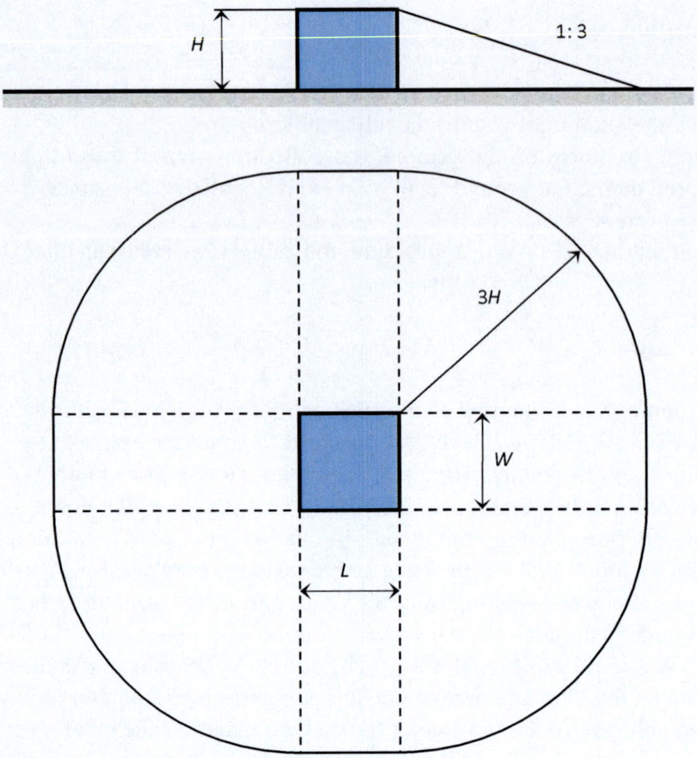

Fig. 1 Collection area A_D of a simple structure

Up to 500 m a lightning flash can create damaging surge inside the structure.

$$A_M = 2 \times 500 \times (L + H) + \pi \times (500)^2 \tag{14}$$

(c) Calculation of N_L

A line may consist of several sections (for example one section is underground when another one is overhead). For each section of line, the value of N_L may be evaluated as follows:

$$N_L = N_G \times A_L \times C_L \times C_E \times C_T \times 10^{-6} \tag{15}$$

where

A_L is the collection area of flashes striking the line (m^2);
C_L is the installation factor of the line;
C_T is the line type factor;

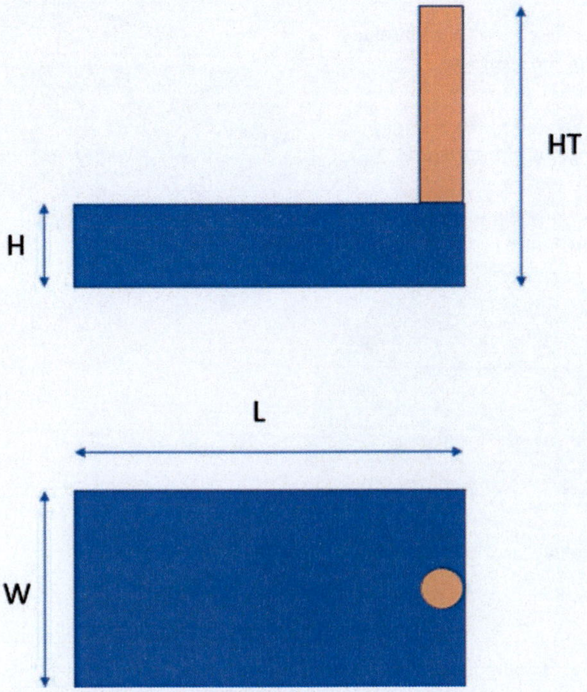

Fig. 2 Structure (in blue) with a tower/chimney (in orange) with height HT

C_E is the line environmental factor.

The collection area for flashes to a line can be calculated as follows:

$$A_L = 400 * L_L \tag{16}$$

L_L is the length of the line section (m).

Where the length of a line section is unknown, $L_L = 1000$ m is to be assumed.

(d) Calculation of N_i

For each section of line, the value of N_I may be evaluated as follows:

$$N_i = N_G \times A_i \times C_L \times C_E \times C_T \times 10^{-6} \tag{17}$$

with the collection area for flashes near a line

$$A_i = 4000 * L_L \tag{18}$$

where the length of a line section is unknown, $L_L = 1000$ m is to be assumed.

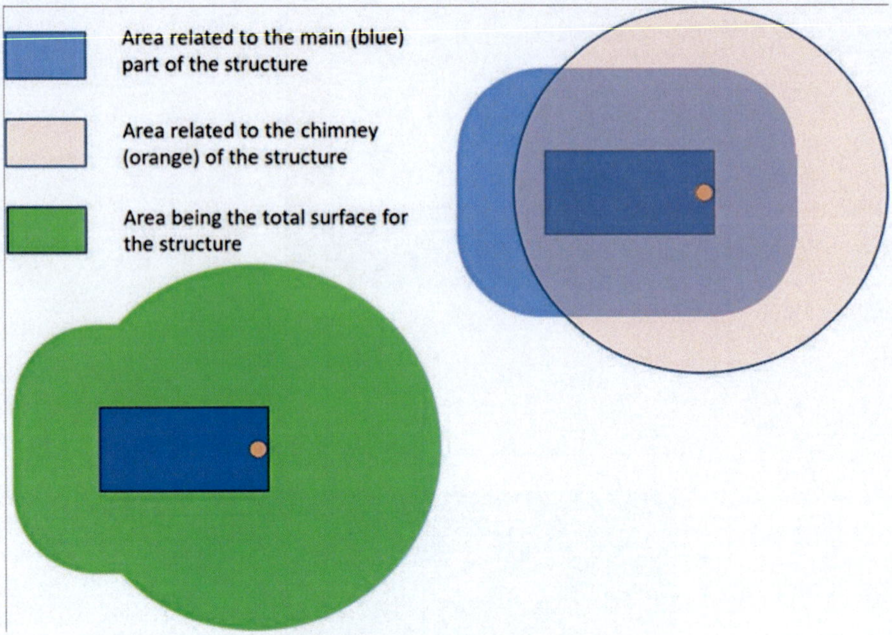

Fig. 3 Collection area being the total surface for the structure

(e) Representation of all the collection areas

Figure 3 is representing a structure connected to an adjacent structure with a line 1 when line 2 is connected to the studied structure but not to an adjacent structure (for example a line longer than 1000 m and in that case the adjacent structure is ignored because the maximum line length is 1000 m, or a line connected to a small structure such as for example a telecom box whose small dimensions can be ignored). When there are many lines connected to the studied structure, the risk components related to lines should be added to calculate the risk (Fig. 4).

ii. Assessment of probability P of damage

(a) P_A

$$P_A = P_{TA} \times P_B \tag{19}$$

where P_{TA} and P_B are given in Table 13.

If more than one provision has been taken, the value of P_{TA} is the product of the corresponding values.

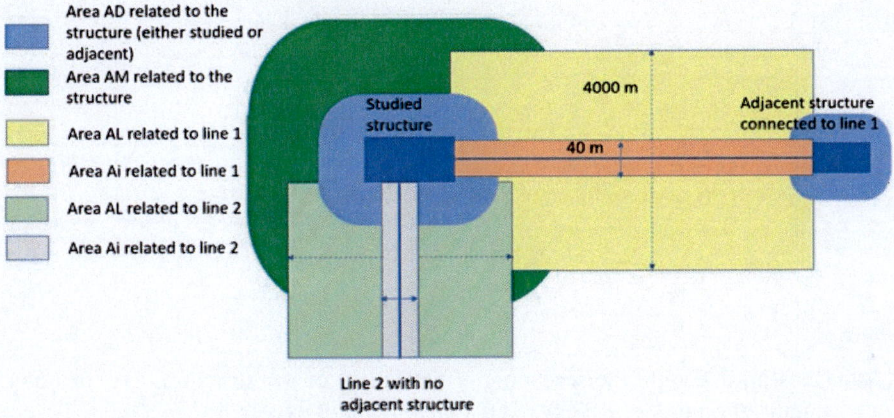

Fig. 4 Collection areas A_D, A_M, A_i and A_L for this structure

(b) P_B

As indicated above, better values than LPL I, are possible.

(c) P_C

$$P_C = P_{SPD} \times C_{LD} \tag{20}$$

P_{SPD} depends on a coordinated SPD system conforming to IEC 62305-4 and to the lightning protection level (LPL) for which its SPDs are designed (i.e. at least the LPL of the SPD at entrance of the installation.

C_{LD} a factor depending on shielding, grounding and isolation conditions of the line to which the internal system is connected.

Once again, better values than LPL I, are possible, as explained previously.

It should be noted that the LPL needed for coordinated SPD system is at least the LPL of the equipotential bonding SPD. The LPL of the equipotential bonding SPD is itself at least the LPL of the LPS.

A coordinated SPD system consists of an equipotential bonding SPD at the entrance of installation (generally in Main Panel Board) and other SPDs (generally in subsidiary panel board but can be also in special cabinet in front of equipment or in the power socket for example), coordinated with equipotential bonding SPD, installed to protect sensitive or specific equipment inside the installation (equipment listed by the lightning risk assessment as being to be protected. It is not possible to protect all equipment in an installation with only a few SPDs).

(d) P_M

$$P_M = P_{SPD} \times P_{MS} \tag{21}$$

The values of P_{MS} are obtained from the product:

$$P_{MS} = (K_{S1} \times K_{S2} \times K_{S3} \times K_{S4})^2 \tag{22}$$

where

- K_{S1} takes into account the screening effectiveness of the structure, LPS or other shields at boundary LPZ 0/1. This factor cannot exceed 1;
- K_{S2} takes into account the screening effectiveness of shields internal to the structure at boundary LPZ ½. This factor cannot exceed 1;
- K_{S3} takes into account the characteristics of internal wiring;
- K_{S4} takes into account the impulse withstand voltage of the system to be protected. This factor cannot exceed 1

For continuous metal shields with thicknesses not lower than 0.1 mm, $K_{S1} = K_{S2} = 10^{-4}$.

If shield thickness is less than 0.1 mm, K_{S1} (or K_{S2}) should be assumed equal to 1.

For spatial grid-like shields factors K_{S1} and K_{S2} may be evaluated as

$$K_{S1} = 0.12 \times w_{m1} \tag{23}$$

$$K_{S2} = 0.12 \times w_{m2} \tag{24}$$

where w_{m1} (m) and w_{m2} (m) are the mesh widths of grid-like spatial shields, or of mesh type LPS down-conductors or the spacing between the structure metal columns, or the spacing between a reinforced concrete framework acting as a natural LPS.

These formulas only apply at a safety distance from the boundary screen at least equal to the mesh width w_m. If equipment or cabling are located nearer to shield boundary the values of K_{S1} and K_{S2} will be greater than calculated. For example, values should be doubled where the distance to the shield ranges from 0.1 w_m to 0.2 w_m. In case of doubt, K_{S1} (K_{S2}) should be assumed equal to 1.

For a cascade of LPZs inside a LPZ2 the resulting K_{S2} is the product of the relevant K_{S2} of each LPZ.

$$K_{S4} = \frac{1}{U_W} \tag{25}$$

where

Table 22 Values of probability P_{Li}

Line type	Withstand voltage U_W in kV				
	1	1.5	2.5	4	6
Power lines	1	0.6	0.3	0.16	0.1
TLC lines	1	0.5	0.2	0.08	0.04

U_w the rated impulse withstand voltage of system to be protected, in kV (this applies to the lowest impulse withstand of equipment in the structure or of the zone inside the structure).

(e) P_U

$$P_U = P_{TU} \times P_{EB} \times P_{LD} \times C_{LD} \tag{26}$$

where P_{TU}, P_{EB} and P_{LD} are given in Table 19.

If more than one provision has been taken, the value of P_{TU} is the product of the corresponding values (Table 22).

(f) P_V

$$P_V = P_{EB} \times P_{LD} \times C_{LD} \tag{27}$$

(g) P_W

$$P_W = P_{SPD} \times P_{LD} \times C_{LD} \tag{28}$$

(h) P_Z

$$P_Z = P_{SPD} \times P_{Li} \times C_{Li} \tag{29}$$

iii. Assessment of losses

(a) Loss of human life (L1) (Table 23).

where

L_T the typical mean relative numbers of victims injured by electric shock;
L_F the typical mean relative numbers of victims by physical damage;
L_O the typical mean relative numbers of victims by failure of internal systems;
n_z the number of persons in the zone;
n_t the total number of persons in the structure;
t_z the time in hours per year for which the persons are present in the zone.

Table 23 Typical losses per zone for L1

Type of damage	Typical loss	Equations
D1	$L_A = L_U = \frac{r_t \times L_T \times \frac{n_Z}{n_t \times t_Z}}{8760}$	(30)
D2	$L_B = L_V = \frac{r_p \times r_f \times h_z \times L_F \times \frac{n_Z}{n_t \times t_Z}}{8760}$	(31)
D3	$L_C = L_M = L_W = L_Z = \frac{r_t \times L_O \times \frac{n_Z}{n_t \times t_Z}}{8760}$	(32)

In case of a TWS associated to procedures to ensure that people are in safe area in stormy periods, t_Z can be multiplied by (1—FTWR). FTWR the Failure To Warn Ratio is obtained based on IEC 62793 [5]) (Tables 24, 25, 26 and 27).

The specific fire load is the ratio of the energy of the total amount of the combustible material in a structure and the overall surface of the structure.

When there is an explosion risk more risk component should be calculated. So either the calculation is done only for explosion risk or, when r_f, is greater for fire risk than for explosion risk, calculation is done first for explosion and then for fire risk and the worst result should be considered (Table 28).

(b) Loss for the Environment (LE)

Table 24 Typical mean values of L_T, L_F and L_O for L1

Type of damage	Typical loss value		Type of structure
D1	L_T	10^{-2}	All types
D2	L_F	10^{-1}	Risk of explosion
		10^{-1}	Hospital, hotel, school, civic building
		5×10^{-2}	Public entertainment, church, museum
		2×10^{-2}	Industrial, commercial
		10^{-2}	Others
D3	L_O	10^{-1}	Risk of explosion
		10^{-2}	Intensive care unit and operation block of hospital
		10^{-3}	Other parts of hospital

Table 25 r_t function of the type of surface of soil (L_A) or floor (L_U)

Type of surface	Contact resistance kΩ	r_t
Agricultural, concrete	≤1	10^{-2}
Marble, ceramic	1–10	10^{-3}
Gravel, moquette, carpets	10–100	10^{-4}
Asphalt, linoleum, wood	≥100	10^{-5}
Asphalt 5 cm thick or a layer of gravel 15 cm thick		0

Table 26 r_p function of provisions taken to reduce the consequences of fire

Provisions	r_p
No provisions or structures with risk of explosion	1
Manual (extinguishers; fixed manually operated extinguishing installations; manual alarm installations; hydrants; fire compartments; escape routes)	0.5
Automatics (fixed automatically operated extinguishing installations; automatic alarm installations if protected against over voltages and other damages and if firemen can arrive in less than 10 min)	0.2

Table 27 r_f as a function of risk of fire or explosion of the structure

Risk	Amount of risk	r_f
Explosion	Zones 0, 20 or solid explosive	1
	Zones 1, 21	10^{-1}
	Zones 2, 22	10^{-3}
Fire	High (with a specific fire load larger than 800 MJ/m^2)	10^{-1}
	Ordinary (with a specific fire load between 800 MJ/m^2 and 400 MJ/m^2)	10^{-2}
	Low (with a specific fire load lower than 400 MJ/m^2)	10^{-3}
Explosion or fire	None	0

Table 28 h_z function of level of panic

Level of panic	h_z
No special hazard	1
Low level of panic (e.g. a structure with a number of persons lower than 100 and limited to two floors)	2
Average level of panic (e.g. structures with a number of participants between 100 and 1000 persons, for example structure designed for cultural or sport events, or structure with more than 2 floors whatever is the number of persons)	5
Difficulty of evacuation (e.g. structures with immobile persons, hospitals) whatever is the number of floors and persons	5
High level of panic (e.g. structures with a number of participants—greater than 1000 persons, for example structure designed for cultural or sport events)	10

When the damage to a structure due to lightning involves surrounding structures or the environment (e.g. chemical or radioactive emissions, toxic fumes, water or soil pollution), additional loss should be taken into account to evaluate the total loss L1 [6]:

$$L_{BT}(orL_{VT}) = L_B(orL_V) + L_{BE}(orL_{VE}) \quad (33)$$

$$L_{CT}(orL_{MT}orL_{WT}orL_{ZT}) = L_C(orL_M orL_W orL_Z) + L_{CE}(orL_{ME}orL_{WE}orL_{ZE}) \quad (34)$$

where

$$L_{BE} = L_{VE} = \frac{r_p \times r_f \times L_{FE} \times t_e}{8760} \quad (35)$$

$$L_{CE} = L_{ME} = L_{WE} = L_{ZE} = \frac{r_p \times r_f \times L_{OE} \times t_e}{8760} \quad (36)$$

L_{FE} being the typical mean percentage of persons outside the structure injured by physical damage; If unknown, the values proposed in Table 29 can be used.

L_{OE} being the typical mean percentage of persons outside the structure injured by failure of internal systems;

Table 29 Proposed values for time of presence of people t_e

Type of surrounding	$t_e/8760$[a]
Inland waterways	0.1
Temporary attendance	0.1
People working inside the site fence	0.25
Railways	0.25
Undeveloped lands and areas with little attendance such as fields, meadows, forests, vacant lots, marshes, horticultural gardens, gardens, vineyards, fishing areas, freight train stations	0.25
Buildings with presence of public	0.5
Areas frequented and very frequented such as parking lots, parks, supervised swimming areas, sports grounds	0.5
Activity zones (industries and other activities not generally open to the public)	0.75
Paths and pedestrian ways	0.75
Site with patrollers or operation of the site with more than one shift	1.0
Residences	1.0
Automobile traffic lanes (local, national, expressways and highways)	1.0

[a]In case of mix environment with different values, the highest value should be used

t_e being the time of presence of people in the dangerous place outside the structure. If values of t_e are unknown, $t_e/8760 = 1$ should be assumed. Otherwise the values proposed in Table 29 can be used (Table 30).

In case of a TWS associated to appropriate procedures, the values for L_{FE} and L_{OE} related to environmental risk remaining inside the site fence, can be multiplied by (1-FTWR) as indicated above.

(c) Loss of service to the public (L2) (Table 31)

where

n_z the number of users served by the zone;
n_t the total number of users served by the structure (Table 32).

(d) Loss of cultural heritage (L3) (Table 33)

Table 30 Typical mean values of L_{FE} and L_{OE}

	Environmental risk remaining inside the site fence		Environmental risk spreading outside of the site fence	
Scenario	L_{FE}	L_{OE}	L_{FE}	L_{OE}
Explosion or overpressure that exceeds 50 hPa	0.25	0.025	0.5	0.05
Thermal flux that exceeds a value of 3 kW/m^2	0.05	0.005	0.1	0.01
Toxic fumes	0.1	0.01	1.0	0.1
Soil pollution	0.1	0.01	0.5	0.05
Water pollution	0.25[a]	0.025[a]	2.5	0.25
Radioactive material[b]	0.5	0.05	5	

[a]Only if pollution can reach the water bed or fresh water or sea/oceans
[b]Not applicable to sealed sources for example used in measuring devices or medical equipment

Table 31 Typical losses per zone for L2

Type of damage	Typical loss	Equations
D2	$L_B = L_V = \frac{r_p \times r_f \times L_F \times n_Z}{n_t}$	(37)
D3	$L_C = L_M = L_W = L_Z = \frac{L_O \times n_Z}{n_t}$	(38)

Table 32 Typical mean values of L_F and L_O for L2

Type of damage	Typical loss value		Type of service
D2	L_F	10^{-1}	Gas, water, power supply
		10^{-2}	TV, telecommunications lines
D3	L_O	10^{-2}	Gas, water, power supply
		10^{-3}	TV, telecommunications lines

Table 33 Typical losses for L3

Type of damage	Typical loss value	Equation
D2	$L_B = L_V = \frac{r_p \times r_f \times 10^{-1} \times c_Z}{c_t}$	(39)

Table 34 Typical losses for L4

Type of damage	Typical loss	Equations
D1	$L_A = L_U = \frac{r_t \times L_T \times c_a}{c_t}$	(40)
D2	$L_B = L_V = \frac{r_p \times r_f \times L_F \times (c_a + c_b + c_C + c_s)}{c_t}$	(41)
D3	$L_C = L_M = L_W = L_Z = \frac{L_O \times c_s}{c_t}$	(42)

(e) Economic loss (L4) (Table 34)

The ratios $\frac{c_a}{c_t}$ and $\frac{(c_a + c_b + c_c + c_s)}{c_t}$ and $\frac{c_s}{c_t}$ have only to be considered only if the full economic risk assessment procedure is conducted. When, in absence of a few parameters, the tolerable risk is used instead, then these ratios have to be replaced by 1.

where

c_a the value of animals in the zone;
c_b the value of building relevant to the zone;
c_c the value of content in the zone;
c_s the value of internal systems including their activities in the zone;
c_t the total value of the structure (sum over all zones for animals, building, content and internal systems including their activities) (Table 35).

Table 35 Typical mean values of L_T, L_F and L_O for L4

Type of damage	Typical loss value		Type of structure
D1	L_T	10^{-2}	All types where animals are present
D2	L_F	1	Risk of explosion
		0.5	Hospital, industrial, museum, agricultural
		0.2	Hotel, school, office, church, public entertainment, commercial
		10^{-1}	Others
D3	L_O	10^{-1}	Risk of explosion
		10^{-2}	Hospital, industrial, office, hotel, commercial
		10^{-3}	Museum, agricultural, school, church, public entertainment
		10^{-4}	Others

Table 36 Environmental factor f_{env}

Environment	f_{env}
Rural and suburban environment	85
Urban environment	850

iv. Simplified methods

a. Electrical installations of buildings
For this simplified method, only the economic risk is concerned and only surges on the lines (induced and direct) are calculated.

In this method, a so-called calculated risk level (CRL) is used to determine if SPDs are required at entrance of the installation.

CRL is given by the following formula:

$$CRL = \frac{f_{env}}{(L_P \times N_G)} \tag{43}$$

where

f_{env} an environmental factor

$$L_P = 2 \times L_{PAL} + L_{PCL} + 0.4 \times L_{PAH} + 0.2 \times L_{PCH} \tag{44}$$

L_{PAL} the length (km) of low-voltage overhead line;
L_{PCL} the length (km) of low-voltage underground cable;
L_{PAH} the length (km) of high-voltage overhead line;
L_{PCH} the length (km) of high-voltage underground cable.

The total length $(L_{PAL} + L_{PCL} + L_{PAH} + L_{PCH})$ is limited to 1 km or by the distance from the first overvoltage protective device installed in the power network (either LV or HV) to the entrance of the installation whichever is the smaller. If the distribution networks lengths are totally or partially unknown then L_{PAL} shall be taken equal to the remaining distance to reach a total length of 1 km (Table 36).

If CRL $\geq$ 1000. No SPD is needed;

If CRL $<$ 1000. SPD is required (Fig. 5).

b. Photovoltaic systems

For this simplified method, only the loss of service to the public risk is concerned and only induced surges on the DC lines are calculated [7].

In this method, risk assessment is based on the evaluation of the so-called critical length Lcrit and its comparison with L.

SPDs shall be installed on the DC side of the installation when L $\geq$ Lcrit.

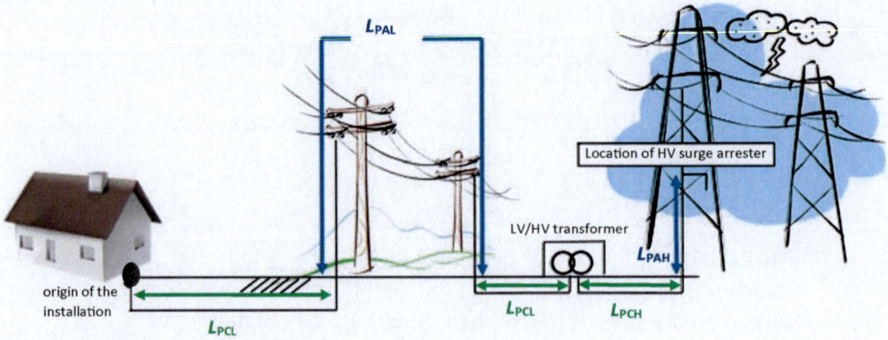

Fig. 5 Illustration of an installation showing the lengths to consider

where

L	the cumulative length between inverters and the furthest point of PV modules in a chain, considering each of the paths. In case of many inverters the length to consider is the sum of the length L for each inverter.
Lcrit (m)	depends on the type of PV installation (Table 37 and Fig. 6).

Table 37 Critical length Lcrit

	PV installation attached to a building	PV installation not attached to a building (e.g. PV farm)
Lcrit (m)	115/Ng	200/Ng

Fig. 6 Method for determining L

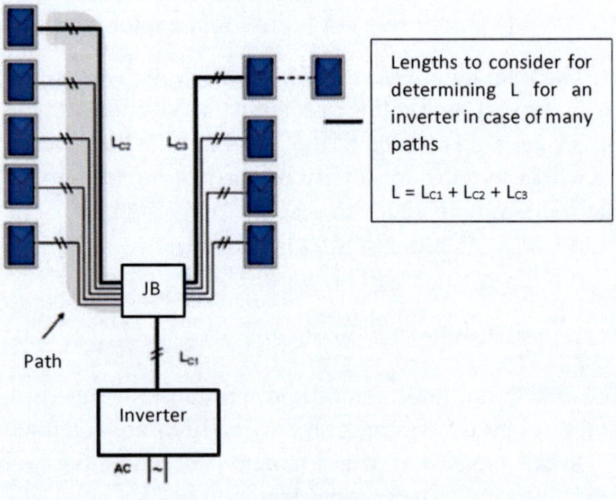

JB: junction box.

References

1. IEC 62305-2 (2010) Protection against lightning—Part 2: risk management
2. IEC 60364-4-44 (2018) Low-voltage electrical installations—Part 4-44: protection for safety—Protection against voltage disturbances and electromagnetic disturbances
3. IEC 60364-7-712 (2017) Low voltage electrical installations—Part 7-712: requirements for special installations or locations—Solar photovoltaic (PV) power supply systems
4. IEC 61400-24 (2019) Wind energy generation systems—Part 24: lightning protection
5. IEC 62793 (2020) Protection against lightning—Thunderstorm warning systems
6. Rousseau A, Kern A (2014) How to deal with environmental risk in IEC 62305-2. In: 2014 international conference on lightning protection (ICLP), Shanghai, China
7. Rousseau A, Guthrie M (2017) Risk assessment of lightning-induced surges on PV systems, CIGRE SC C4. In: International colloquium on lightning and power systems, Ljubljana, Slovenia

Protection of Buildings and Structures

Alexis Barwise

Abstract So often standards and norms do not spell out exactly what it recommends, leaving the reader unsure about where to start, what the next steps are in order to design and install a lightning protection system. The spirit of this chapter is one of cookbooks, containing cost-effective recipes together with clear step-by-step guides to practically cook up a lightning protection system for a building and or structure, which is both safe and effective.

Keywords Structural protection · Rolling sphere method · Protective angle · Air termination · Down conductor

1 Damages Caused by Lightning to Buildings

A lightning strike can cause physical damage to buildings as well as internal damages to the contents and or occupants of these buildings. The type, scale and cost of the damages depend on the characteristics of the building itself and the characteristics of the lightning strike.

i. The characteristics of buildings

The main characteristics of a building that contributes to lightning damages are.

- The occupants of buildings
- The type of roof construction
- The type of building construction
- The type of utility lines entering the building
- The contents in the building.

A. Barwise (✉)
South African National Standards, TC81 Mirror Committee Chairman, Lightning Protection Concepts, Pretoria, South Africa
e-mail: alexis@lpconcepts.com
URL: https://lpconcepts.com/

C. Gomes (ed.), *Lightning*, Lecture Notes in Electrical Engineering 780,
https://doi.org/10.1007/978-981-16-3440-6_4

a. The occupants of buildings

The probability that a building or structure can present a lightning risk (e.g. agricultural soil used as a floor) and the total time spent by the occupant in this area are the most important factors contributing to the risk of losing lives. For example, animals hiding under trees during a lightning storm (Fig. 1) are at higher risk than workers are at the office.

b. The type of roof construction

Buildings consist of many different types of roofing materials (e.g. thatched, steel sheets, or concrete tiles) each of which changes the risk of lightning damage to the building. For example, if the roof consists of dried grass (thatch) then the risk of lightning starting a fire is high which can lead to the destruction of the building as is shown in Fig. 2.

Fig. 1 Lightning fatalities (animals) (*Source* lpconcepts.com)

Fig. 2 Roof damage (thatched) (*Source* lpconcepts.com)

c. The type of building construction

Depending on the country and or local region, buildings are constructed using different methods and materials (e.g. wood, brick and mortar, reinforced concrete or steel). The different materials and construction methods change the risk of lightning damage to the building. For example, parts of a concrete building may not be damaged although the rest of the building was completely destroyed (Fig. 3).

d. The type of utility lines

Utility lines such as power, telecommunication, gas and or sewage all bring with them an increased risk of lightning damage to a building. For example, the risk of lightning damage from an unshielded overhead power line in a rural community (Fig. 4) is far greater than from a shielded underground urban power line.

e. The contents in the building

During the last few years, buildings have been equipped more and more with sensitive electrical and electronic equipment which are more prone to lightning damage. As a

Fig. 3 Structural damage (concrete) (*Source* lpconcepts.com)

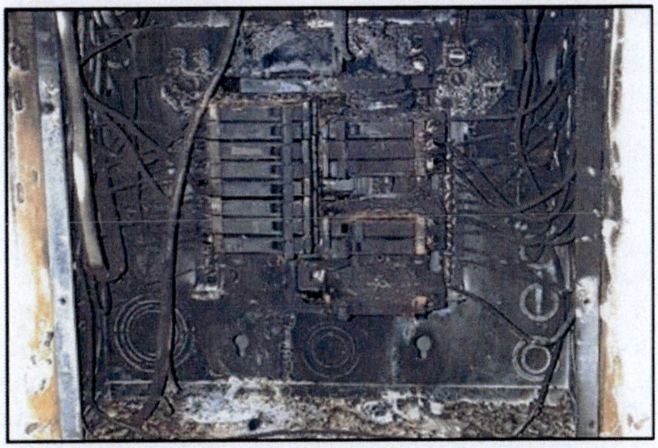

Fig. 4 Utility line damage (power) (*Source* strikecheck.com)

Fig. 5 Content damage (computer) (*Source* strikecheck.com)

result, insurance companies have experienced an increase in electrical and electronic equipment claims, in particular to electronic equipment such as computers and TVs (Fig. 5).

ii. The characteristics of lightning currents

In order to protect a building/structure, it is important to first understand the characteristics of a lightning strike for these determine many of the requirements of an effective and safe lightning protection system. The main characteristics which will be discussed are.

- The peak value of a lightning current
- The specific energy of a lightning current
- The mechanical force of a lightning current.

a. Peak value of a lightning current

It is understood that a lightning discharge can be considered to be a current source, which means that when lightning strikes an object, a potential gradient will be generated along with the object. The voltage, $V(t)$ along a unit length of the object is given by

$$V(t) = R\,i(t) + L\frac{di}{dt}$$

where

R: Resistance per unit length.

L: Inductance per unit length.

This potential gradient gives rise to a potential difference along with the object as well as that between the current passage and the nearby objects. If parts of the body of a person are in contact with two points at different potentials a partial current will flow through the person. This hazard is termed the touch potential.

As the lightning current passes to the ground, the potential at the point of injection of current rises to a large value. The potential decreases radially as the lightning distributes radially in the soil. The maximum potential (φ) at a particular distance (r) away from the point of current injection depends on the peak lightning current and the earth's ability to conduct these currents (i.e. soil-resistivity). Figure 5, graphically depicts this phenomenon.

If a living being (person or animal) stands radially along the potential gradient it will be subjected to a potential difference which is termed step potential. For a person with feet separation s along the potential gradient, at a distance r away from the point of current injection, the maximum step potential (U) is given by

$$U = \frac{\rho I s}{2\pi r(r + s)}$$

where

ρ: soil resistivity.

I: Peak lightning current.

The equation shows that as s increases the step potential increases. Hence, an animal with a larger feet separation has a higher risk of getting subjected to a larger step potential (Fig. 6).

The minimum and maximum peak lightning currents values are used to define the dimensioning and interception criteria for each of the four (4) lightning protection levels. The lightning protection level required is dependent on the lightning risk assessment as per IEC 62305-2. The lowest level of protection is LPL 4 and the highest is LPL 1. The maximum lightning current values are used to define the size, thickness and mechanical requirements of lightning protection system components, whereas the minimum lightning current values are used to determine where to place

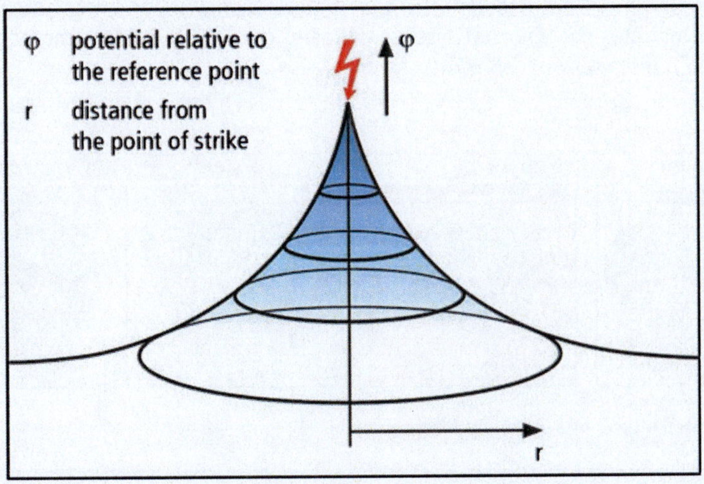

Fig. 6 Potential distribution of a lightning strike to homogenous soil [1]

Table 1 Assignment of lightning current parameters [2]

Lightning protection level (LPL)	Interception criteria (min. lightning current) (kA)	Dimensioning criteria (max. lightning current) (kA)	Interception efficiency (%)
LPL 1	3	200	98
LPL 2	5	150	95
LPL 3	10	100	86
LPL 4	16	100	79

the above-mentioned lightning protection system components. Table 1 summarizes the above-mentioned requirements.

b. The specific energy of a lightning current

The specific energy of a lightning current is used to calculate the temperature rise in conductors carrying lightning currents. This is done to determine whether the conductors/components of the lightning protection system will be able to withstand the temperature rise without causing hots-spots, punctures and or molten metal. The specific energy defines the thermal stress which causes the deformation of components of a lightning protection system. The specific energy can be calculated using the equation below.

$$W = R \times \int i^2 \times dt$$

R: Low-frequency resistance of the conductor (temperature-dependent).

It is important for the lightning protection designer to calculate the temperature rise of conductors if the risk from fire or explosion has to be taken into account. Figure 7 illustrates the thermal effects of a lightning strike on different types of metals with a thickness of 0.5 mm.

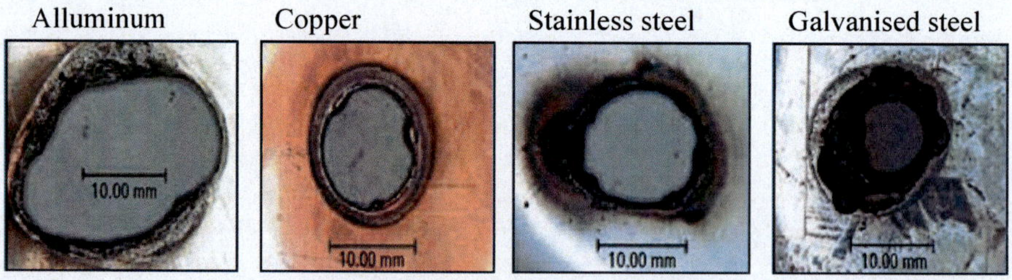

Fig. 7 Thermal effects of a strike on different metals [1]

c. The mechanical force of a lightning current

During the interception and whilst conducting a lightning current to ground, the lightning protection system (i.e. the components) are subjected to large electrodynamic forces (F) which can be calculated as an approximation using the following equation:

$$F = \mu_0/2\pi \times i^2 \times l$$

μ_0 Magnetic field constant in air.

l Conductor length.

The force is proportional to the square of the current in the lightning protection components (e.g. clamp, conductor). Thus, the specific energy of the short-stroke defines the stress which causes reversible or irreversible deformation of components and arrangements of a lightning protection system. These effects are considered in the test setups of the product standards concerning the requirements made on lightning protection components for lightning protection systems.

Annex D of IEC 62305-1 describes in detail in which way the lightning current parameters relevant to the point of strike are important for the physical integrity of the system.

2 Lightning Protection (LP)

The IEC standards describe lightning protection as a complete system used to protect buildings/structures and their occupants and contents against the harmful effects of lightning. In general, this protection is offered through the combination of lightning protection systems (a system used to reduce the physical damage) and surge protection measures (measures taken to protect internal systems) as is shown in Fig. 8.

However, when considering the safety and behavior of human and living beings, lightning protection should include further mitigation measures such as lightning

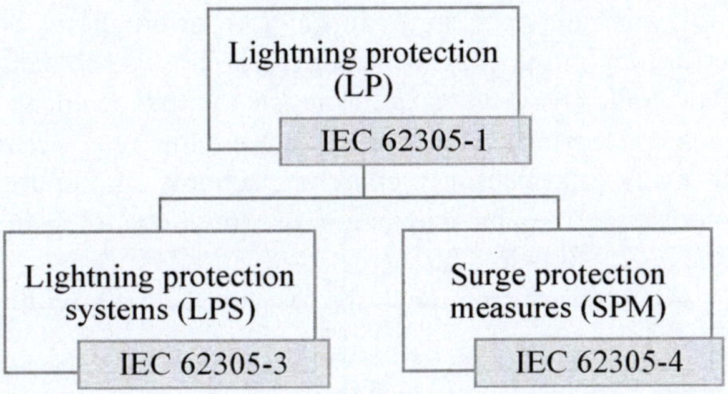

Fig. 8 Lightning protection definition [3]

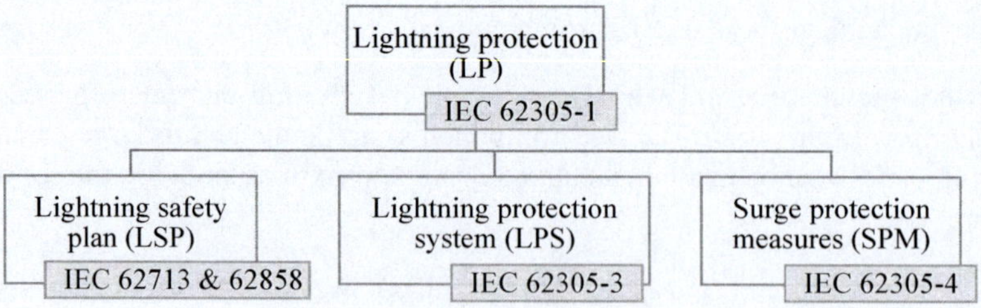

Fig. 9 A more comprehensive lightning protection definition

awareness, lightning detection systems, lightning warning systems and evacuation procedures. All of these and other items should form part of a lightning safety plan (LSP) which should be incorporated into the lightning protection of the building/structure, see the proposal in Fig. 9.

3 Lightning Safety Plan (LSP)

Everyone knows, or at least, everyone should know that there is no place outdoors that is safe, hence no person is safe when active with outdoor activities such as sport (e.g. Golf, hiking, rugby and swimming). Through recorded lightning fatality statistics, we gain further insight into the fact that not only are most fatalities occurring outdoors but that the leading cause of these fatalities is step & touch potentials. Therefore, it is imperative to develop a lightning safety plan/procedure to ensure the protection of human beings during high-risk lightning activities.

A lightning safety plan should include the following items.

- Lightning awareness campaign (pre-lightning activity)
- Lightning detection and warning systems (pre-and during-lightning activity)
- Lightning procedures (during lightning activity)
- Lightning safe structures (pre-lightning activity).

i. Lightning awareness campaign

An awareness campaign can range from brochures being handed out at school, to informing employees or training scholars on the dangers of lightning in a manner that would answer their questions so nothing technical. The United States of America have successfully implemented such a campaign *"When Thunder Roars, Go Indoors"* to many industries e.g. churches, schools, and sports field. The results speak for themselves they have managed to bring down their lightning-caused fatalities from above 300 to below 50.

For more information (https://www.weather.gov/ind/LightningSafetyAwareness).

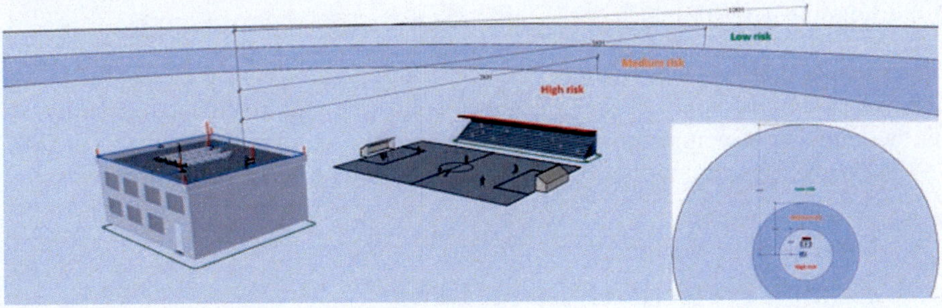

Fig. 10 Typical low, medium and high-risk distances

ii. Lightning detection and warning systems

In all crowd protection schemes, the use of a lightning detection device/network is considered paramount. It should be able to indicate the advance of lightning strike fronts, and by the time strikes reach an appropriate distance from a location all required protection strategies should be activated.

The person monitoring this device should be invested with full authority to activate the protocol. A risk assessment is needed to determine the necessity for an action plan to be executed based on data from the device/network. The protocol should incorporate clear procedures for each risk level (e.g. low, medium and high) and these risk levels will be determined by the distance the storm is away from the protected area (Fig. 10). It is commonly known that a stand-alone lightning detection device often gives false alarms for it cannot accurately pinpoint the strike location nor can it track the direction of the cell, therefore it is recommended to rather sign up with a lightning detection network service provider (eg. www.blitz-detect.com) than investing in a stand-alone device.

iii. Lightning procedures

If thunder can be heard, the lightning strike is 10–15 km away and it is time to take shelter. The lightning detection device should also come with a loud siren/alarm to ensure that its warning can be heard, nowadays these alarms can also be linked to a messaging system, public announcement system, mobile apps and building management systems. Once the alarm is triggered people should seek shelter in either a substantially enclosed building or a fully metal-enclosed vehicle, pavilion or similar. The escape to these locations should be planned in the initial planning and changes of plan.

Once there has been no thunder or no visible lightning for 30 min, it is generally safe to resume activities. If possible when the risk of a venue (e.g. a concert) being struck becomes large, the crowd should be evacuated to a safe location. The factors to consider are the ease and speed of crowd movement, and the availability of a protected area for the size of the crowd, in the provided examples a school building and a sports pavilion will be used.

iv. Lightning safe structures

The principle that no place outdoors is safe from lightning injury guides the basic dictum that shelter should be sought immediately when the risk is recognized, and that building, or structure must satisfy the minimum criteria as stipulated in IEC 62305. The next section will cover the requirements with practical perspectives.

4 Lightning Protection System (LPS)

The main purpose of a lightning protection system is to make a building lightning safe to protect the occupants. A lightning protection system (LPS) consists of both an external and an internal lightning protection system as illustrated by the graphic below.

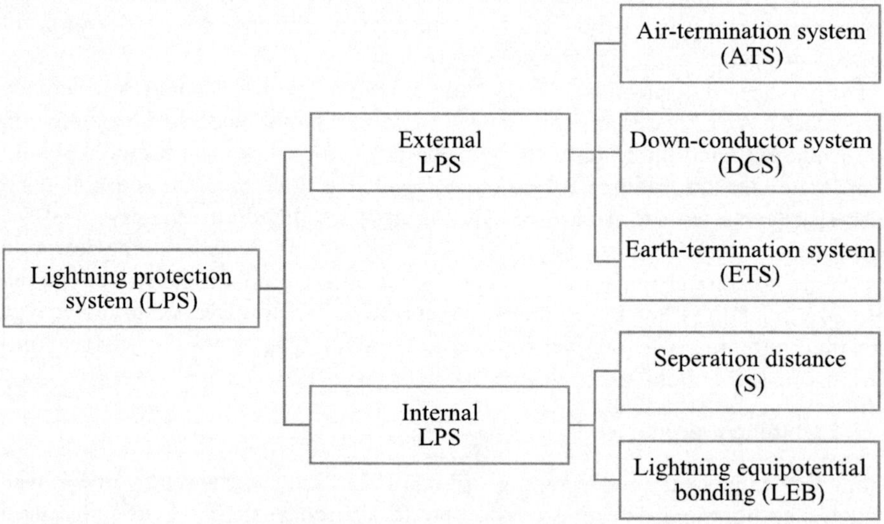

The functions of an external lightning protection system are.

- To intercept, perhaps a better term would be to catch, the direct lightning strikes and or side strikes to a building by means of an air-termination system (ATS)
- To safely and controllably conduct the intercepted lightning current down to the earth by means of a down-conductor system (DCS)
- To dissipate the lightning currents into earth by means of an earth-termination system (ETS) whilst minimizing any potentially dangerous conditions.

The function of an internal lightning protection system is

- To prevent uncontrollable flashovers and or dangerous sparking between conductive elements or metal parts of the building/structure by keeping a separation distance (S) between these parts or through establishing lightning equipotential bonding (LEB) between these parts.

5 External Lightning Protection System

In order to design a safe and effective system a lightning protection designer not only has to take into consideration the requirements of the lightning protection standards but also be cognizant of the associated costs and the aesthetics of such a system. Therefore, it is important that the designer finds ways to design safe, practical and cost-effective lightning protection systems by making use of any and all attributes and materials each building, and structure has to offer (Fig. 11).

i. Air-termination system (ATS)

The purpose of the ATS is to intercept the direct lightning strikes, which is probably why parts of the world refer to this system as a lightning-termination system instead. The German lightning protection community calls an air-termination system a "Fang-system" which when translated means a catching system, which is exactly what the system should do. Therefore, the lightning protection designer should consider the placement of air-terminals as if they are placing catchers in strategic positions to

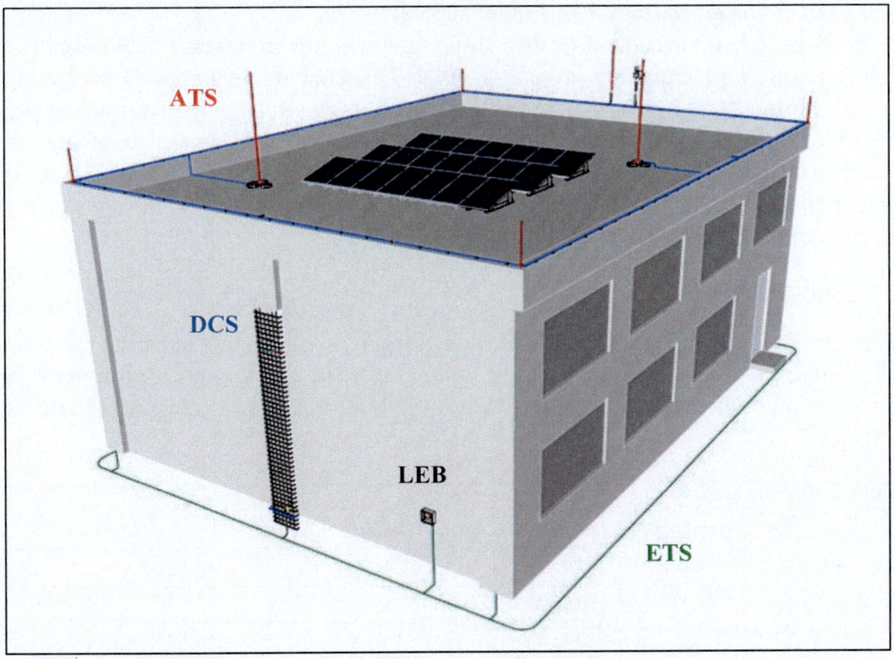

Fig. 11 Typical sub-systems of a lightning protection system

ensure that the direct lightning strike is intercepted (i.e. caught). The most suitable method to use to determine the placement of air-terminals is the rolling sphere method, which is derived from the strike distance of a lightning strike.

a. Strike-distance of a lightning strike

In most cases, a lightning strike moves from a negatively charged cloud to a positively charged earth via a downward step-leader. It is termed a stepped leader for the downward leader moves in a step-by-step movement, where the length of the steps is mainly determined by the amplitude of the lightning current (the larger the lightning current the larger the steps are).

Once the downward leader gets within a couple of hundred of meters from the earth upward streamers begin to grow towards the downward leader. Upward streamers often launch from sharp edges and corners for here the electric field intensity is at its highest for charges commonly accumulate on sharper objects. The distance of the last step of a downward leader, before terminating with the upward streamer, is termed the final striking distance and this distance (i.e. step) depends on the amplitude of the lightning strike.

The relationship between the lightning current and the radius of the rolling sphere (center is the head of the downward leader) is captured in the equation below.

$$r = 10 \times {I_p}^{0.65}$$

r: striking distance (radius of rolling sphere).

With the help of this equation, the rolling sphere radius values (see Table 2) can be calculated for each lightning protection level. This enables the lightning protection designer to apply the rolling sphere method to a building/structure to determine where possible interceptions (between the downward leader and the upward streamer) can take place. Using the rolling sphere method the lightning protection designer will be able to place the air-terminals (i.e. catchers) in strategic positions to ensure that the direct lightning strike will be intercepted.

b. Rolling sphere method

The concept of the rolling sphere method is captured in the name meaning a certain sized sphere is rolled over the building and or structure needing protection in every direction and wherever the sphere touches the building and or structure would be

Table 2 Rolling sphere radius values [2]

Lightning protection level	Min. peak lightning current (kA)	Rolling sphere (radius) (m)
LPL 1	3	20
LPL 2	5	30
LPL 3	10	45
LPL 4	16	60

considered a possible point of termination, the figures below illustrated how to apply the rolling sphere method (Figs. 12 and 13).

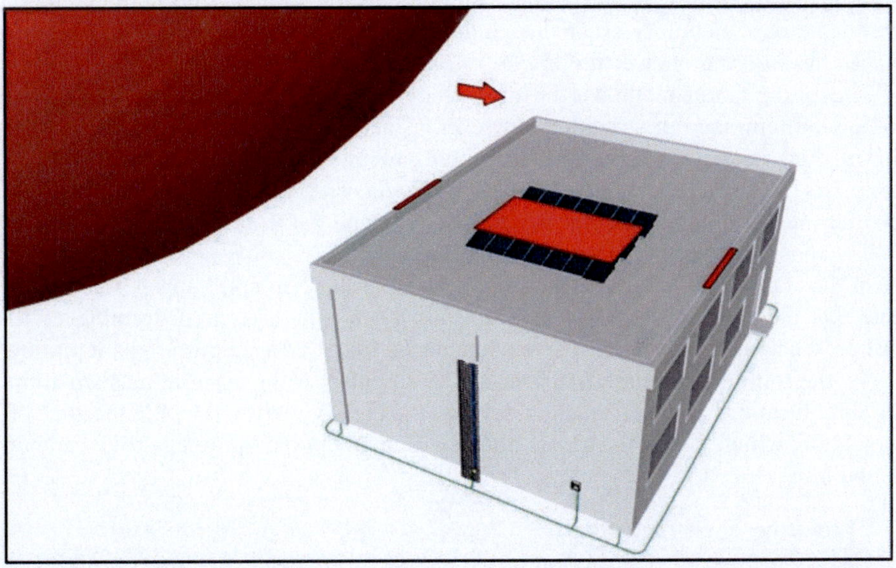

Fig. 12 Rolling the sphere along the x-axis

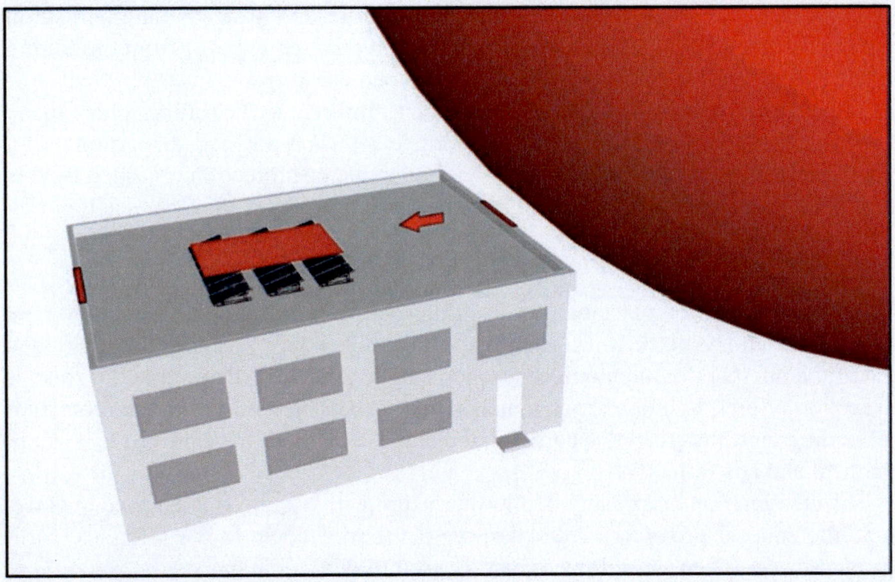

Fig. 13 Continuing the sphere along the y-axis

Placing air terminals in the highlighted areas would ensure the interception of the lightning strikes. One could almost say it is like having the playbook of your competition and knowing where to place your catchers. Once the air-terminals have been placed the rolling sphere method should be applied again to ensure that there are no interception points left on the building or structure, this is referred to as having placed the building or structure in the zone of protection (i.e. covered area).

The rolling sphere method is the most effective interception method and is suitable for any building and or structure. However, as can be seen from the figures above, this method requires computer-aided programs to run and not all lightning protection designers will have access to such programs. To this end, Table 3 has been developed to offer the lightning protection designers a practical guide as to the sag (m) of the rolling sphere between two air-terminals (Fig. 14).

In the above example the air-terminals are spaced 15 m apart, using Table 3 we note that the sag for LPL 3 is 0.63 m, thus the height of 1 m air-terminals will suffice. Once the air-terminals have been placed in the areas exposed to a lightning strike, the rolling sphere method needs to be repeated. If the air-terminals are all in the right locations and of correct length then the building will fall within the zone of protection, which means the air-terminals will intercept the lightning strike and not the building (Fig. 15).

c. Protective angle method

The rolling sphere method most often require computer-aided programs for complex building structures and not all lightning protection designers have access to such programs. At times, a single air-terminal may be enough to cover the structure; for e.g. a small hut, security guard house etc. Therefore, the protective angle method (a simplified and practical method), has been derived (see Fig. 16) from the rolling sphere method and is applicable for simple shaped buildings.

The concept of the protective angle method is different to the rolling sphere in the sense that the lightning protection designer needs to first place an air-terminal of a certain height, then use Table 4 to look up the angle of protection provided by this terminal and derive the cone-shaped zone of protection across the area (see Fig. 17).

d. The placement/positioning of air-terminals

When determining the placement or position of air-terminals it is important to note that these should be placed on corners, exposed points/edges first (e.g. 1 m high) then one of the interception methods mentioned can be applied to determine the zone of protection offered by these air-terminals. However, when it comes to the protection of buildings and structures the combination of these methods have proved to be more practical and cost-effective.

The combination is applied as follows; the protective angle is to be used to determine the zone of protection from the top of the roof down to the reference earth (ground-level). Then the rolling sphere is used to determine the sag between these terminals (i.e. the spacing between them). It is important to note that we should ensure that the diagonal sag requirements are met.

Table 3 Sag distance between two (2) air-terminals using the rolling sphere method

Height between air-terminals	LPL 1	LPL 2	LPL 3	LPL 4
	Sag (m)	Sag (m)	Sag (m)	Sag (m)
1	0.01	0.00	0.00	0.00
2	0.03	0.02	0.01	0.01
3	0.06	0.04	0.03	0.02
4	0.10	0.07	0.04	0.03
5	0.16	0.10	0.07	0.05
6	0.23	0.15	0.10	0.08
7	0.31	0.20	0.14	0.10
8	0.40	0.27	0.18	0.13
9	0.51	0.34	0.23	0.17
10	0.64	0.42	0.28	0.21
11	0.77	0.51	0.34	0.25
12	0.92	0.61	0.40	0.30
13	1.09	0.71	0.47	0.35
14	1.27	0.83	0.55	0.41
15	1.46	0.95	0.63	0.47
16	1.67	1.09	0.72	0.54
17	1.90	1.23	0.81	0.61
18	2.14	1.38	0.91	0.68
19	2.40	1.54	1.01	0.76
20	2.68	1.72	1.13	0.84
21	2.98	1.90	1.24	0.93
22	3.30	2.09	1.37	1.02
23	3.64	2.29	1.49	1.11
24	4.00	2.50	1.63	1.21
25	4.39	2.73	1.77	1.32
26	4.80	2.96	1.92	1.43
27	5.24	3.21	2.07	1.54
28	5.72	3.47	2.23	1.66
29	6.23	3.74	2.40	1.78
30	6.77	4.02	2.57	1.91

If the entire building and or its components are not covered by the first air-terminals, then additional air-terminals should be added until the zone of protection covers the entire building and its components. It is also very good practice to try and interconnect the air-terminals at roof level, this will ensure that the intercepted lightning current be divided into partial lightning currents (Fig. 18).

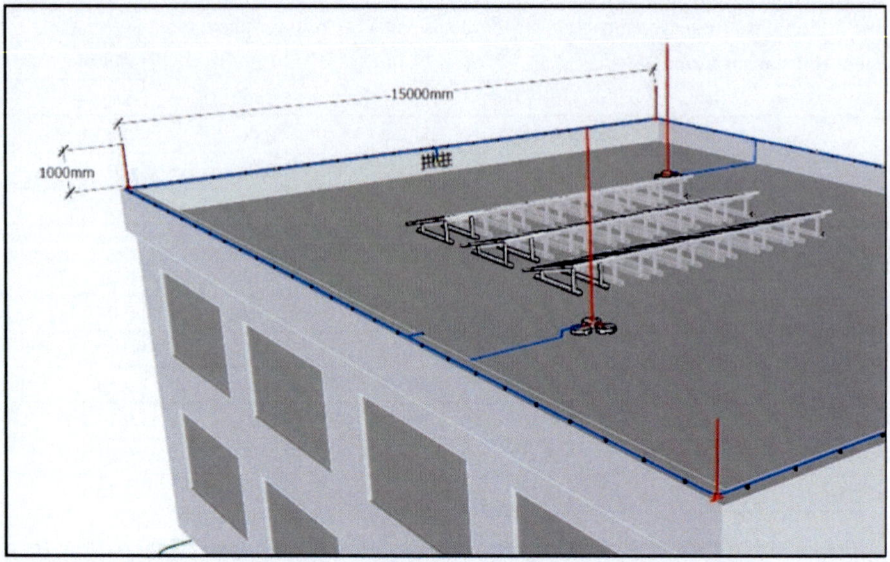

Fig. 14 Placing of air-terminals in exposed areas

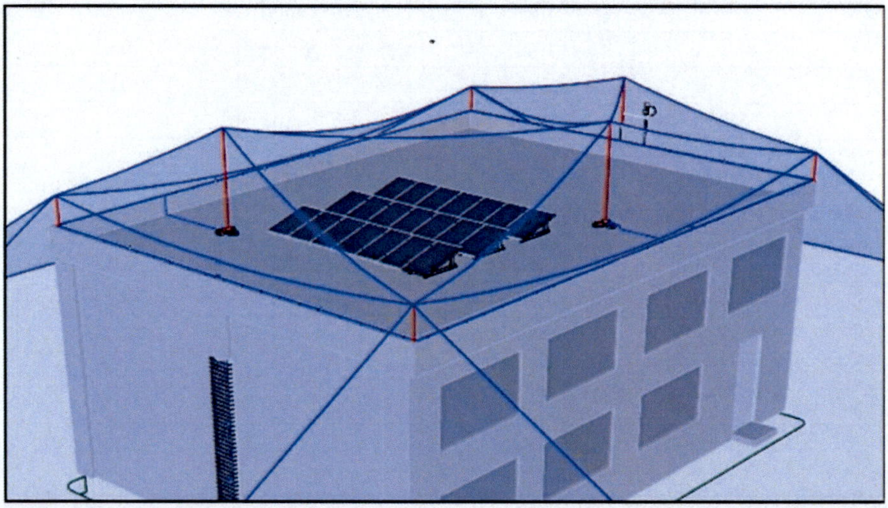

Fig. 15 Zone of protection indicates that the building is protected

e. Natural air-terminals

The use of natural air-terminals as an air-termination system (ATS) is highly recommended, for it offers a more economical and aesthetical solution seeing as buildings and structures commonly comprise of conductive elements such as railing,

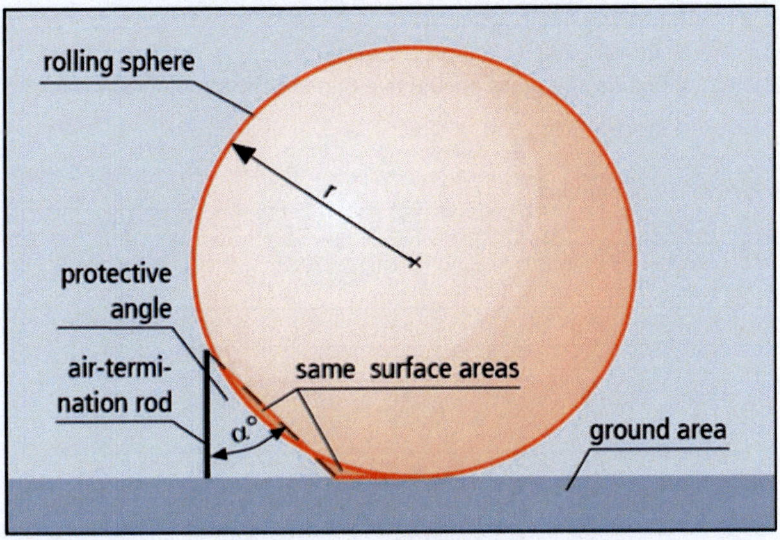

Fig. 16 Relationship between rolling sphere and protective angle methods [1]

claddings, facades and corrugated metal sheets. However, before using these natural air-terminals consideration needs to be made in terms of the required thickness (see Table 5) of these elements and once this is confirmed then the electrical continuity needs to be tested to ensure a permanent continuous electrical connection exists. If there is no electrical connection, these elements must be additionally connected by means of bridging braids, cables, welding, pressing, screwing or riveting.

If the thickness of the natural air-terminal is not less than the value (Thickness B) and if melting at the point of strike or the ignition of flammable material under the air-terminal does not have to be taken into account, such air-terminals can be used as an air-termination system.

It is important to note that the IEC standards do not accept any air-terminals which artificially enhance the interception (i.e. lengthening the catcher) or which disrupts the electromagnetic fields, these systems are yet to be proven effective and at the current moment are still considered unsafe (Fig. 19).

ii. Down-conductor system (DCS)

Ever heard of the expression *'less is more'*? Well, when it comes to down-conductors the total opposite is true. Down-conductors are the electrically conductive path between the ATS and the earth-termination system (ETS) whereby the intercepted lightning currents will flow. Hence, the more down-conductors there are, the more the lightning current will divide (i.e. Kirchhoff's rule) and with this division, the lightning effect (i.e. damage probability) is reduced.

The IEC standard requires down-conductors to be arranged in a way that.

- several parallel current paths exist
- length of the current paths is kept to a minimum
- equipotential bonding to other conductive parts of the structure is performed.

Table 4 Protective angle method

Height of air-terminal	LPL 1		LPL 2		LPL 3		LPL 4	
	Angle (deg.)	Distance (m)	Angle (deg.)	Distance (m)	Angle (deg.)	Distance (m)	Angle (deg.)	Distance (m)
1	71	2.90	74	3.49	77	4.33	79	5.14
2	71	5.81	74	6.97	77	8.66	79	10.29
3	66	6.74	71	8.71	74	10.46	76	12.03
4	62	7.52	68	9.90	72	12.31	74	13.95
5	59	8.32	65	10.72	70	13.74	72	15.39
6	56	8.90	62	11.28	68	14.85	71	17.43
7	53	9.29	60	12.12	66	15.72	69	18.24
8	50	9.53	58	12.80	64	16.40	68	19.80
9	48	10.00	56	13.34	62	16.93	66	20.21
10	45	10.00	54	13.76	61	18.04	65	21.45
11	43	10.26	52	14.08	59	18.31	664	−16.31
12	40	10.07	50	14.30	58	19.20	62	22.57
13	38	10.16	49	14.95	57	20.02	61	23.45
14	36	10.17	47	15.01	55	19.99	60	24.25
15	34	10.12	45	15.00	54	20.65	59	24.96
16	32	10.00	44	15.45	53	21.23	58	25.61
17	30	9.81	42	15.31	51	20.99	57	26.18
18	27	9.17	40	15.10	50	21.45	56	26.69
19	25	8.86	39	15.39	49	21.86	55	27.13
20	23	8.49	37	15.07	48	22.21	54	27.53
21	Not applicable		36	15.26	47	22.52	53	27.87
22			35	15.40	46	22.78	52	28.16
23			34	15.51	45	23.00	51	28.40
24			32	15.00	44	23.18	50	28.60
25			30	14.43	43	23.31	49	28.76
26			29	14.41	41	22.60	49	29.91
27			27	13.76	40	22.66	48	29.99
28			26	13.66	39	22.67	47	30.03
29			25	13.52	38	22.66	46	30.03
30			23	12.73	37	22.61	45	30.00

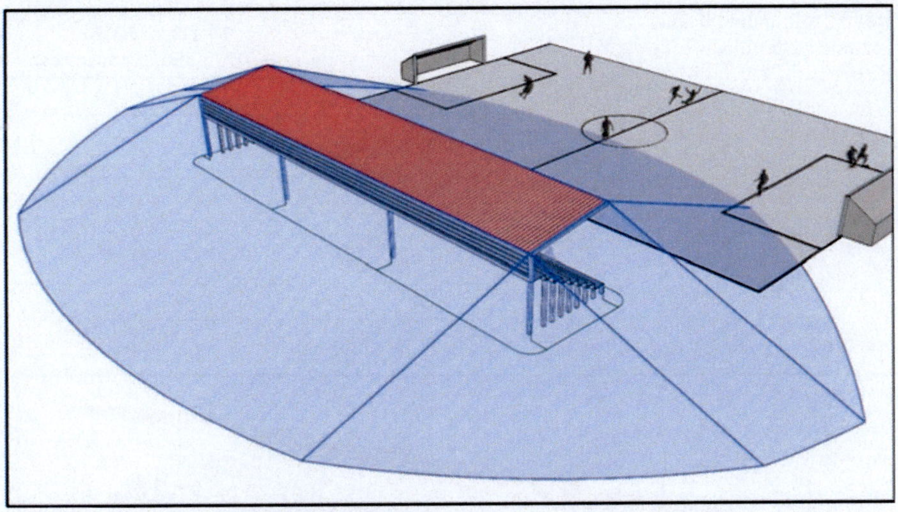

Fig. 17 Cone-shaped zone of protection (protective angle method)

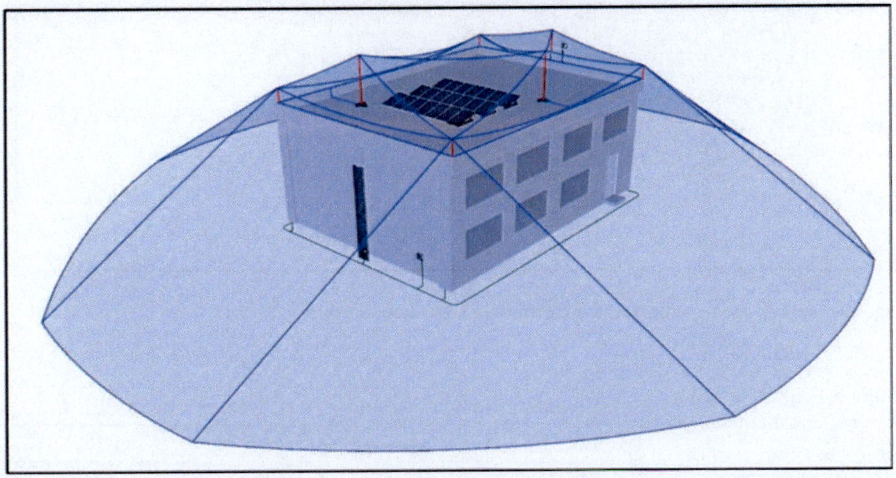

Fig. 18 Combination of the interception methods (most practical)

a. Several parallel current paths

It is advised to install as many down-conductors as possible, at equal spacing (see Table 6) around the perimeter and be interconnected by ring conductors at roof and ground level. The table values are guidelines, in many cases, the buildings or structures offer different distances, for e.g. reinforced columns of a building may only be 4 m apart so then it means by utilising every 3rd column the spacing becomes 12 m or the I-beams of a structure might be spread 6 m apart then the spacing becomes

Table 5 Minimum thickness of natural air-terminals [2]

Material	Thickness A (prevents punctures)	Thickness B (allows punctures)
Lead	–	2.0
Steel (stainless, galvanised)	4.0	0.5
Titanium	4.0	0.5
Copper	5.0	0.5
Aluminium	7.0	0.65
Zinc	–	0.7

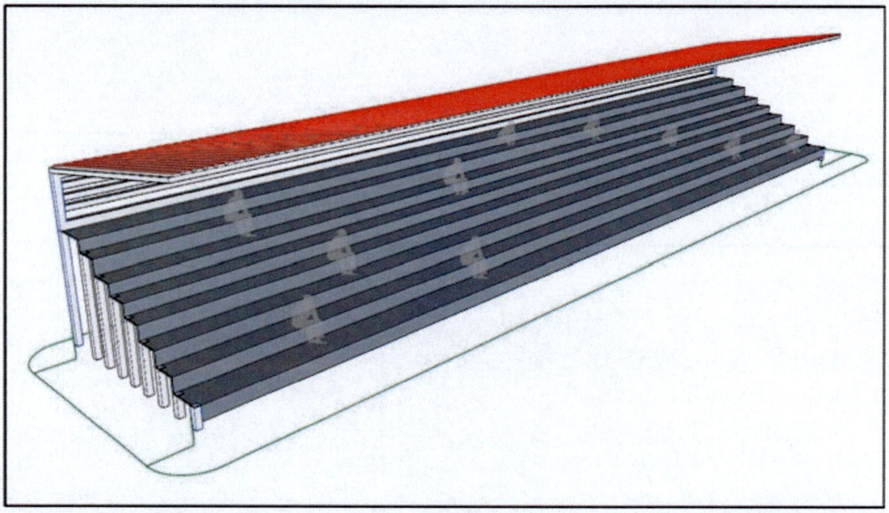

Fig. 19 Natural air-terminals in the form of an IBR sheeted roof

Table 6 Typical spacing between down-conductors [2]

Lightning protection level	Average spacing (m)
LPL 1	10
LPL 2	10
LPL 3	15
LPL 4	20

6 m, either one is acceptable for the goal is to achieve multiple paths to the ground to ensure current sharing.

The requirements of the IEC are fulfilled in metal framework structures and in reinforced concrete structures in which the interconnected steel is electrically continuous. Buildings frequently comprise conductive sections such as metal framework structures and steel pile type structures and these can be used as a natural DCS if it can be proven that they are electrically continuous.

For structures with steel pillars e.g. I-beams the continuity can visually be verified, but if, for e.g. the reinforced columns of buildings are used, a continuity test should be conducted from roof to ground level to ensure an overall resistance of 0.2 Ω is achieved, if this is not the case then external dedicated down-conductors needs to be installed.

b. Length and shape of down-conductors

We have all seen the movie where the train comes off the railway tracks because it is running too fast for the bends, in a similar way this is what happens when a lightning current travels through the down-conductors. Therefore, it is imperative to keep them as short and as straight as possible with supporting holders for every meter.

Achieving this is quite difficult on modern-day buildings for architects are designing more and more ridges, edges, and curve-like facades so it is imperative to work closely with architects and construction companies (Fig. 20).

c. Test joints

It is considered good practice to install a test joint, capable of being opened with a tool, at the end of each down-conductor for measuring (i.e. electrical continuity

Fig. 20 Typical dedicated external down-conductor

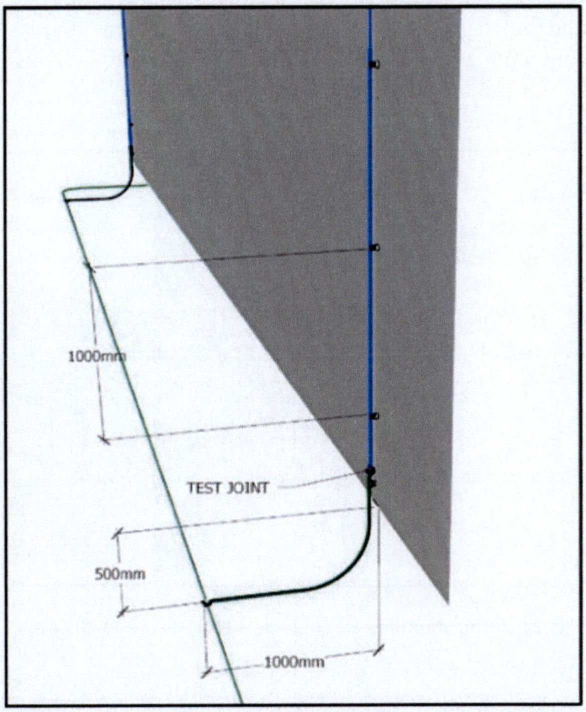

Table 7 Galvanic corrosion between metals [1]

	Galvanized Steel	Aluminum	Copper	Stainless Steel
Galvanized Steel	Yes	Yes	**No**	Yes
Aluminum	Yes	Yes	**No**	Yes
Copper	**No**	**No**	Yes	Yes
Stainless Steel	Yes	Yes	Yes	Yes

tests) and maintenance purposes. These test joints also give the lightning protection designer the opportunity to transition from one metal type to the next without having galvanic corrosion between these metals. Table 7 gives an indication of which metals can be connected without the need for a transition metal.

In reinforced columns, I-beams and other natural down-conductors, a connection point will not always be possible and should in these cases rather be disregarded and other measurement methods should be adopted to ensure electrical continuity (Figs. 21 and 22).

iii. Earth-termination system (ETS)

Most laymen have heard and understood the saying *"the path of least resistance"* which is true for the flow of low frequency (50–60 Hz) currents, however when it comes to designing an earth-termination system capable of dispersing high frequency (MHz-GHz) lightning currents the resistance value is not as important as the shape and dimensions of the system is. However, if at all possible, one should strive towards achieving a low earthing resistance (i.e. R_E<10 Ω) as well.

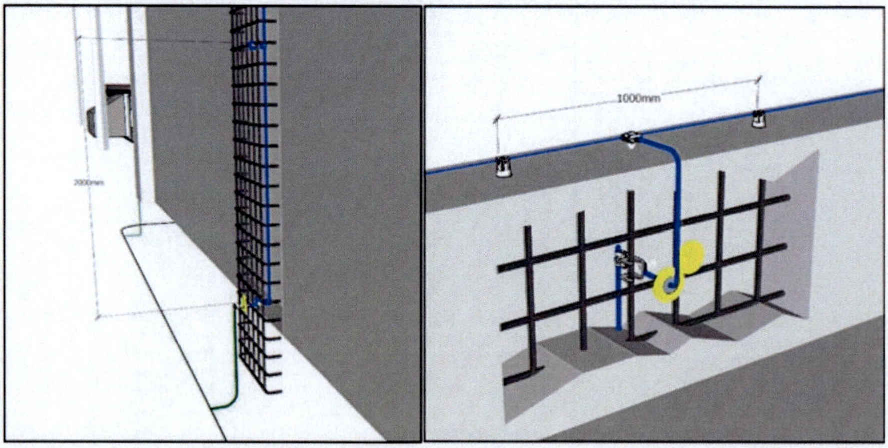

Fig. 21 Typical reinforced concrete (right) down-conductor

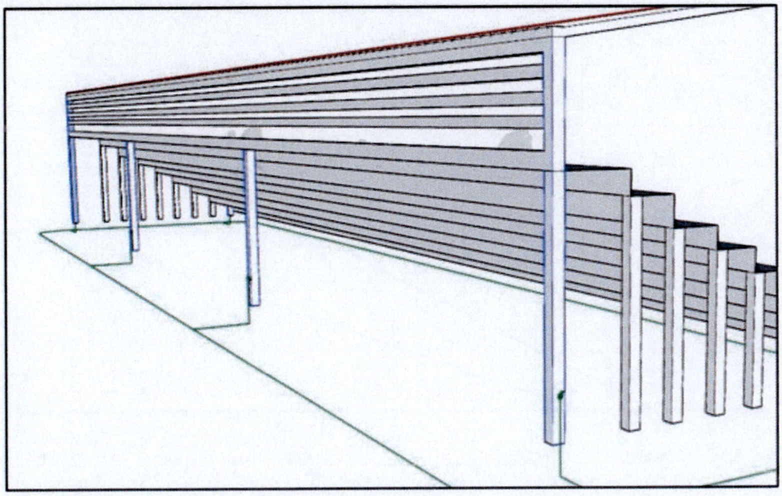

Fig. 22 Natural down-conductors in the form of I-beams

a. Type of earthing arrangements

There are two types of earth-termination arrangements (Type A and Type B), each having different requirements for they are used in different applications. Type A arrangement is intended for simple structures for e.g. telecommunication towers, where the presence of human beings isn't as high as for e.g. local retail stores, for these higher-risk areas a single integrated earth-termination system (i.e. Type B) is preferred, sometimes better known as a ring earth electrode or earth matt.

There are both dimensional and length requirements (l_1) for these two arrangements, the latter of which is defined as a function of the soil resistivity, see Fig. 23. The graph clearly shows that LPL 3 and LPL 4 systems are independent of soil resistivity.

b. Type A arrangement

This type of arrangement describes either a single vertical or horizontal earth electrode, each being connected to a dedicated down-conductor and will be installed outside the structure to be protected. The minimum length requirements for these arrangements are.

l_1 for horizontal electrodes, or

$0.5 \times l_1$ for vertical (or inclined) electrodes.

Taking these equations, the absolute minimum length for a horizontal earth electrode is 5 m and for a vertical earth electrode is 2.5 m. It is recommended that both of these electrodes be buried at least 0.5 m deep and for the vertical earth electrodes to be installed at least 1 m away from the building as shown below in Fig. 24.

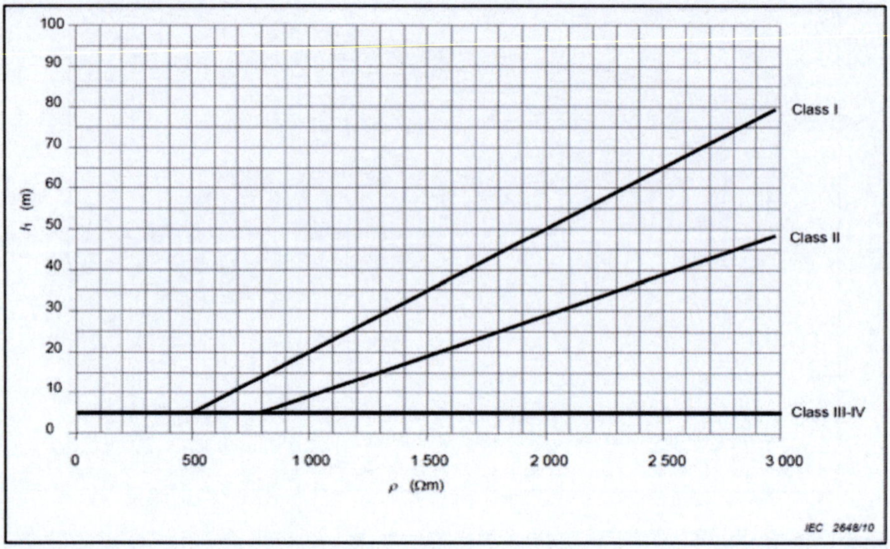

Fig. 23 Minimum length of each earth electrode [2]

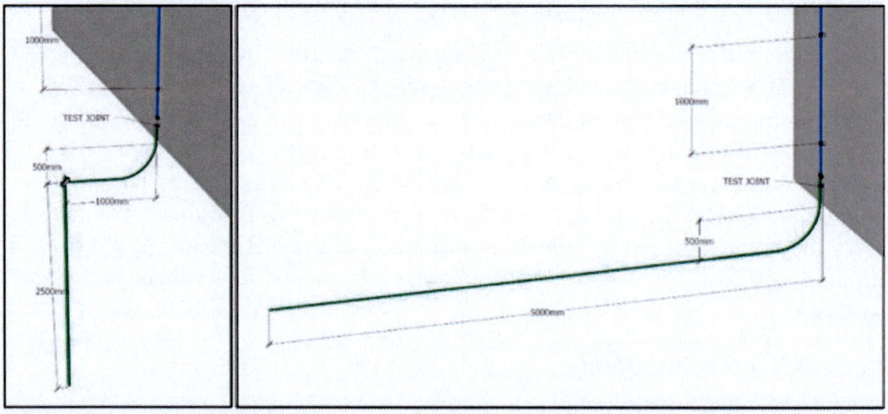

Fig. 24 Typical type A, vertical (left) and horizontal (right) arrangements

c. Type B arrangement

For the ring earth electrode (or foundation earth electrode), the mean radius (r_e) of the area enclosed by the earth electrode shall not be less than the value of l_1 and at least 80% of the electrode must be in contact with soil. Every down-conductor must be connected to the ring earth electrode via testing joints.

$r_e \geq l_1$ for ring earth electrodes.

It is recommended that the ring earth electrode be installed at least 0.5 m deep and be kept at least 1 m away from the building's walls. It is important to note that a single integrated earth-termination system would have to fulfill the requirements of different systems e.g. the lightning protection system, low-voltage power system and even the telecommunication systems (Fig. 25).

d. Soil-resistivity factor

The most used and practical electrical sounding array is the *"Wenner"* method, where the spacing between the test probes are equal. Four probes are used, and the current is passed through the two outer probes (C_1, C_2) and the voltage drop is measured with the two inner probes (P_1, P_2).

The soil-resistivity, at a depth equal to the spacing of the probes, can then be calculated, therefore a practical approach would be to space the probes e.g. 1.5 m apart and to keep increasing the spacing by 1.5 m outwards for standard earth electrodes are 1.5 m long and as such it would be easy to match the reading taken at 3 m by installing 2 × 1.5 m earth electrodes (Fig. 26).

Based on the soil-resistivity values the type of soil i.e. the corrosiveness of the soil can also be determined. This is helpful for the lightning protection designer should consider the corrosiveness of the soil when selecting the type of material used in the ground (Table 8).

In soil with resistivity higher than 3000 Ωm, the use of type B earth electrodes with earthing enhancing compounds is recommended.

e. Overall earthing resistance

In order to measure the earth resistance of an electrode using the same instrument used for the soil-resistivity measurement. Connect C_1 and P_1 to the earth electrode

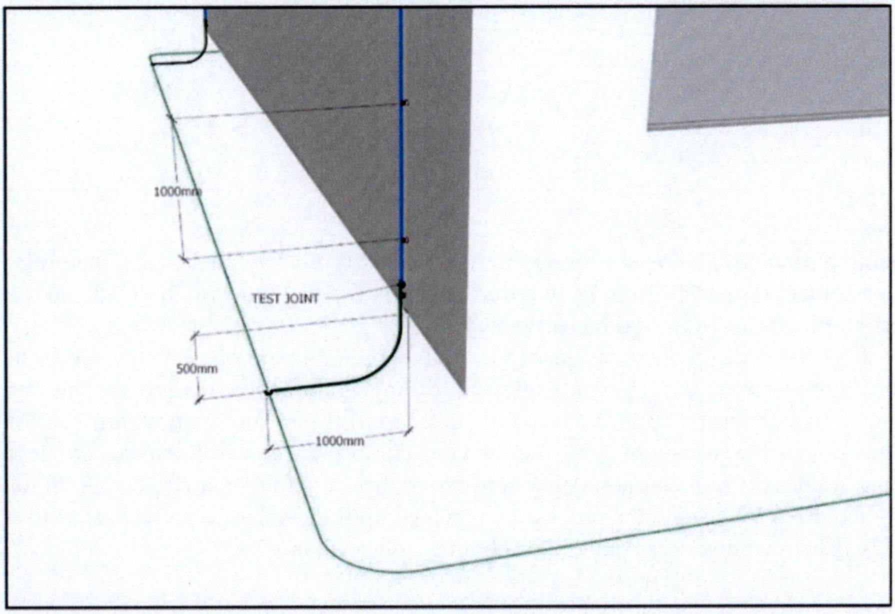

Fig. 25 Typical type B (ring earth) arrangement

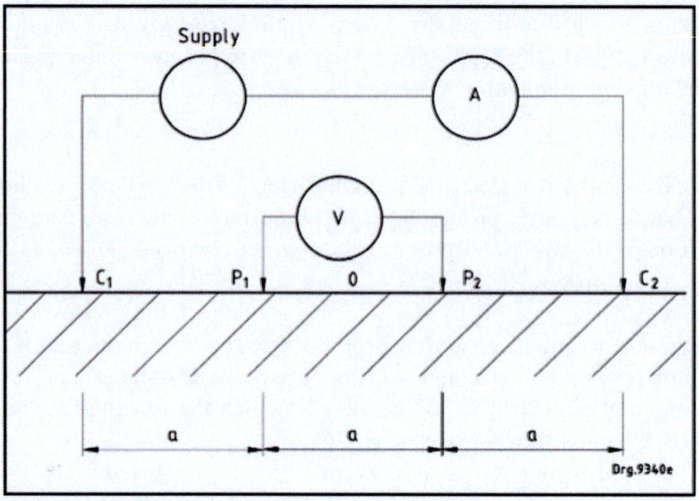

Fig. 26 Wenner sounding array [4]

Table 8 Type of soil according to soil-resistivity

Type of soil	Soil resistivity (Ωm)	Corrosiveness
Marsh, soil in a humid environment	30	Moderate—severe
Loamy soil, clay soil, farmland	100	Moderate—severe
Sandy clay	150	Mild
Sandy soil (dry)	1000	Mild
Gravel (humid)	500	Mild
Gravel, agricultural sand	1000	Mild
Stony soil, back-fill, compacted soil	3000	Nor corrosive

being measured, then space the probe P_2 at least 20 m away, then space the probe C_2 another 20 m away from P_2 to ensure that this is done in a straight line from the geometric center of the earth electrode (Fig. 27).

Take the measurement and record the value, then move probe P2 10% closer to the earth electrode e.g. 18 m and take the reading again, repeat the process but this time move the probe 10% closer to C_2 e.g. 22 m. If these values are within 10% of one another the measured value can be considered the true earth-resistance, but if these values differ then the spacing between probes should increase to e.g. 30 m for P_2 and 60 m for C_2 and the process be repeated until the values obtained are within 10%. The recommended value to be obtained is less than 10 Ω.

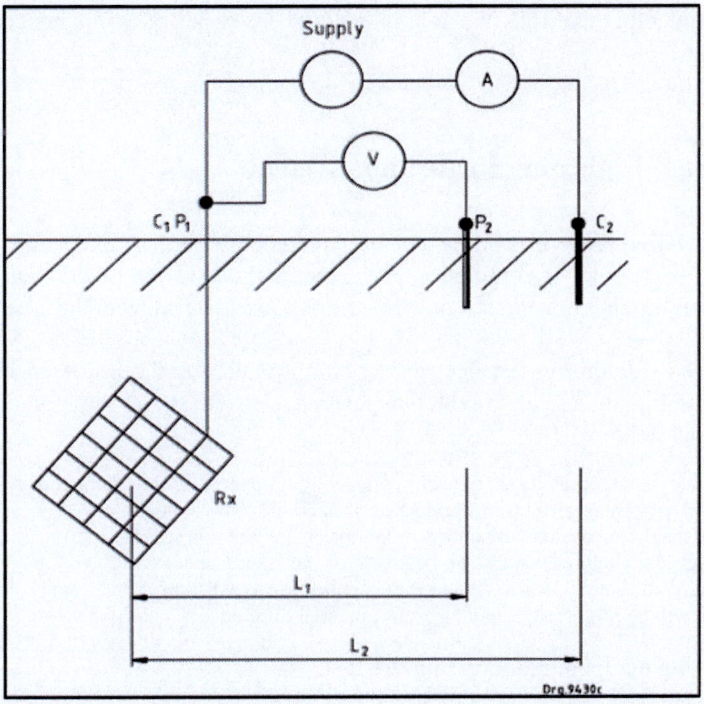

Fig. 27 Earth-resistance measurement [4]

iv. Material dimension and configuration

It is important to note that there are minimum cross-section requirements for different types of metals and configurations of both air-termination and down-conductor system components. Table 9 showcases some of the most common metals and configuration requirements, for more details see IEC 62305-3 Table 6. It is imperative that a lightning protection designer not only take the minimum cross-section into consideration but also to ensure that all the lightning protection components used (e.g.

Table 9 Material requirements for lightning protection system components

Material	Configuration	Air-terminals and Down-conductors (mm^2)	Earth-termination system
Copper	Solid	50	50 mm^2
	Stranded	50	50 mm^2
Aluminium (alloy)	Solid	50	Do not use
	Stranded	50	Do not use
Steel (stainless, galvanised)	Solid	50	50 mm^2
	Stranded	70	70 mm^2

clamps, conductors and rods) are compliant with the series of component standards in IEC 62561.

6 Internal Lightning Protection System

The internal lightning protection will be covered in detail in a separate chapter. However, for the sake of completion of the external protection, which has no strict boundary of separation from internal protection, a brief overview of the latter is given below.

Wikipedia's definition for electrical bonding highlights the importance of electrically bonding any and all conductive parts of a system to the same potential (i.e. equipotential bonding).

> the practice of intentionally electrically connecting all exposed metal items not designed to carry electricity in a room or building as protection from electric shock. If a failure of electrical insulation occurs, all bonded metal objects in the room will have substantially the same electrical potential, so that an occupant of the room cannot touch two objects with significantly different potentials. Even if the connection to a distant earth ground is lost, the occupant will be protected from dangerous potential differences.

i. Lightning equipotential bonding (LEB)

This concept is definitely one of our industry's tongue twisters, but to try and simplify the concept we use the example of a cold beer (i.e. own potential) being placed outside on a hot day (i.e. high potential) then condensation happens as a result of the energy (i.e. heat) being transferred due to the fact that there exists a large temperature difference. The exact opposite is proven when a glass of red wine is placed in a room then we notice no condensation for there is no temperature difference. During a lightning strike potential differences (i.e. a hot day, cold beer) are created either through ground potential rise (GPR) from a direct strike or through magnetic excitation from a distant strike. This gives rise to uncontrolled flashovers and or dangerous sparking (i.e. condensation in the previous example) inside the building between the external lightning protection system and other conductive elements such as gantries, facades, walkways and handrails, but the same goes for internal systems such as electrical and electronic systems thus the need to establish lightning equipotential bonding (i.e. create a condition where the temperatures are the same, like the glass of red wine in the room).

All conductive parts, which have no other potential, can directly be bonded to the main earth bar (i.e. common potential in the building) using for e.g. bonding conductors. However, internal systems that have their own potential (e.g. 230 V), such as an electrical distribution board, can only be bonded indirectly to the main earth bar. Indirect bonding is made possible by installing surge protection devices (SPD) (Fig. 28).

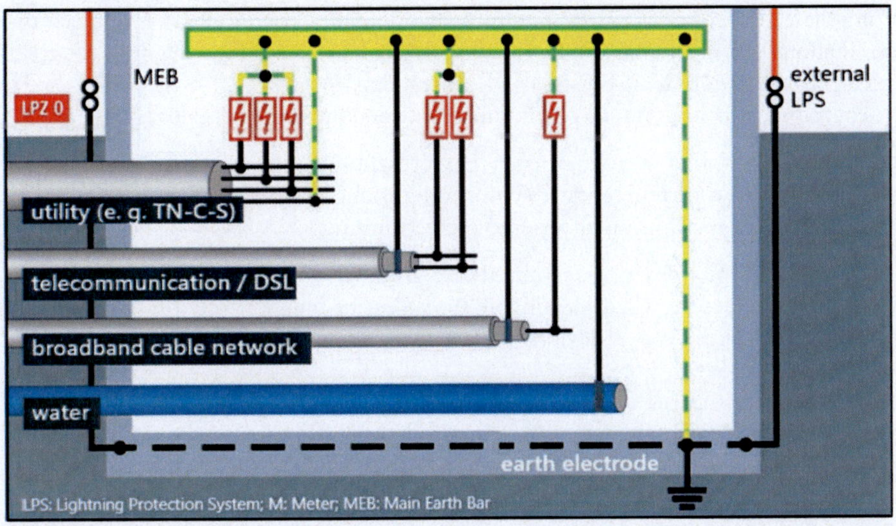

Fig. 28 Typical lightning equipotential bonding [1]

By establishing lightning equipotential bonding, it is important to note that parts of the lightning current can pass through both the direct and indirect bonding methods and this effect shall be considered in terms of the sizing of materials/components.

a. Direct bonding

The minimum values of the cross-section of the bonding conductors connecting different bonding bars and of the conductors connecting the bars to the earth-termination system and the minimum values of the cross-section of the bonding conductors connecting internal metal installations to the bonding bars are listed in Table 10.

b. Indirect bonding

It is a common misunderstanding that SPDs are used to only limit overvoltages conditions and although these devices can do this as well, the main purpose of

Table 10 Material requirements for bonding conductors

Material	Bonding to ETS (mm^2)	Bonding of Internal metals (mm^2)
Copper	16	6
Aluminium (alloy)	25	10
Steel (stainless, galvanised)	50	16

fitting these devices is to ensure lightning equipotential bonding between the earth-termination system and the conductors (i.e. live conductors) of the internal electrical and electronic equipment (Fig. 29).

Therefore, it is important to distinguish between the three types of SPDs.

- Type 1—Lightning current arrester (Direct lightning current)
- Type 2—Surge current arrester (Induced lightning current)
- Type 3—Voltage limiting arrester (Voltage limiting).

The use of Type 1 devices is required for lightning equipotential bonding whereas the Type 2 devices are used to mitigate the effect of magnetic coupling caused by distant strikes and Type 3 devices are used to limit overvoltages caused by e.g. switching operations.

For a building or structure with an external LPS, it is imperative to fit a combination type 1 + 2 surge protection device in the main electrical distribution board. Thereafter type 2 SPDs can be fitted in sub-distribution boards, when they are located more than 10 m away from the main distribution board, and finally for finer protection a Type 3 can be fitted on the terminal device itself (Fig. 30). Note, only fitting a Type 3 SPD is considered inadequate, an upstream and coordinated Type 1 or Type 2 SPD must be fitted as well.

The figure above illustrates wiring details that needs to be taken into consideration by the lightning protection designer and or electrical contractor. It is important to ensure that a connection from the SPD is made to both the electrical earth (main earth bar) as well as the local earth electrode.

Furthermore, please note that the protection level of SPDs decrease when the wiring length (the length of a wire from a live conductor to the closest earth) increase, therefore it is recommended to keep the total wire length below 1 m. In order to avoid nuisance tripping, ensure that SPDs are wired upstream from any earth-leakage or RCCD devices (Fig. 31).

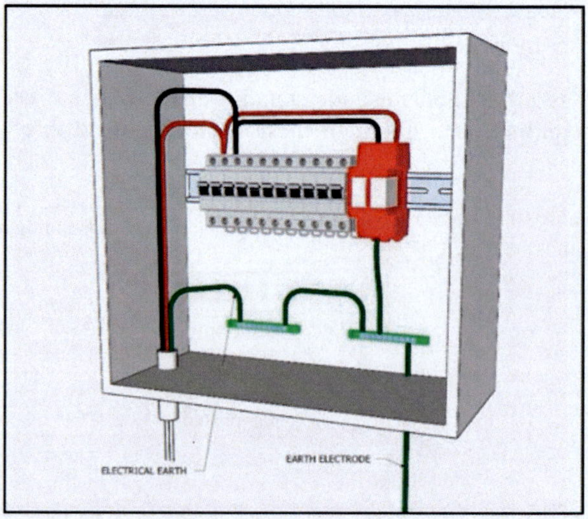

Fig. 29 A typical indirect lightning equipotential bonding method

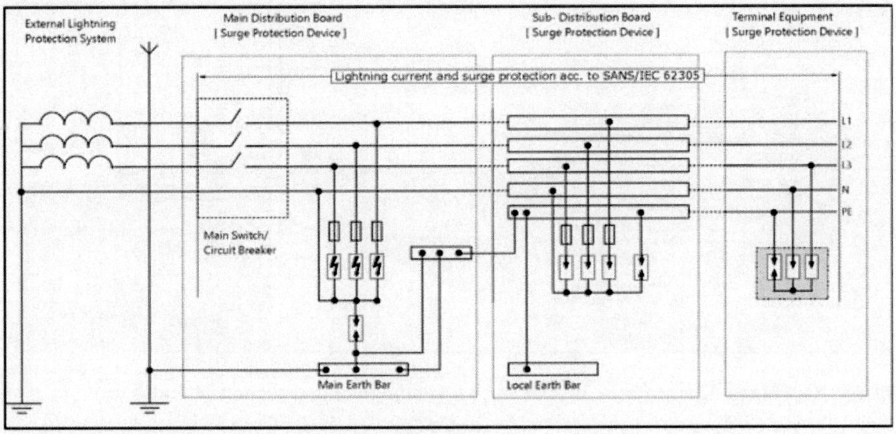

Fig. 30 Coordination of surge protection devices

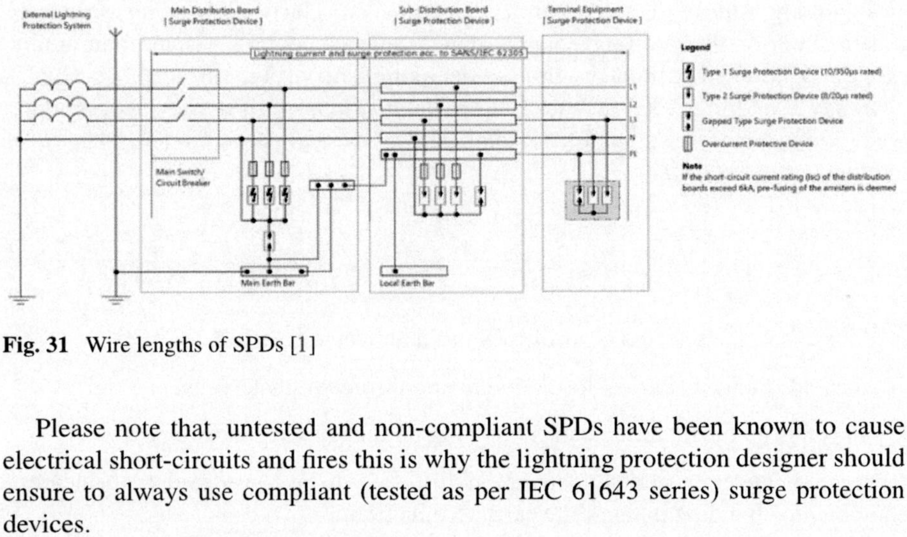

Fig. 31 Wire lengths of SPDs [1]

Please note that, untested and non-compliant SPDs have been known to cause electrical short-circuits and fires this is why the lightning protection designer should ensure to always use compliant (tested as per IEC 61643 series) surge protection devices.

ii. Separation distance (S)

In order to avoid uncontrolled flashovers between the external LPS and other conductive parts or electronic systems, a separation distance (air gap) between these parts needs to be kept. An isolated external LPS should be used to achieve this separation distance in order to ensure that the flow of the lightning current into bonded internal conductive parts do not cause damage to the structure or its contents. This is achieved by placing air-terminals or masts adjacent to the conductive parts or electronic systems needing protection.

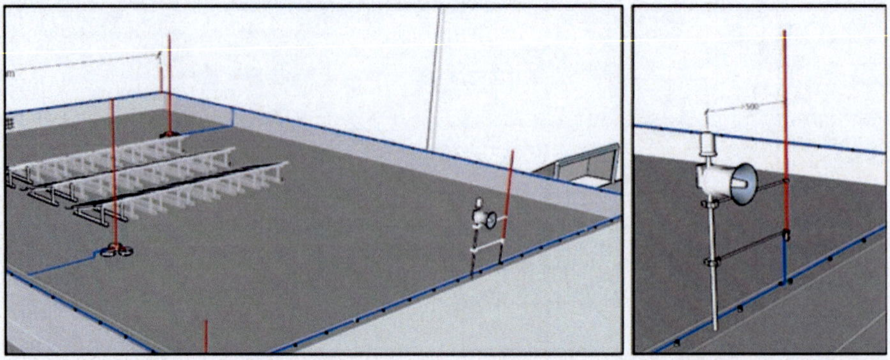

Fig. 32 Separation distance between external LPS and lightning detection system

An isolated LPS is frequently used when a roof is covered with flammable material (e.g. thatched roof) or also for systems located in hazardous areas (e.g. tanks) but for most buildings, these are used to protect electrical and electronic systems which are situated in or on the roof (e.g. solar systems, cameras, cooling systems, and in this case the lightning detection system) see the figure below (Fig. 32).

Annex E of IEC 62305-3 describes the detailed approach on how to calculate the separation distances, but for the safest and most practical method the simplified approach should be taken.

$$S = k_i \frac{k_c}{k_m} l$$

k_i: depends on the selected class of LPS (assume worse case LPL1).

k_m: depends on the electrical insulation material (predominately air).

k_c : depends on the lightning current flowing on the (single conductor)

l : length, in meters, where the separation distance is to be considered, to the nearest equipotential bonding point or the earth termination.

Then a safe and practical equation can be redefined as

$$S = 0.08 \times l$$

References

1. DEHN + SÖHNE (2014) Lightning Protection Guide (3rd updated edn.)
2. Standards International Electrotechnical Commission (2018) Physical damage to structures and life hazard, IEC 62305 Part 3

3. Standards International Electrotechnical Commission (2018) General principles, IEC 62305 Part 1
4. Standards South Africa (2010) The design and installation of earth electrodes. SANS 10199

Protection of Low-Voltage Equipment and Systems

Hélio Eiji Sueta, Sergio Roberto Santos, and Ruy Alberto C. Altafim

Abstract This chapter describes the main aspects related to the protection of electrical equipment and low-voltage systems. It initially addresses the way lightning surges can occur in such systems—they can be induced by lightning strikes inside the clouds, or between different clouds; those conducted by low-voltage network conductors due to direct lightning strikes; surges from lightning strikes on medium-voltage networks; discharges that reach points near networks and are, therefore, induced in low-voltage systems; discharges that reach the LPS (Lightning Protection Systems) and return to the systems by the MEB (Main Earthing Busbar), and those induced in low-voltage systems by lightning strikes through the LPS conductors are also analyzed. The chapter details the main surge protection measures, such as earthing and bonding, shielding, routing, surge protection devices coordination, and isolating interfaces. The chapter defines the concept of Lightning Protection Zone (LPZ) for the positioning of Surge Protection Devices (SPD) and specified and detailed, types and characteristics of these devices. The chapter also covers grounding concepts, resistance and resistivity measurements, and describes the main elements and use of the earth-termination system.

Keywords Surge Protection Devices · Earthing · Overvoltage protection · Surge Protection Measures · Lightning Protection Zone

H. E. Sueta (✉)
Planning, Analysis and Energy Development Scientific Division of the Energy and Environment Institute (IEE-USP), University of São Paulo, São Paulo, Brazil
e-mail: sueta@iee.usp.br

S. R. Santos
Lambda Consulting, São Paulo, Brazil
e-mail: sergio@lambdaconsultoria.com.br

R. A. C. Altafim
University of São Paulo, São Paulo, Brazil

Federal University of Paraiba, João Pessoa, Brazil

R. A. C. Altafim
e-mail: altafim@usp.br

C. Gomes (ed.), *Lightning*, Lecture Notes in Electrical Engineering 780,
https://doi.org/10.1007/978-981-16-3440-6_5

1 Surge Protection Scenario

Low-voltage electro-electronic equipment and systems are quite vulnerable to lightning surges (Lightning ElectroMagnetic imPulses—LEMP). While lightning strikes are megajoule-energy phenomena, electro-electronic equipment can be damaged by only a few millijoules. All such damages from LEMP may be caused by either effect of electromagnetic fields radiated directly to the equipment, or conducted or induced surges transmitted to the equipment by connecting metal conductors (power cables and/or metal signal cables).

Lightning surges can affect installations, damage electronic equipment, and cause dangerous sparks that may start a fire, explosion, or electric shock, as described below.

i. Surges in low-voltage systems due to lightning strikes

a. Surges caused by cloud-to-cloud or intracloud discharges

Approximately 75% of lightning strikes occur between clouds or within them [1]. Such discharges typically last between 200 and 500 ms [2], and are especially important for studies on aircraft protection [3]. According to aviation agencies, commercial aircrafts are hit by lightning at least once a year. The electromagnetic fields generated by these lightning currents can induce surges in electrical networks. Because of the large distance between such discharges and networks, the intensity of surges caused by them is extremely low. The number of studies on this type of discharge is small, in comparison to those on cloud-to-ground discharges, whose harmful effects are much more significant [4].

b. Direct lightning in low-voltage networks

If a lightning directly strikes a low-voltage network, an extremely high voltage surge may penetrate into the facility's electrical installation, thus damaging equipment, or generating sparks that can start a fire, explosion, or electric shock. The relatively short length of such networks and especially those "shielded" by the primary network (Medium Voltage), nearby structures (e.g., buildings, houses, trees), and, in many cases, underground, substantially reduces the possibility of damage. In low-voltage rural grids, those lengths may not be so short, and lines may be more exposed, therefore, when directly hit by lightning, the damage can be severe, especially if such lines are not adequately protected.

c. Direct lightning in medium voltage networks

If a lightning directly strikes a medium-voltage network (e.g. 13,8 kV), the current injected into the struck conductor splits into two waves that propagate in directions opposite the point of impact. The value of each portion of the current multiplied by the line surge impedance, i.e., approximately 400–550 Ω [5], produces over voltages that will shutdown the power grid through disruptive discharges at some points in the line. Such disruptive discharges that occur between phase conductors and/or between phase conductors and ground will cause several voltage and current waves reflections and a reduction in the effective grounding impedance. The electrical charges (consumers near the lightning impact point) will suffer a voltage drop during the short-circuit and there will be a small interruption of the electrical energy due to the actuation of the protection system for the elimination of the failure [6]. If a lightning strikes primary networks, part of its current will be injected into a neutral conductor, thus causing overvoltages in low-voltage networks [4]. A potential rise results from the current flow through the ground resistance, which follows the operation of the lightning arrester for the protection of the transformer and/or occurrence of flashover of insulators in the medium-voltage network. Such transferred surges (medium- to low-voltage) have a wide range and depend on several factors, such as location of incident, observation point, network configuration, lightning current amplitude and waveform, and characteristics of both transformer and protective devices [4].

d. Induced surges in electrical networks

A lightning striking near electrical networks will cause induced surges, which can directly damage equipment, whether or not followed by hazardous sparks, if they occur in low-voltage networks.

Such overvoltages in low-voltage conductors can be subdivided according to the following components [7]:

- Voltages transferred from primary networks;
- Voltages associated with the lightning current portion intercepted by the neutral conductor grounding points, and;
- Voltages directly induced to low-voltage cables due to the electromagnetic coupling between the LV line and the lightning discharge channel.

In the case of voltages transferred from the primary network, we also emphasize those transferred to low-voltage networks by distribution transformers [8].

Regarding voltages associated with the current portion intercepted by the neutral grounding points, experiments performed at ICLRT (International Center for Lightning Research and Testing) in Camp Blanding, Florida, USA, must be highlighted, since lightning rockets were used [9]. According to the study, when lightning strikes a few dozen meters from the line, a substantial portion of the total current can enter the system through grounding points of the neutral conductor. For 60, 40 and 19 m distances between the line and the impact site, peak values of the currents injected into the system range from 5 to 18% of the total current of the lightning discharge.

Among the studies on surges directly induced in low-voltage networks [10] conducted experiments in which lightning induced by rockets and simultaneous measurements of induced voltages in a 210 m long low-voltage overhead line with twisted conductors showed maximum values of phase-to-ground and neutral-ground voltages between 2 and 12 kV. The line was connected at one end to a transformer and at the other to a 60 m long underground cable terminated by low-voltage surge arresters. Some launches were made and the lightning hit a tower near the cable termination and also 50 m from this point. Induced voltages were measured at the low-voltage terminals of the transformer, where current measurements were also varied from 4 to 50 kA.

Piantini and Janiszewski [11] studied the influence of several parameters on induced voltages by indirect discharges for a 300 m long single-phase line. In 2005 [12], Silva Neto and Piantini analyzed the case of multiplexed line and observed the main difference between multiplexed cable and conventional-type network was the former was characterized by a much stronger coupling between much closer conductors. The greater the mutual surge impedance between phase and neutral conductors, the smaller the amplitude of induced voltages. In the case of multiplexed cables, the conductors are braided and their impedance varies along the line. However, as the distance between the conductors is shorter than their height from the ground, the impedance variation is extremely small and can be neglected in practical cases.

All those studies confirmed the substantial impact of induced voltages on low-voltage networks [4].

If lightning strikes some point near a medium-voltage network, the surges may be transferred to low-voltage networks, thus harming the installation and probably shutting it down.

The knowledge of induced voltages in a primary distribution transformer and its frequency-related behavior is fundamental, as highlighted by Piantini [7]; Piantini [13]; Piantini and Malagodi [14]; De Conti and Visacro [15]; Borghetti et al. [16]; Piantini et al. [17].

e. Lightning in Lightning Protection Systems (LPS)

If lightning strikes LPS, the lightning current must flow through the conductors towards the earth-termination system; part of the current flows to the ground, whereas the other portion returns to the MEB. Some portions flow through MEB equipotentialized metallic conductors, which may be low-voltage cables, phone cabling, or computer networks, TV cabling, metal pipes, among others.

IEC 62305-1: 2010 [18], Protection against lightning, Part 1: "General principles", presents a calculation methodology for verifying the portion of surge currents that will flow through external conductive parts and lines connected to structures when lightning reaches the LPS of the structure in Annex E, "Surges due to lightning at different installation points".

The portion of the current that returns to the MEB will be part of the lightning current expected to strike the structure, and depends on the number of parallel paths, conventional grounding impedances of buried conductive parts (Z_1), or their grounding resistances to the aerial parts (Z_2) connected to other buried parts and the conventional grounding impedance of the earth-termination system (Z).

This IEC annex provides two formulas for buried installations and aerial installations, which calculate both current portions as a function of impedances (Z, Z_1 and Z_2) and total number of external parts of buried (n_1), or overhead (n_2) lines, and a table of the conventional grounding impedance (Z and Z_1) values as a function of ground resistivity and Lightning Protection Level (LPL) considered for protection. The data enable the estimation of the lightning current portion that will flow through the lines or metal pipes that enter the structure. After such values have been obtained, the impulse discharge current of the Surge Protection Devices (SPD) can be specified for each line, for example.

f. Surges induced in low-voltage installations due to lightning currents circulating in the LPS

The same current flowing through LPS conductors toward the earth-termination system will induce surges in the facility's internal conductors. Due to the proximity between the down-conductors and the metallic conductors inside the building, the intensities of the induced surges are high, thus threatening both their installation and structures.

IEC 62305-1: 2010 [18] in Annex E presents Tables E.2 "Conventional surge currents due to lightning flashes in low-voltage systems" and Table E.3 "Conventional surge currents due to lightning flashes in telecommunication systems". The last columns for 8/20 μs waveform for induced currents in the internal systems of the tables show surge values from the currents flowing in the down-conductor system. 10, 7.5, and 5 kA currents are estimated for Lightning Protection Levels I, II and III or IV, respectively, which enable the specification of the nominal discharge current values (8/20 μs) of the SPD to be installed in internal systems.

g. Surges in equipment and components caused by a direct effect of lightning electromagnetic fields

Electromagnetic fields generated by lightning can directly damage the electrical equipment and its components. Electromagnetic Compatibility Studies should be performed for structures containing sensitive electro-electronic equipment.

The same IEC 62305-1 [18] Tables (E.2 and E.3) show values estimated for surges induced in internal systems due to lightning strikes near the structure. 0.2, 0.15, and 0.1 kA currents are estimated for each conductor for Lightning Protection Levels I, II, and III or IV, which enables the specification of the nominal discharge current values (8/20 μs) of the SPD to be installed in internal systems closest to the equipment.

ii. Surge Protection Measures (SPM)

Surge Protection Measures (SPM) should be taken for avoiding the harmful effects of LEMP on both structures and their contents' protection, including people.

The main SPM are earthing, bonding measures, electromagnetic shielding, conductor shielding, line routing, Coordinated Surge Protection Devices system, and isolating interfaces.

a. Earthing and bonding measures

Earthing and equipotential bonding are essential for the effectiveness of SPM. The earth-termination system conducts and disperses lightning currents in the soil. Equipotentialization is a set of measures that minimize potential differences between parts of the installation and elements of the LPS and reduce both magnetic field within structures and risk of electric shock.

The Brazilian standard [19] emphasizes this concept applies only to direct current systems or, approximately, low frequencies, which is not exactly the case of lightning strikes. Such connections should be as short as possible for higher frequencies, for a voltage reduction effect between the points of equipotential bonding.

The ring earth electrode around the structure buried in the ground, the natural electrode formed by the foundation of the building and reinforced concrete, and meshed electrodes are good earth-termination system options for protection against lightning and their surges.

Main Equipotentialization Busbars (MEB) and Local Equipotentialization Busbars (LEB) are important elements for installations, since they reduce the voltage between service conductors that feed the structure (e.g., energy, telecommunication, network, cable TV, etc.), protective conductors of the electrical installation, metallic components of internal systems (e.g., enclosures and racks), and any magnetic shielding that defines, for example, a Lightning Protection Zone (LPZ) on the periphery or within the structure.

b. Shielding

Shielding, which can be a space shield in a room, or even an entire shielding building, attenuates induced surges. The space shield can be a mesh formed by conductors or metal sheet that covers the environment, or comprise "natural components" of the structure itself, such as steel from reinforced concrete columns or steel for the constructions of slabs and floors.

The conductors can be shielded by metal covers in the form of shielded cables or embedded in properly grounded metal conduits.

In general, space shields define protected zones, which can cover the entire structure, part of it, a room, or only the enclosure of the equipment to be protected.

Space shielding should preferably be defined in the early design stages of a structure, since natural components are better used and the necessary connections are more appropriately made, which reduce costs.

The shielding of internal lines, e.g. cables used in metal ducts, shielded cables and equipment in properly grounded metal enclosures, is also a good practice.

Space shields can mitigate risks calculated in accordance with IEC 62305-2: 2010 [20]. Risk management must take into consideration the widths of the grid-like shielding, distances between down-conductors, or spacing between the reinforced concrete structures acting as natural LPS.

c. Routing

The loops formed by the installation of the conductors result in surges due to the coupling with electromagnetic fields. Hazardous sparks can occur at some points of those loops, thus damaging equipment and components, and causing fire, explosion, or electric shock.

In equipment whose power is supplied with a routing and signal is fed by a different one, the route can be damaged when the electromagnetic field is generated by lightning coupling with the loop formed by the conductors, and a sparking or surge may also occur.

According to IEC 62305-2:2010 [20] and for risk analysis purposes, an unshielded cable whose installation was not regarded for the avoidance of loops can be considered an "Unshielded cable—no routing precaution to avoid loops", for loop conductors with different routings in large buildings. In this case, loops may occur with areas of the order of 50 m^2. If the looping conductors are routed in the same conduit, or if the installation is a small building of an approximately 10 m^2 loop area, "routing to avoid large loops" is considered. If loop conductors are routed on the same cable (e.g., parallel cable, three-pole cable, etc.) with a 0.5 m^2 loop area, then "routing to avoid loops" is taken into consideration, according to Table B.5 of standard [20]. Such considerations significantly influence the risk considered.

d. Coordinated Surge Protection Devices (SPD) system

Surge Protection Devices (SPD) are the main facility's components for reduction in surges. They must be installed in a coordinated manner and can reduce surges to levels withstood by the equipment with no damage caused to it. They will be defined and specified in an appropriate item.

The concept of Lightning Protection Zones (LPZ), described below, is fundamental for the establishment of locations SPD in the installation. In this respect, Annex E (informative) of IEC 62305-1 [18] (Surges due to lightning at different installation points), and the several parts of IEC 61643 standards [24] (Low-voltage surge protective devices) shall be used for power and signal systems.

e. Isolating Interfaces

Isolating interfaces are surge protection measures. In telecommunication and computer networks, the use of optic fibers is fundamental for a higher data transmission rate, since they are immune to LEMP.

Isolating interfaces are especially used for a proper separation of an existing installation from a new one in the same structure. Interference can be avoided through the use of insulating interfaces, such as class 2 insulated equipment (e.g., double insulated, without a PE conductor), isolation transformers, optic fiber cables with no metal components, and optocouplers.

2 Zonal Concept

i. General

The concept of Lightning Protection Zones (LPZ) is important for the definition of the effects of lightning on structures and their surroundings, which can be direct discharges, electromagnetic fields, and surges driven by the facility's metal conductors.

ii. Zonal concept

The concept of lightning protection zones is fundamental for the definition of Surge Protection Measures (SPM), especially for the SPD (Surge Protection Devices) location in the facility. It enables adaptations of environments with different LEMP exposures to the supportability of electronic systems. Therefore, LPZ can be defined according to the characteristics necessary for protection against lightning of an electric-electronic system (Table 1).

Zone 0, represented by light blue and green in Fig. 1, is outside the structure. It is divided into Zone 0_A and Zone 0_B. The former (light blue color, Fig. 1) is outside the structure and the protection volume given by the LPS. Objects, including people, are exposed to direct lightning strikes and their total electromagnetic field. Zone 0_B (green color, Fig. 1) is also external to the structure; however, the objects are protected against direct lightning, but not against effects of the electromagnetic fields.

The other Lightning Protection Zones are located in the structure. Zone 1 (yellow color, Fig. 1), admits no direct lightning strikes, and their electromagnetic field is attenuated by the metal in the walls of the structure.

Some structures can be divided by other LPZs internal to zone 1 (e.g., LPZ 2, 3,… n, and an LPZ 2 as a computer room with an elevated floor above the grounding screen

Table 1 Characteristics and protection conditions of an LPZ

LPZ	Characteristics	Protection conditions
LPZ 0_A	Zone of a possible direct impact from a lightning strike and no attenuation of the electromagnetic field generated	LPZ 0_A is a totally unprotected environment
LPZ 0_B	Zone of a highly improbable direct impact from a lightning strike and no attenuation of the electromagnetic field generated	LPZ 0_B is an environment protected against direct lightning, but unprotected against the effects of lightning electromagnetic impulses (LEMP)
LPZ 1	Zone of no direct impact of a lightning strike. The electromagnetic field generated is reduced by space shields created with metallic materials separating LPZ 0_B from LPZ 1	LPZ 1 is an environment protected against direct lightning strikes and effects of LEMP. Shields, equipotentialization, isolating interfaces and surge protection devices installed at the boundary between this zone and the previous zone are the SPM used for protection
LPZ n	Zone of no direct impact of a lightning strike. The electromagnetic field generated is successively reduced by the space shields created with metal structures separating LPZ 0_B/LPZ 1/LPZ 2/…LPZn	LPZ n is an environment protected against direct lightning strikes and LEMP effects. Shields, equipotentialization, isolating interfaces and surge protection devices installed at the boundary between this zone and the previous one are the SPM used for protection

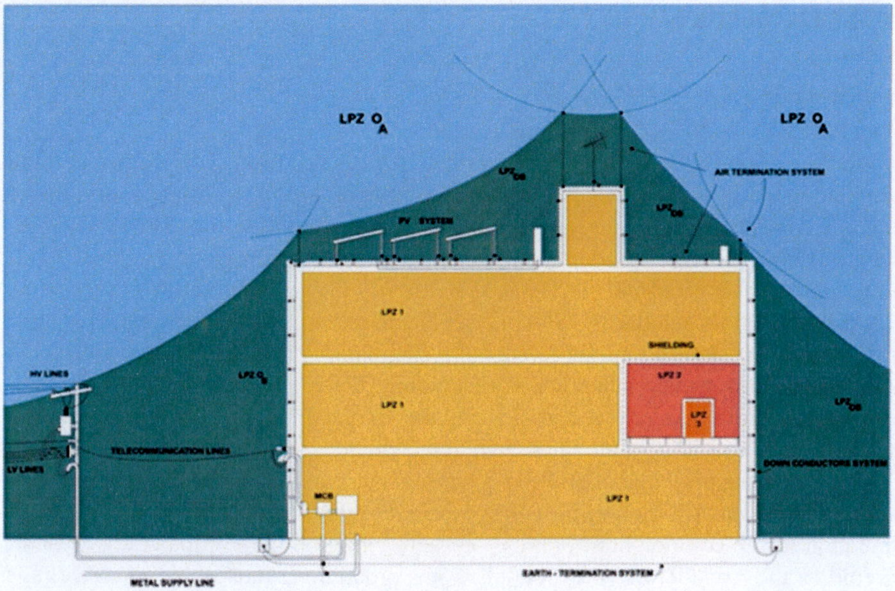

Fig. 1 Lightning Protection Zones (LPZ)

and shielding by the walls). In this case, the electromagnetic field of the lightning is further attenuated (pink color, Fig. 1).

Zone 2 might have a Zone 3 inside it (e.g., a computer network metal rack whose electromagnetic field is further attenuated (orange color, Fig. 1)).

iii. SPD positioning in LPZ

SPD should be positioned at the boundary, between the LPZs. Typically, Type 1 SPD are close to the Main Equipotentialization Busbar, at LPZ O_B and LPZ 1 interface. Type 2 SPD can be positioned at the interfaces of LPZ 1 and 2 (for example, in a Distribution Board), whereas Type 3 SPD can be located at the interfaces of LPZ 2 and 3 (for example, when entering a metal rack or entering equipment to be protected). SPD types represent the characteristics of SPD regarding the way they were tested.

Type 1 SPD, tested with a lightning surge current (10/350 μs waveform), shall be installed at the interface of LPZ O_B and 1 and subject to total or partial lightning currents reaching lines, and those that reach LPS and return to the lines through MEB.

Type 2 SPD, tested with a nominal discharge current (8/20 μs waveform), can be installed on internal LPZ interfaces and shall be subject to lightning-induced currents.

Type 3 SPD, tested with a combined wave of 1.2/50 μs waveform open circuit voltage and 8/20 μs short-circuit current of a combined wave generator with a relationship between them of 2 Ω, shall be installed at the equipment.

3 SPD Specifications

i. General

Surge Protection Devices (SPD) are electrical devices that protect an entire electrical installation, parts of it, or specific electrical equipment against transient overvoltages or surge currents. Their use is part of the Surge Protection Measure (SPM), as established by international standard IEC 62305-4:2010 [21], Protection against lightning—Part 4: Electrical and electronic systems within structures and their respective equivalent national standards. SPD protect electrical and electronic systems against transients overvoltages and surge currents originated from lightning or switching operations. They prevent electrical withstanding (dielectric strength) of an installation element from being exceeded, limiting the voltage in the system points to levels withstood by the components [22].

The operating principle of an SPD is based on a change in its internal impedance, which decreases with increasing voltage at its terminals promoting a surge current deviation to the equipotential bonding system, common mode protection, or short circuiting the conductors, differential mode protection, thus avoiding a voltage increase above that withstood by an electrical installation element, i.e. equipment electrical supportability [23].

Regarding applications, SPDs are divided into three classes or types (1, 2 and 3) arranged throughout the installation, from its origin to the equipment to be protected, and following the LPZ concept. Type 1 SPD protects the entire installation against the effects of a direct atmospheric discharge in either the building, or the utility distribution network and the electrical earthing system. Type 2 SPD are installed in switchboards or electrical panels to protect the circuits originated from these elements against residual Type 1 SPD overvoltage or installation-induced overvoltage caused by remote lightning. Type 3 SPD protects electronic devices against overvoltage originated in the installation itself and caused by voltage variations from motor starting, circuit breaker tripping or other types of switching. The specification of an SPD requires, initially, the understanding that SPD depends on its correct positioning throughout an installation according to the concept of lightning protection zones (LPZ).

Since SPDs protect power and signal lines, some parameters are common to both applications, while others are specific to power or signal. They can also be divided into two groups. The first is related to the operation of the SPD, and their parameters, called here performance parameters, refer to the protection effectiveness. The second group is comprised of compatibility parameters, related to the non-interference of SPD with the normal operation of the systems.

According to standard IEC 61643-11:2011 [24], Low-voltage surge protection devices—Part 11: Surge protective devices connected to low-voltage power systems—Requirements and test methods, the most important parameters are:

a. **Cut-off frequency (f_G)**, which describes the behavior of an SPD as a function of frequency. It causes an insertion loss (aE) of 3 dB under specific test conditions

(IEC 61643-21). Unless otherwise specified, the specified cut-off frequency refers to a 50 Ω system.

b. **Lightning impulse current (I_{imp})**, an impulsive current standardized by waveform 10/350 μs used in SPD class 1 test. Its parameters (charge, peak value, specific energy) simulate the stress caused by real lightning currents.

c. **Maximum continuous operating voltage (U_C)**, i.e., the root mean square (rms), or D.C voltage, value of the maximum voltage in SPD input terminals during their operations. It is the maximum voltage applied to SPD in the non-conductive state, and its value depends on the nominal voltage of the system where SPD will be installed.

d. **Nominal discharge current (I_n)**, which is the peak value of the current conducted by SPD. It has an 8/20 μs waveform and classifies the test of type 2 SPDs.

e. **Temporary Overvoltage (TOV)**, which describes temporary power–frequency surges in SPDs and results from faults in high-, medium- and low-voltage systems. The SPD must have a TOV withstand capability.

f. **Voltage protection level (U_P)**, i.e., the maximum instantaneous value of the voltage on SPD terminals that represents the SPD's ability to limit the overvoltage in the protected installation.

ii. Types of SPD components and their characteristics

The SPD components can be classified into two different families, namely voltage switching SPD and limiting voltage SPD, according to their nonlinear behavior. The manufacture of an SPD involves different components and technologies. The main component is its nonlinear element, responsible for the most important SPD characteristics. The most used nonlinear components are Spark gaps, Gas discharge tubes (GDTs), Metal-oxide varistors (MOV) and suppressor diodes.

The technology defines the characteristics of the nonlinear components and the way SPD will be manufactured. Performance characteristics of an SPD and its constructive aspects interfere with its efficiency as a surge protector, its life cycle, maintenance and selling price.

As addressed elsewhere, SPDs are divided into 3 different classes or types of protection, namely 1, 2 and 3, of specific purposes in the surge protection. Although any nonlinear component can be used in the manufacture of an SPD of any type, spark gaps, varistors, and diodes are typically used in type 1, 2 and 3, respectively. Variations in their use are considered exceptions, rather than rules, at the time of publication of this book.

a. Surge Protection Device Switching type

The SPD switching type has an extremely high impedance when the voltage in its terminals is the normal, but to chance suddenly its impedance to a low value when the voltage increase. Typical examples of switching components in the manufacture of an SPD switching type are spark gaps, gas tubes, thyristors and triacs. Such types of SPDs are also called crowbar types.

b. Surge Protection Device Voltage Limiting type

The SPD voltage limiting type also has extremely high impedance when no voltage surge occurs and the voltage in its terminals is normal; however, it is continuously reduced when the voltage starts to increase. The most common examples of components used in the manufacture of an SPD voltage limiting type, also called SPD clamping type, are varistors and suppressor diodes. A clamping device absorbs much more transient energy internally than a similarly rated crowbar device [25], i.e., a crowbar device is more appropriate for conducting an energy of a surge.

c. Spark-gap

Studies on electric discharges in gases date from the late nineteenth century [26], and have led to the development of the first spark gap devices and their use as electric switches. Spark gaps were the first option for the manufacture of an SPD, and consist, basically, in an arrangement of two electrodes separated by a certain distance by an insulating medium, usually a gas, such as air, calculated in a way a sparking can occur from a specific voltage value between the two terminals. When the potential difference between the conductors exceeds the dielectric stiffness of the insulating medium, the sparking occurs and an electrical current flow until the path of the ionized gas is either broken or reduced to below a minimum value. When the voltage between the two electrodes reaches a value higher than the dielectric strength of the air, i.e., approximately 3 kV/mm under specific conditions, an electric arc is created. In comparison to the impedance under normal conditions in the order of giga-ohm, the impedance of the electric arc is very low, and the spark gap changes its state to a closed electrical switch, where the drop voltage between the electrodes is the residual voltage of the spark gap.

An advantage of the spark gap technology is its exceptionally low impedance, which promotes the passage of a high surge current in the SPD without dissipating much energy. The spark gap characteristic is especially useful when discharging lightning currents. The spark gap can discharge extremely high lightning currents (10/350 μs) of the several tens of kilo Amperes. Because the lower the residual voltage of the spark gap, the lower the energy input to electrical installation protected. Due to the extremely fast change in the internal impedance of the spark gap and the behavior of the voltage difference across their electrodes, the spark gap is known as voltage-switching device.

Spark gaps suffered from two serious drawbacks in the past. The first was the electrical arc caused the release of hot gases, which might damage the SPD and the surrounds. The other, and much more significant drawback was the short circuit between the line and the ground caused by the electrical arc was not spontaneously interrupted after the lightning current had finished.

To avoid the first drawback, the new Spark Gaps are encapsulated in enclosures, so that in discharge situations, no ionized hot gases created by the electric arc escape into the environment and burn the electrical components around the SPD. Another important characteristic of modern spark gaps are their triggering devices, which

Fig. 2 Spark gap symbol

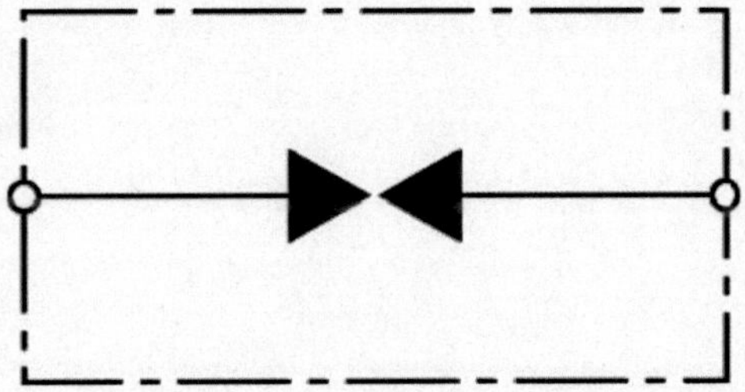

decrease their response time to limits the voltage protection level to a very low value, lower than the voltage of older spark gap based on the dielectric strength of the air.

Regarding the second drawback, modern spark gaps have built-in follow current interruption systems that interrupt these currents that might pass through the SPD after it has conducted surge currents. Figure 2 shows a spark gap symbol.

d. Varistors

Varistors (voltage-dependent resistors) are nonlinear resistors, whose resistance decreases when the voltage magnitude increases in the terminals, thus exceeding the maximum continuous operating voltage. The correlation between the voltage in the varistor terminal and the current it conducts is given by:

$$I = K \cdot V^{\alpha} \quad (1)$$

where α is the non-linearity coefficient of the varistor and represents a V/I ratio. The greater its value, the better the varistor surge protection. Because of their characteristics, varistors are known as voltage limiters, or clamping devices.

Due to their attributes, MOVs are currently the most widely used components in SPD manufacture. They are composed of a thin disk wafer of a material (metal oxide) of a known voltage breakdown characteristic and developed using zinc oxide because this chemical compound has better nonlinear characteristics (α), far superior to older silicon carbide varistors.

At low voltages, an MOV conducts a very weak electrical current (micro-amperes); however, when the voltage in its terminals approaches the breakdown value, it begins to conduct a much stronger current, which enables the surge current to be shunted to the equipotential bonding and avoids overvoltage in the electrical installation. When the varistor is subjected to a voltage higher than its rated voltage, the interfaces between zinc oxide grains are affected by the electric field. The main effect is a reduction in the electrical resistance and conduction of a surge current from an indirect lightning, for example. When the surge ceases, the electric field decreases, the electrical resistance between the grains rises, and the only current flowing through the varistor becomes again the leakage current.

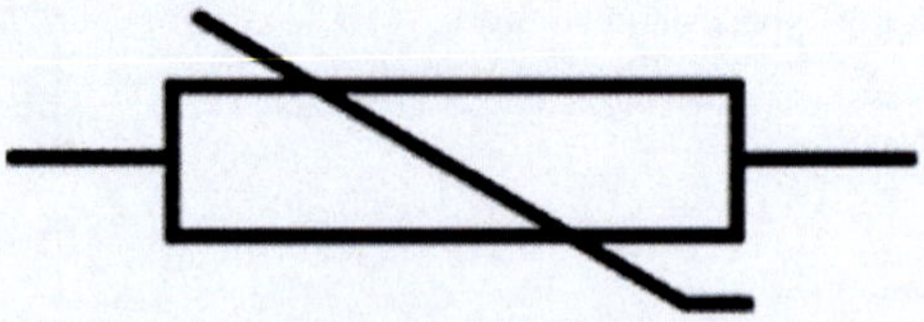

Fig. 3 Varistor symbol

Fig. 4 Diode suppressor symbol

An important characteristic of varistors is their aging since they have a "service life" and should not be installed without proper supervision. MOVs have a large, but finite capacity to absorb energy; therefore, after a given number of operations, they suffer a failure and must be replaced. Their operations succeed when the component is new, however, when subjected to large or repetitive surge currents, the high temperature resulting from the passage of electrical current may fuse the zinc oxide grains, thus reducing the insulation between them and between the electrodes of the varistor. The fusion phenomenon increases the steady-state leakage current, which increases over time and may even result in a short circuit of the component. Figure 3 shows a varistor symbol.

e. Suppression diode

A transient-voltage-suppression (TVS) diode is an electronic component also used in the manufacture of an SPD. Standard and Zener diodes can also be used for surge protection, however, they are designed specifically for rectification and voltage regulation, and are not as reliable or appropriate as TVS diodes.

A TVS diode is a clamping device, suppressing the overvoltage and always come back to its previous state when the overvoltage is finished. It normally responds to overvoltage in less time than other SPD components (e.g., spark-gap or varistors), i.e., in picoseconds, and, consequently, is used mainly in type 3 SPD.

Figure 4 shows a diode suppressor symbol.

4 Earthing Concerns

In LPS, earthing systems are elements that enable lightning or fault currents to safely flow into the ground and protect individuals near grounded installations against critical electrical shocks or other electrical injuries. The soil normally has high electrical resistance concentrated next to the point of contact, between the electric system and

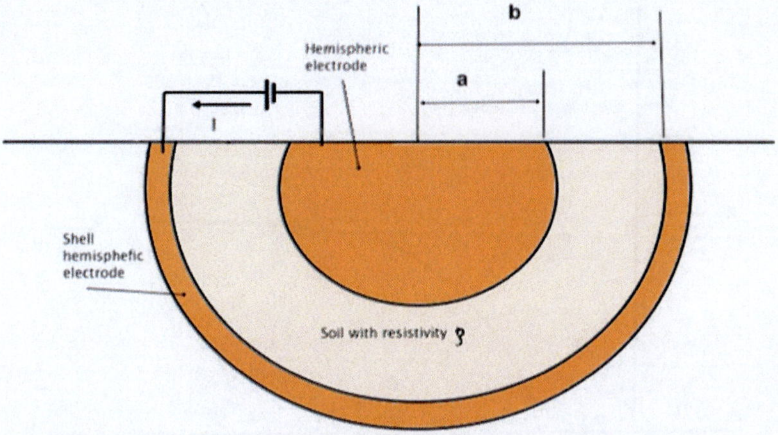

Fig. 5 Electrical Resistance (R_e) between a hemispheric electrode and a shell hemispheric electrode filled with a soil of resistivity (ρ)

the soil. Such resistance, called Earth Resistance, can be significantly reduced by the introduction of metallic elements, such as natural electrodes, rods, cables or wires, properly interconnected and defined in a specific earthing system project. However, the resistivity of the soil and existence of other metallic materials or water in its vicinity must be known for a project of this system. Such features are best described in the following topics.

i. Earthing resistance

The earthing resistance can be better understood by considering the electrical earthing resistance between a conductive hemisphere of radius (a) and a hemispheric shell of radius (b), separated by a soil of resistivity (ρ), as illustrated in Fig. 5, calculated as:

$$R_e = \frac{\rho}{2\pi}\left(\frac{1}{a} - \frac{1}{b}\right) \tag{2}$$

If the radius of the hemispherical shell (b) tends to infinity, the electrical resistance of this hemispherical system becomes only

$$R_e = \frac{\rho}{2\pi}\left(\frac{1}{a}\right) \tag{3}$$

This resistance (R_e) is called earthing resistance of the hemispheric electrode, and as it can be seen only a function of local parameters, such as soil resistivity and the hemisphere radius.

Figure 6 displays it as a linear region, which can be also determined in the field, as shown in Fig. 7. In this case, an electrical voltage (V_1) is applied between the earthing system and an auxiliary electrode away (h1), and a current (I) circulates in

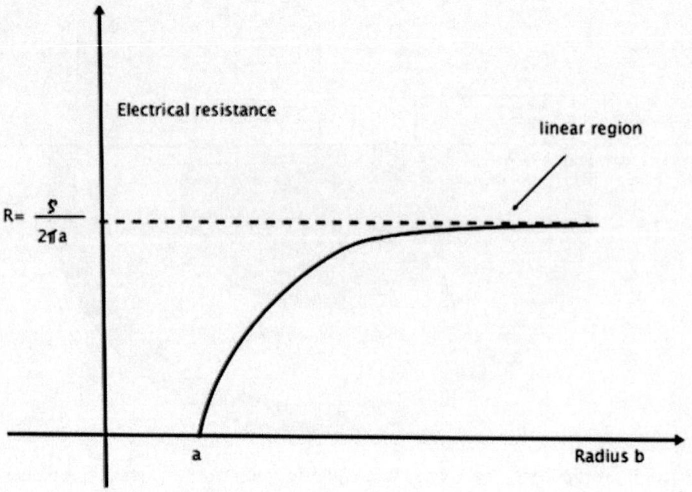

Fig. 6 Resistance of a metallic hemispheric electrode of radius (a) in the soil with resistivity (ρ)

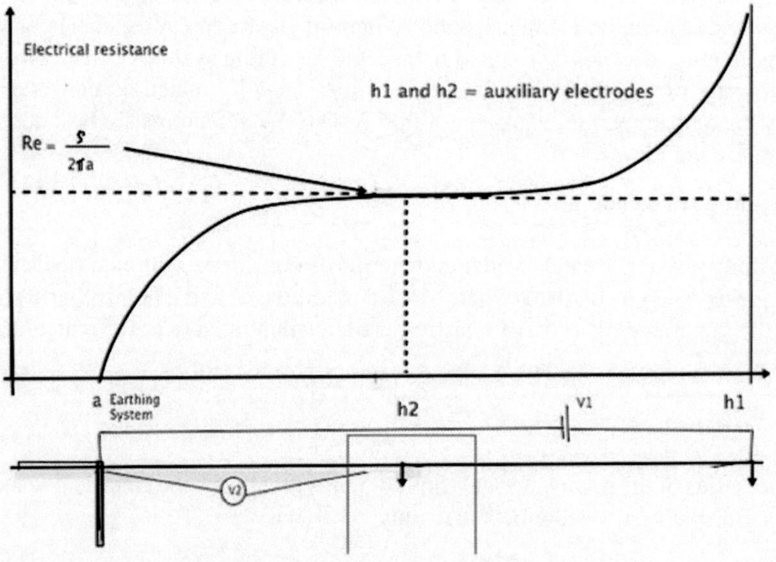

Fig. 7 Measurement of the earthing system resistance in the field

this circuit. The resistance (R_e) is given by the measurement a voltage (V_2) between some point of the earthing system and another auxiliary (h2) sticked in a middle point of the earthing system and the auxiliary electrode (h1), as

$$R_e = \frac{V_2}{I} \tag{4}$$

Special care must be taken regarding the distances of electrodes (h1) and (h2) towards the avoidance of their interference with the measurement, and presence of the linear region.

Resistance (R_e) depends on the materials of the electrodes, cables and their connections, types of contact between the electrodes or cables (whether exothermic welds or pressed connectors), soil characteristics, including temperature and humidity, and geometry of the arrangement of the electrodes on the ground. The first two factors, i.e., materials and types of contact, are often ignored, especially due to their low resistivity and low influence on resistance (R_e). However, they may increase over time because of the corrosion effect of the electrodes connections involved. The other two factors, i.e., soil resistivity and geometry, effectively influence the resistance value (R_e).

The earthing resistance of other electrode configurations can be calculated analytically or digitally by several methods. However, one of the first calculations was proposed by Dwight in 1936 [27] and has been widely applied. For instance, the earthing resistance of a cylindrical rod driven vertically into the soil of resistivity (ρ), driven length (L), and radius (r) is given by

$$R_e = \frac{\rho}{2\pi L}\left(ln\frac{4L}{r} - 1\right) \tag{5}$$

For a buried straight wire of length (2L), we have

$$R_e = \frac{\rho}{4\pi L}(ln\frac{4L}{r} - 1) \tag{6}$$

Another useful earthing system, such as two cylindrical rods driven vertically in parallel at a distance longer than 3 m, or groups of such rods we can find the formulas in the same reference [27].

ii. Soil resistivity

Soil composition, temperature, water-retention capacity and seasonal variations (Table 2) [28] strongly influence the soil resistivity value, measured in [Ω·m]. Regarding moisture retention capacity, one of the most important characteristics of the soil is, basically, a function of its porosity or compaction. In fact, according to earthing system designers, "the best soil is water". Such a statement carries a logical meaning, because in very humid soils, soil resistivity is very low and high electrical currents rapidly tend to disperse, whereas in dry or moderately wet soils, the drainage of such currents may be hampered by water evaporation and further increase in soil resistance.

Table 2 shows typical resistivity values; however, according to ANSI/IEEE std 80 [29], Loboda [30], such values can dramatically change (see Table 3), thus requiring on-site measurements for the obtaining of effective resistivity values. Soils of heterogeneous mineral and geological formation and whose resistivities vary from one point to another are beyond the scope of this chapter. However, the determination of an equivalent electrical resistivity for those soils can be found in reference [31].

Table 2 Typical soil resistivity of various types of soil

Type of soil or water	Typical resistivity (Ω·m)
Sea water	2
Clay	40
Ground well and spring water	50
Clay and sand mix	100
Shale, slates, sandstone	120
Peat, loam and mud	150
Lake and brook water	250
Sand	2,000
Morane gravel	3,000
Ridge gravel	15,000
Solid granite	25,000
Ice	100,000

Table 3 Variation of soil type resistivity

Soil type	Resistivity (Ω·m)
Wetlands, slime, humus, mud	Up to 150
Arable, clay-sandy soils	50–500
Clay	40–5,000
Sand	1,000–8,000
Limestone	500–5,000
Granite and sandstone	100–10,000
Basalt	10,000–20,000

iii. Earthing elements

Earthing systems also comprehend three other important components, namely earthing electrodes, cables and connectors. Cables and metallic rods of different shapes and dimensions are normally used as earthing electrodes. Figure 8 shows one of such electrodes, i.e., one of cylindrical shape of length (L) and diameter (D). Electrodes can be inter-connected by cables and connectors in different geometries, thus forming the earthing system. Each geometry is used to reduce the earthing resistance and applied to a specific soil. For instance, in rocky soils, the widely used geometry is the interconnection of cables that extend horizontally over long distances. However, numerous metal elements (e.g., pillars and beams) can also be used as earthing electrodes in a building, and, when properly connected, metal structures can be an excellent earthing system.

Another solution with good results is to encapsulate the electrodes in the concrete. In contact with the soil, concrete may show approximately 3,000 [Ω·m] resistivity at 20 °C, which is generally lower than the soil itself. However, regardless of the electrodes or metal structure used, all metal elements must be interconnected for

Fig. 8 Cylindrical metallic rod with connector to the cable

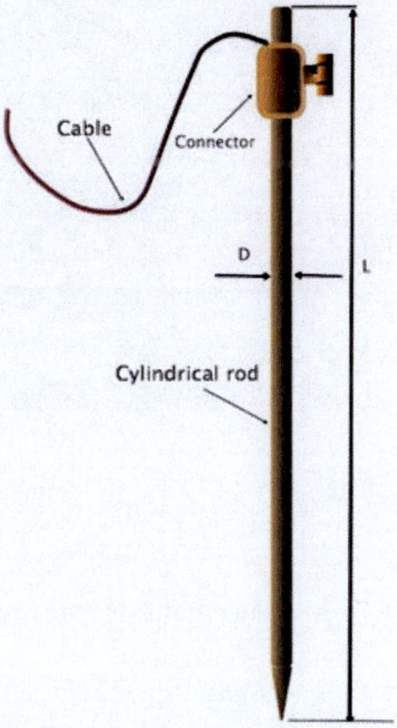

ensuring the grounding system becomes equipotential, with an earthing resistance below 0.25 Ω. Bare copper cables of 50 mm^2 (1/0 AWG) minimum dimensions and 7 wires of 3.00 mm diameter each are normally used for such connections. In rocky or sandy terrain, the soil tends to be very dry and show high resistivity, and foundations, metal strips or horizontally buried cables are the most suitable solutions both technically and economically. Mechanical connectors, exothermic welds, and pressure connectors are important elements for connecting cables to electrodes or metal structures. While mechanical connectors are of easy installation and disconnection of part of the grounding system during inspections and ground resistance measurements, they may suffer from corrosion or poor contact problems, which can be avoided by some special protection (e.g., paints, enamels, oxides and other metals). Crimp wire connectors are also of easy installation, have low contact resistance and face no corrosion problems; however, they enable no disconnection from the grounding system. Finally, exothermic welds promote a permanent connection between two metal components, thus preventing corrosion between them and contact-resistance problems. This technique, called aluminothermic chemical reaction, consists in an exothermic chemical reaction at 900 °C average temperatures, involving a heavy metal oxide (e.g., copper oxide) and a reducing agent (e.g., aluminum oxide). The copper oxide is reduced by aluminum, thus resulting in a copper alloy that forms permanent molecular bonds between welded materials. The corrosion process of such bonds is similar to that of the joined components. Figure 9 shows an example of this type of weld.

However, the grounding system designer decides on the most suitable type of connector to be used.

Fig. 9 **a** An electrode and a cable being prepared for exothermic welds. **b** Final results of the weld

iv. Measurement of soil resistivity

Wenner's method [32] (Fig. 10) is the main method for measurements of soil resistivity. It consists in measuring the soil resistivity of four metal rods grounded to a depth (b) and equally spaced by a distance (a). The electrical current (I) is measured by the connection of a voltage source to the two external rods located at points 1 and 4. This current (I) flowing through the ground between these electrodes produces a potential difference between points 2 and 3, called voltage (V_{23}). The relationship between voltage (V_{23}) and current (I) is the resistance (R_t) or $R_t = \frac{V_{23}}{I}$. With the resistance (R_t) known, we can have

$$\rho = \frac{4\pi a R_t}{1 + \frac{2a}{\sqrt{a^2+4b^2}} - \frac{2a}{\sqrt{4a^2+4b^2}}} \tag{7}$$

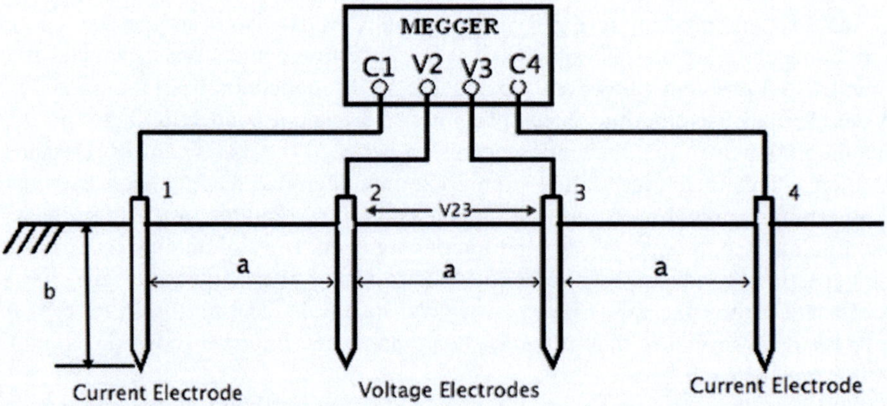

Fig. 10 Measurement of earthing resistivity

In case the parameter (b) is large in comparison with the parameter (a),

$$\rho = 4\pi a\ R_t \tag{8}$$

and in case the parameter (b) is small in comparison with the parameter (a)

$$\rho = 2\pi a\ R_t \tag{9}$$

Variations in the distances (a) between the electrodes (e.g., 1, 2, 4, 8, 16, 32, 64 m) provide data for soil stratification in two or more soil layers [31].

v. Earth-termination system for protection of structures

For lightning-protection purposes in buildings, the earthing system may be the foundation itself, a mesh system or a grounding ring around them, but also connected to the MEB inside the power distribution board to be the earth reference for the protection of internal systems and electrical equipment. An important premise to be established is all connections to the earthing system should be as short and straight as possible. It is important to reduce the effects of current reflections, since lightning currents have high-frequency components. Figure 11 shows an example of a MEB.

Fig. 11 Example of a Main Equipotentialization Busbar (MEB)

Fig. 12 Earth joint box and connections

As an effective Surge Protection Measure (SPM), all non-energized metallic elements of a building and points of earthing of some equipment, especially SPD, must be connected to the same earthing system for guaranteeing equipotentialization, since the separation distance was not achieved [33]. In the case of SPD, their grounding shall be as short as possible, i.e., at most 0.5 m.

Large buildings require the installation of Local Equipotentialization Busbars (LEB) (on the same or different floors), also connected to the MEB and always referenced to the same earthing system. A single earthing system is recommended for lightning-protection purposes and, when adopted, it must include the LPS, the electrical system, and the signal system (telecommunication and computer networks). However, if a building has more than one non-connected earthing system, spark arrestors must be used to connect them [34]. Figure 12 shows a test joint box and connections to the down-conductor system and grounding ring.

References

1. Rakov VA (2007) Lightning phenomenology and parameters important for lightning protection. In: Proceedings SIPDA 2007, pp 541–564
2. Cooray V (2003) The mechanism of the lightning flash. In: IEE power engineering series, 34, IEE, cap 4, pp 127–239

3. Piantini A (2020) Lightning interaction with low-voltage overhead power distribution networks. In: Piantini A (ed) Lightning interaction with power systems, vol 2: applications, 1st ed, IET Press, London, pp 173–226
4. Piantini A (2019) Sobretensões atmosféricas em redes de baixa tensão, Revista O Setor Elétrico, julho de 2019, ano 14, edição 162, pp 64–69 (in Portuguese)
5. Anderson RB, Eriksson AJ (1979) Lightning parameters for engineering applications. Electra 69:65–102
6. Piantini A (2008) Lightning protection of overhead power distribution lines. In: Proceedings ICLP 2008, Uppsala, p 1-1-1-1-1-29 (invited lecture 4)
7. Piantini A (2010) Lightning protection of low voltage networks. IET Power and Energy Series, 58, Lightning protection. IET 2010, Cap 12, pp 553–634
8. Piantini A, Kanashiro AG (2002) A distribution transformer model for calculating transferred voltages. In: Proceedings ICLP 2002, Cracow, pp 430–434
9. Rakov VA, Uman MA (2003) Artificial initiation (triggering) of lightning by ground-based activity. Lightning: physics and effects. Cambridge University Press, Cap 7, pp 265–307
10. Clement M, Michaud J (1993) Overvoltages on the low voltage distribution networks. Origins and characteristics. Consequences upon the construction of Electricite de France networks. In: Proceedings CIRED, IEE, pp 2.16.1–2.16.6
11. Piantini A, Janiszewski JM (1999) Lightning induced overvoltages on low-voltage lines. In: Proceedings SIPDA 1999, pp 234–239
12. Neto AS, Piantini A (2005) Induced overvoltages on LV lines with twisted conductors due to indirect strokes. In: Proceedings SIPDA 2005, pp 234–239
13. Piantini A (1997) Tensões induzidas por descargas atmosféricas em linhas aéreas, rurais e urbanas, considerando diferentes métodos de proteção – modelagens teórica e experimental e aplicação ao cálculo de interrupções. Tese de doutorado EPUSP. Universidade de São Paulo, 316 p (in Portuguese)
14. Piantini A, Malagodi CVS (1999) Voltages transferred to the low-voltage side of distribution transformers due to lightning discharges close to overhead lines. In: Proceedings SIPDA 1999, pp 201–205
15. De Conti A, Visacro S (2005) Evaluation of lightning surges transferred from medium voltage to low-voltage networks. IEE Proc Gener Transm Distrib 152(3):351–356
16. Borghetti A, Morched AS, Napolitano F, Nucci CA, Paolone M (2009) Lightning-induced overvoltages transferred through distribution power transformers. IEEE Trans Power Delivery 24(1):360–372
17. Piantini A et al (Aug 2013) Lightning protection of low-voltage networks, CIGRÉ, 2013, 81 p, CIGRÉ working group C4.408, Technical brochure 550
18. IEC 62305-1: 2010 (2010) Protection against lightning, part 1: general principles. International electrotechnical commission, p 137
19. ABNT NBR 5419: 2015 (2015) Proteção contra descargas atmosféricas, 4 partes (in Portuguese)
20. IEC 62305-2: 2010 (2010) Protection against lightning, part 2: risk management. International electrotechnical commission, p 171
21. IEC 62305-4: 2010 (2010) Protection against lightning, part 4: electrical and electronic systems within structures. International Electrotechnical Commission, p 178
22. Uman MA (2008) The art and science of lightning protection. Cambridge University Press, Cap 6, pp 99–110
23. Cooray V (2010) Lightning protection. The Institution of Engineering and Technology
24. IEC 61643-11:2011 (2010) Low-voltage surge protection devices—part 11: surge protective devices connected to low-voltage power systems—requirements and test method, International electrotechnical commission, p 201
25. Kularatna N et al (2019) Design of transient protection systems: including supercapacitor based design approaches for surge protectors. Elsevier Inc
26. Standler R (1989) Technology of fast spark gaps. The Pennsylvania State University Department of Electrical Engineering

27. Dwight HB (1936) Calculation of resistance to ground. Electrical engineering, pp 1319–1328
28. Lacroix B, Calvas R (Mar 2002) Earthing system in LV. Schneider electric's Cahier's technique no. 172
29. ANSI/IEEE std 80-1986, IEEE Guide for safety in AC substation grounding
30. Loboda M, Marciniak R (Nov 2004) Practical aspects of earthing rods application for power system grounding. In: Proceedings 1st LPE international conference on lightning physics and effects, pp 285–290
31. Russell A, Blumentritt BS (Aug 1969) A theoretical earth resistivity study with applications on the LLano Estacado. A thesis in geology at Texas Technological College-Master of Science
32. Wenner F A (1915) Method of measuring earth resistivity. Bulletin of the bureau of standards, pp 469–475
33. IEC 62305-3: 2010 (2010) Protection against lightning, part 3: physical damage to structures and life hazard. International electrotechnical commission, p 313
34. NBR 5410 (Mar 2008) Electrical installation of buildings - low voltage (in portuguese)

Lightning Protection of High-Risk Installations: Petrochemical Plants

Arturo Galván Diego

Abstract Petrochemical plants are outdoor facilities that hold and produce vast quantities of chemicals. Critical stages of production deal with toxic, flammable and explosive substances. The petrochemicals industry sources raw materials from refining and gas-processing and converts these raw materials into valuable products using a variety of chemical process technologies. Historically the industry evolved out of technological innovation in the developed industrialised economies. Until the last quarter of the twentieth century production of petrochemicals was concentrated in Western Europe, the United States and Japan. Over the last few decades, however, production in areas with competitively priced feedstocks has increased dramatically. New production capacity has been built in the Middle East and Asia, and lightning strikes can be a major hazard.

Keywords Fire hazard · Galvanic effect · Bound charge · Shielding failure · Spark and lightning protection

Lightning can be responsible for damage of two types in any modern plant: one, the destruction resulting from a direct stroke, and the other, the impairment of equipment due to surges in the electric/electronic system, caused either by direct strokes to power lines at some distance from the plant, or by electrostatically induced voltages [1]. When a chemical plant gets hit directly by lightning in critical stages of production, what usually follows are explosions, process upsets, power outages, and potential injuries to personnel, and both direct and indirect lightning strikes can be a major hazard. All lightning protection standards stress that no single lightning protection measure can get rid of lightning strikes 100%, but it is possible to reduce the likelihood of damage from lightning strikes applying suitable measures combined.

There are three main reasons why to implement lightning protection measures in petrochemical installations, in order of importance.

1. Protect people (staff).

A. Galván Diego (✉)
National Institute of Electricity and Clean Energy (INEEL), Cuernavaca, Mexico
e-mail: agalvan@ineel.mx

C. Gomes (ed.), *Lightning*, Lecture Notes in Electrical Engineering 780,
https://doi.org/10.1007/978-981-16-3440-6_6

2. Prevents damage to systems.
3. Prevents damage to structures and buildings.

The philosophy of lightning protection is, according to author's experiences and standards criteria, as follow:

1. Avoid or get rid of lightning hit upon tall metallic structures by lightning protection terminals. The taller a structure is, the higher the chance it will be struck by lightning.
2. Conduct safely the lightning current to ground. This measure should get rid of the voltage drop due to the inductance of the lightning conductor.
3. Apply equipotential bonding measures in each particular system joined with the general grounding grid. This measure will avoid dangerous potential differences.
4. Apply comprehensive provisions of equipment and electrical system grounding, like those recommended in local or relevant standards, such as IEC and NEC. This provides safe working conditions for personnel.
5. Implement local grounding grids connected to a general grounding grid in order to ensure equipotential surfaces and low ground resistance.
6. Install surge protection system for electrical and electronic systems. This measure will guarantee trouble-free operation of complex electrical and electronic systems without consequential damage.

The complexity of petrochemical plants is due to the diversity of systems involved in the operation of the plant, that deserve properly lightning protection measures:

1. Process installations, with tall process metallic towers, metallic chimneys and flare stacks, process containers/vessels.
2. Oil storage tanks farm.
3. Power substations.
4. Distribution lines.
5. Structures and buildings.
6. Low voltage system.
7. Telecommunication Systems and control rooms.

This chapter is devoted to analyze and recommend measures in order to reduce damage from lightning direct hit and indirect lightning effects in all areas of petrochemical installations, according to the six lightning protection points and the seven types of systems referred above.

1 Direct Lightning Protection

In order to implement a measure for direct lightning protection, the designer has to decide into three valid options [2–4].

1. Self protection.
2. Non-isolated protection.

3. Isolated protection.

For self-protection, lightning stepped leader is allowed to intercept with the structure, thus total lightning current will be injected into the metal structure. The first return stroke of negative lightning strikes, the commonest among cloud-to-ground flashes, has impulse currents with a peak value, Ip, of 30 kA on average, a peak current derivative, (di/dt), of 30 kA/μs on average and continuing current of several hundred of amperes and times of 0.5 s. Thus, interception of lightning leader and consequent passage of impulse current may generate highly localized hotspots, large potential gradients, and mechanical stresses in the material due to the flow of fast varying currents. In such situation, strong walls are essential to prevent puncturing at the point of strike, joule heating close to flammable materials and mechanical collapse due to magnetic forces. Not only real failures but even the sense of the risk of failure may strongly influence the smooth operation of the system [2]. Figure 1 depicts this protection scheme.

For non-isolated protection, lightning stepped leader is allowed to intercept with the air terminal bonded to the structure, thus total lightning current will be injected into the metal structure, but hotspot is put away from the metal structure of the structure to the tip of the air terminal. Effects by the current passage on the metal structure of the tank are similar to those given for self-protection scheme. Figure 2 depicts this protection scheme.

For isolated protection, isolated metal masts (with overhead grounded networks in some cases) are installed to protect the structure (process container, oil tank or small buildings). These conductors are installed with a minimum separation, S, from the structure to be protected to avoid arcing between structure and mast in the event of lightning strike to the mast. The minimum separation is calculated by the pertinent

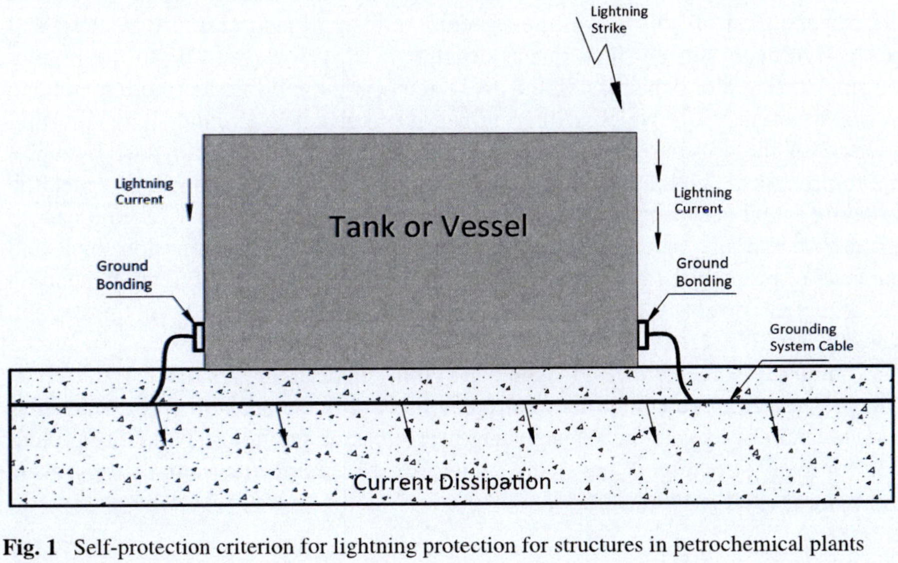

Fig. 1 Self-protection criterion for lightning protection for structures in petrochemical plants

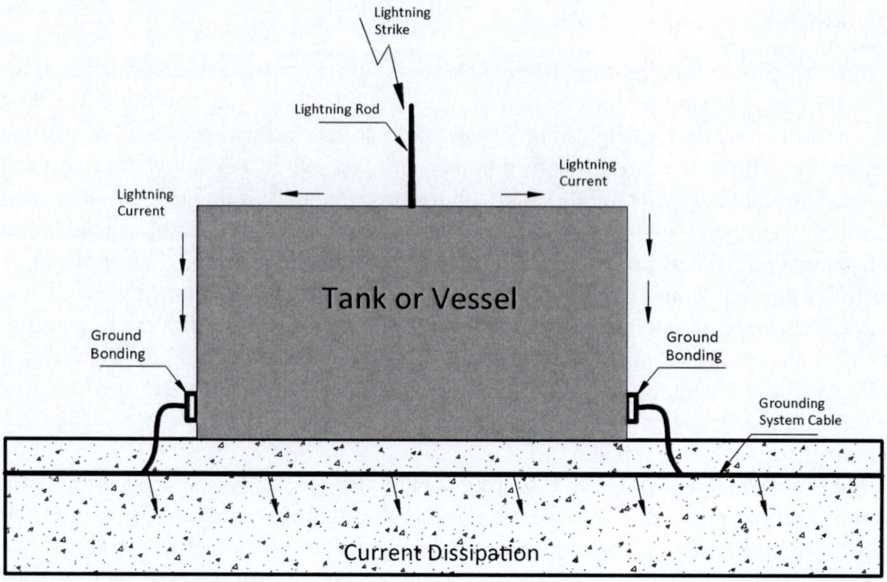

Fig. 2 Non-isolated criterion for lightning protection for structures in petrochemical plants

equation given in [3]. The masts will act both as the air-termination and the down conductor, and in the case of a metallic grid over the tip of the structure will act as the air-termination system. A suitable grounding system should be installed at the earth-termination of the mast in order to disperse the lightning current readily into the soil masses, see Section IV. The structure walls and the mast should be electrically bonded at the ground level to avoid surface arcing. Such isolated LPS will prevent the development of hotspots in the structure walls and also ensure that there will be no significant current flow in the structure walls. However, due to the rapidly varying current along the mast in the event of a lightning will induce certain voltage in the structure walls which will in turn drives current to ground. It is whorthly mentioning that this current will be much smaller than the current injected by direct lightning strikes. Lightning bulk current is managed by air terminals and metallic structure is exposed only to induction voltages and currents. This scheme can be termed "shielding", as employed in lightning protection of transmission lines, and the level of protection (and the number of air terminations or masts) will depend on the accepted current to generate shielding failure. Figure 3 depicts this protection scheme.

Contrary to some believes, Isolated and Non-Isolated systems made of air terminals increases the probability that a strike will occur in the proximity of the installation, but a properly designed system can offer substantial damage avoidance protection for many structures. Conversely, an improperly designed system can aggravate lightning-related problems [2, 5].

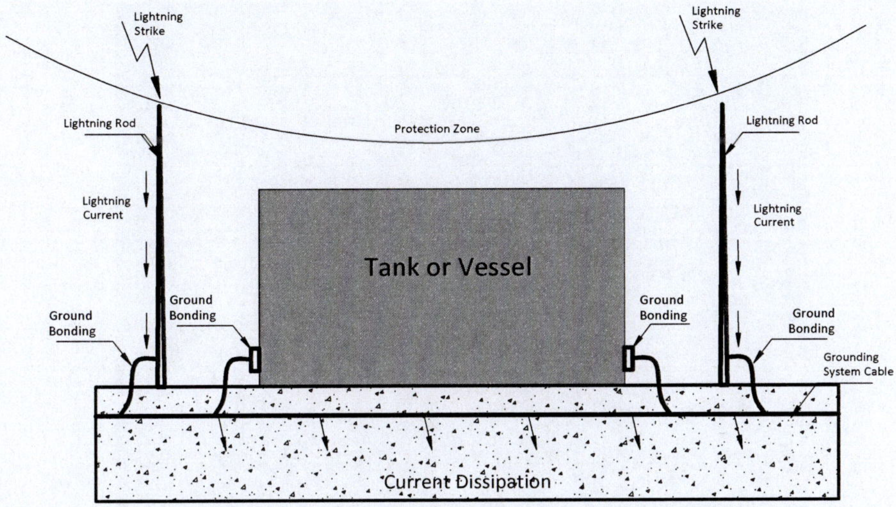

Fig. 3 Isolated criterion for lightning protection for structures in petrochemical plants

1.1 High Metallic Structures

For very tall structures (like destilling towers or flare stacks) or very wide structures (like oil-fuel tanks with big diameters), isolated protection is an impractical measure to adopt. For structures like cracking containers, process tanks and blending containers, isolated protection is more suitable to be adopted, and for buildings non-isolated protection should be adopted [2].

It is probable that some of the higher grounded steel structures, do receive occasional lightning strokes with no appreciable damage to itself. They maybe considered as self-protected against lightning strokes. Such high structures create a zone of protection about their bases [1] (Fig. 4).

For the case of destilling towers, the minimum thickness walls are greater than 6 mm. So, a direct lightning hit theoretically would not produce puncture, but the local hot-spot with an ignition hazard possibility should be evaluated according to local lightning current performance (density, rate of rise of current, charge and long tail current). In this case, non-isolated protection scheme could be a suitable protection measure.

Metal chimneys and flare stacks need no protection against lightning other than that afforded by their construction, provided they are properly grounded. Metal flare stacks are usually remotely located from process areas and rarely fall within a zone of protection from other structures. In such stacks, it is also desirable that metal guy wires and cables be grounded at their lower ends. Guy wires attached to steel anchor rods set in the earth may be considered as sufficiently well grounded. Guy wires set in concrete or attached to buildings or nonconducting supports should be provided with additional grounding facilities. Vent stacks situated at the top of process

Fig. 4 Petrochemical plant inherently selfprotecting for tall structures. Non-isolated protection are normally applied in order to divert the striking point from metal structure to be protected. Adapted from Tompson [6]

structures can usually be considered as adequately grounded through the vessel or supporting structural steel. Special precautions must be taken with chimneys and stacks of nonconducting materials such as brick, tile, concrete, or similar material liable to damage by lightning.

It is customary practice to install air terminals (lightning rods) uniformly about the top rim of the stack at intervals not greater than 8 ft. The air terminals should be connected together by means of a metal ring or band forming a loop about 2 ft below the top of the chimney. The air terminals must then be connected to ground by at least two down conductors installed on opposite sides of the stack. Where stacks have a metal lining, the lining should be connected to the air terminals at its upper end and grounded at the bottom. Reinforced concrete stacks should be treated as above, but in addition, the reinforcing metal should be electrically connected together and bonded to the down conductors on the outside of the stack at the top and bottom of the concrete [1, 7]. A typical arrangement of air terminals on a stack is shown in Fig. 5.

1.2 Oil Storage Tanks or Vessels

IEC Standard [3] establishes that the material of metallic pipes and tanks can be considered itself as air termination (natural components) if thickness meets the value

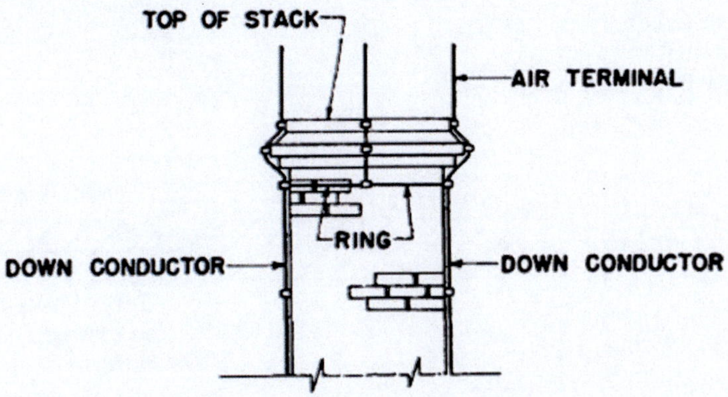

Fig. 5 Non-isolated air terminal arrangement for a stack [1, 7]

Table 1 Risk conditions of lightning-related hotspot on the metal structure

Scheme	Hotspot at structure metal	Bulk lightning current at structure metal
Self-protection	High	High
Non-isolated	Low	High
Isolated	Low	Low

of 4.8 mm for steel (lesser than 3/16″ required by [8] and [9]) and the temperature rise of the inner surface at the point of strike does not constitute a danger.

Table 1 shows the risk conditions for each protection scheme. For self-protection scheme, where the risk of hotspot is very high, it is important to take steps to reduce damage as the result of a lightning strike, side-flash or corona discharge.

In 2009, a study on lightning effects on above-ground storage tanks was carried out [10], especially in terms on puncturing effects when stainless and ferrous steel were exposed to a long duration lightning current. Results of this study indicated that for a 5 mm thick samples, with both 100 C and 400 C (that is, 200 A or 800 A in 0.5 s) the molten pool volumes were 64 mm^3 for 100 A and 400 mm^3 for 400 A. None of the samples in this case were punctured. However, there was a depth of penetration of the molten pool: 2.1–2.5 mm for 100 C and 3.5 mm—3.7 mm for 400 C. According to [10], a threat of 200 C is considered as an adequately severe level against which protection is required in the event of lightning direct strikes. Testing concludes that for 200 C steel can be punctured at 2.5 mm thickness, but at 3.5 mm there is no possibility of puncture. It is then clear that there is no possibility of a 3/16″ (4.8 mm) steel shell being punctured, and with a safety margin of at least 1 mm to allow for reduction in thickness due to corrosion.

As it is shown in Fig. 6 [10], when lightning arc attaches to any part of the tank (or structure), a certain amount of erosion of the metal will occur on the metallic surface, causing hot-spots on the back of thicker metal sheets (inner surface becomes

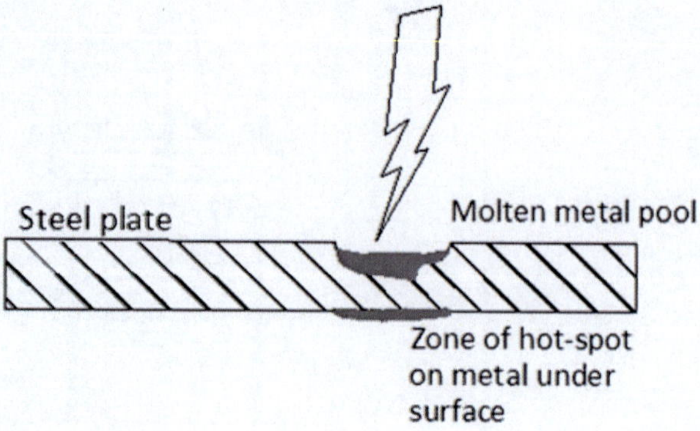

Fig. 6 Effects at the arc attachment point over and under the steel plate surface [10]

quite warm). It is recognized that even if puncture does not occur, the local hot-spot could itself pose an ignition hazard.

All lightning protection measures given above rely on the strict fulfillment of the operational conditions, material thickness and reliable bonding. If some of them are not fulfilled (change in operation conditions, erosion of material, and corrosion in bonding connections), the risk of fire and explosion due to a lightning current can rise to unaccepted levels. For example, the oil industry in some countries establishes a replacement criterion of steel sheets of oil storage tanks when thickness is reduced up to 100 mils (2.54 mm), which represents a good parameter for mechanical integrity of the tank [2]. However, such criterion could not be sufficient for preventing metal puncturing or the development of hotspots. The proposal is upsizing the replacement thickness for better safety, to a value of 3.5 mm, which can be considered as a suitable thickness for lightning protection.

According to Denov and Zoro [11], for areas of very high lightning activity (200 thunderstorm days/year) and long tail lightning, like in Indonesia, 4.8 mm thickness of wall metallic structure of an oil storage tank could become insufficient for lightning self-protection. In such a case, the recommended thickness of steel plate should be greater than 7.1 mm, and if corrosion is of concern, the minimum thickness should be at least 10 mm. Certainly, this lightning protection measure increase considerably the cost of the oil storage tank or vessel.

Properly grounded steel vessels and tanks are essentially self-protecting or maybe considered protected if located in a zone of protection as defined above. Certain definite precautions should be observed, however, in the protection of the vessel and its contents. The following procedures are of value in this respect [1–4]:

1. Flammable liquids and gases should be stored in all metal structures, essentially gastight.
2. Vapor or gas openings to the atmosphere should be protected against the entrance of flame.
3. Positive metallic contact of piping to vessels to eliminate spark gaps at points where there may be an escape or an accumulation of flammable vapors must be assured.

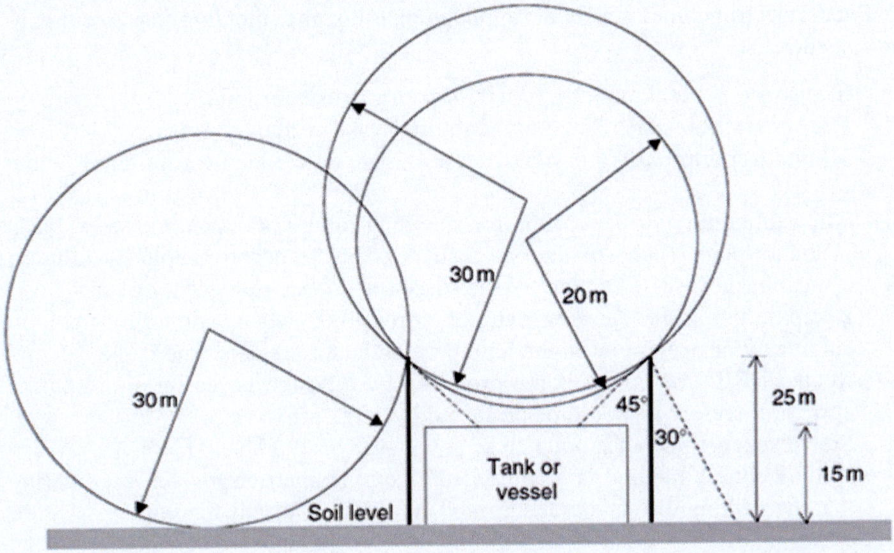

Fig. 7 Direct lightning protection pf storage tanks and vessels, according to [3, 4, 8]

4. In extreme cases, it may be necessary to establish zones of protection through the use of grounded masts and overhead grounded networks of wires or other similar provisions making up a protective cage. In addition, it is recommended that all parts of steel tanks be in metallic contact as provided by riveted and caulked or welded construction and that all fixtures be in intimate electrical contact or bonded. It should be noted that steel tanks with wooden or other nonmetallic roofs have a relatively poor record of safety.

According to [3, 4], the rolling sphere method is suitable for positioning air terminals in all cases, including structures with risk of fire and explosions, and protection zone of levels I and II (20 and 30 m of rolling sphere's radii), are recommended to be used in [4] and [8] respectively, as shown in Fig. 7.

1.3 Power Substations

According to IEEE recommendation [12], there are two methods of direct lightning protection (shielding) of substations: (a) classical empirical methods—fixed angles and empirical curves, and (b) electrogeometrical method. For fixed angle protection, the angle decreases as the height of the air terminal (including ground wires and lightning masts) increases in order to maintain a low failure rate. Using the rolling sphere method, the equivalent angle of protection should be 45° for air terminals up to 15 m height, 30° from 15 to 25 m height and 20° from 25 to 50 m height.

For the electrogeometrical model application, following shielding concepts should be considered:

1. Electrogeometrical model is based on return stroke current.
2. The strike distance is related to the return stroke current.
3. Minimum return stroke current of 2 kA is recommended for voltages below 115 kV.
4. Allowable return stroke current for shielding failure is related to the (a) basic insulation level (BIL) of the bus insulators or the negative polarity impulse critical flashover (CFO), and (b) the surge impedance of the station bus.
5. Decrease the allowable return stroke current by 2 when a direct hit upon an equipment or open point in the high bus conductor are concerned.
6. Reduced BIL equipment is not protected by a design based on return stroke current, therefore it should be protected by surge arresters.
7. Shield spacing becomes quite close at voltages of 69 kV and below.
8. Rolling sphere method is a simplified electrogeometrical model, considering that the striking distance to the ground, a mast or a wire is the same.

1.4 Overhead Lines

Many plants are connected to overhead power lines operating either at utilization voltages or at higher potentials; such a distribution system invites lightning strikes. A high degree of protection for such lines can be achieved by the installation of lightning arresters on the line side of substations fed from overhead lines and at the terminus of an overhead line. Further protection for more extensive overhead systems may be obtained by the use of the overhead ground wire system. This method of shielding live conductors on pole lines has been very successfully employed through the years; its effectiveness, however, depends upon several factors. Ground resistance must be low. The installation must be so designed that the live conductors are not only widely spaced from the ground conductor but, for maximum protection, they must be within a shielded area below the ground wire [1].

According to IEEE recommendation [13], a shielding angle (as shown by Fig. 8) of 45° or less is recommended so that most lightning flashes will terminate on the grounding wire rather than on the phase conductors, being valid for lines less tan 15 m tall with conductor spacing less than 2 m. Taller lines require smaller shielding angles [13]. Figure 8 illustrates pole line shielding.

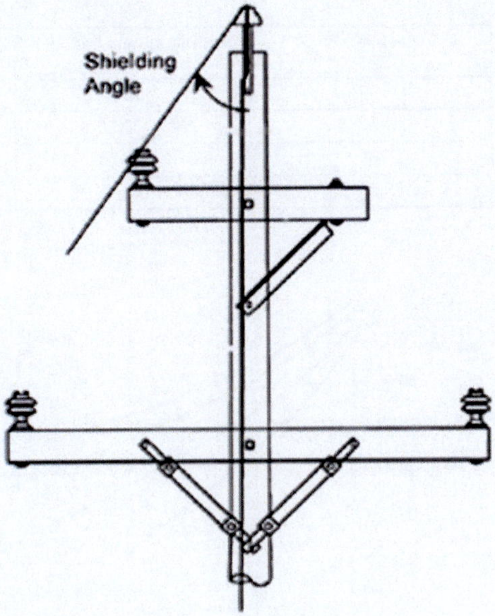

Fig. 8 Direct lightning protection of pole lines. Adpated from [13]

1.5 *Structures and Buildings*

IEC Standard [3] recommend the following protection levels (Fig. 9):

Following recommendations should be implemmented [3]:

1. Acceptable methods to be used in determining the position of the air-termination system include: (a) the protection angle method; (b) the rolling sphere method; (c) the mesh method. The rolling sphere method is suitable in all cases.
2. An isolated external LPS should be considered when the thermal and explosive effects at the point of strike, or on the conductors carrying the lightning current, may cause damage to the structure or to the contents. Typical examples include structures with combustible covering, structures with combustible walls and areas at risk of explosion and fire. An isolated external LPS may also be considered when the susceptibility of the contents warrants the reduction of the radiated electromagnetic field associated with the lightning current pulse in the down-conductor.
3. Natural components made of conductive materials, which will always remain in/on the structure and will not be modified (e.g. interconnected steel-reinforcement, metal framework of the structure, etc.) may be used as parts of an LPS.
4. Air-termination components installed on a structure shall be located at corners, exposed points and edges (especially on the upper level of any facades).
5. The type and location of an LPS should be carefully considered in the initial design of a new sructure, thereby enabling maximum advantage to be taken of the electrically conductive parts of the structure.

Class of LPS	Protection method		
	Rolling sphere radius *r* m	Mesh size w_m m	Protection angle $\alpha°$
I	20	5 × 5	See Figure 1 below
II	30	10 × 10	
III	45	15 × 15	
IV	60	20 × 20	

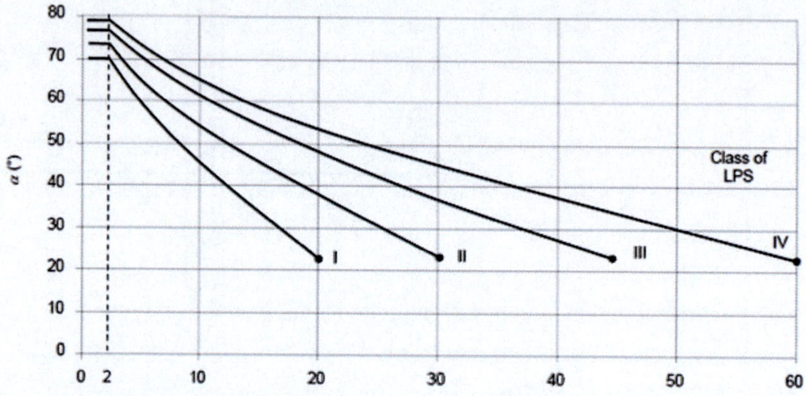

Fig. 9 Class of Lightning protection system according to IEC [3]

6. Regular consultation between LPS designers and installers, architects and builders is essential in order to achieve the best result at minimum cost.

2 Lightning Current to Ground

Generally speaking, metal building of petrochemical plants are inherently self-protected, Fig. 4 [6]. Normally, tall structures and smokestacks are protected with either self-protected scheme or with air terminals bonded to the structure metal (non-isolated protection) and very frequently bonded to down conductors with suitable transversal section according to IEC recommendation [3]. If the down conductor is conventional one (cylindrical stranded copper) the inductance is in the order of 1 μH/m. Figure 10 shows the potential rise of various one-meter long down conductors for a 10 kA 8/20 μs waveshape lightning current impulse. The voltage drop due to lightning current flowing in a conductor is [3, 4, 6]:

$$V = L \cdot \frac{di}{dt} \tag{1}$$

where L is the inductance [μH/m] of the down conductor and di/dt [kA/μs] the rate of rise of lightning current.

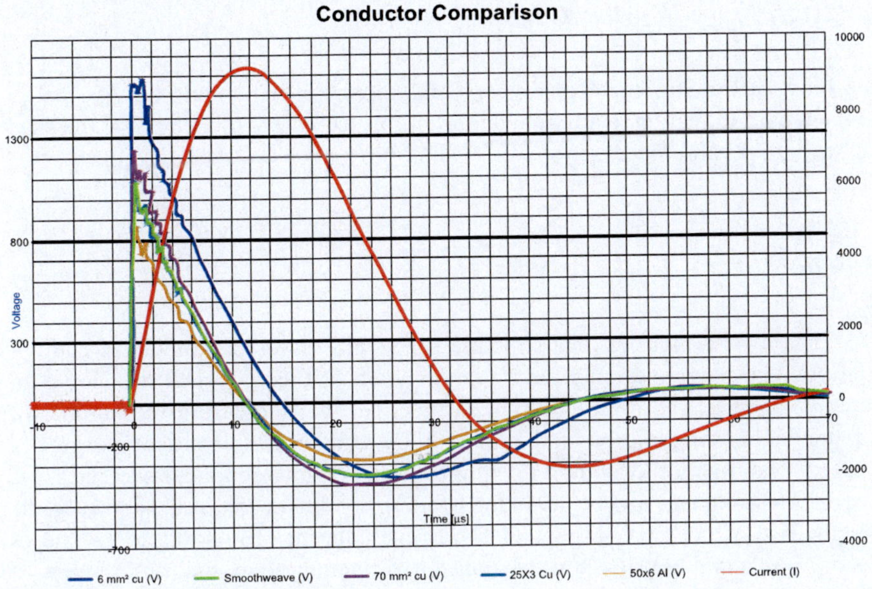

Fig. 10 A 10 kA 8/20 μs impulse applied to various conductors. Adapted from Tompson [6]

The maximum rate of rise is at the start, from zero, of the impulse and here the inductive voltage is the greatest. At the peak of the current curve, di/dt is zero and the voltage due to resistance of the cable is apparent (apart from the 6 mm^2 conductor, it is almost zero). Thus for typical types of downconductor about 1 kV/m/10 kA is expected. So for the plant as shown in Fig. 4, it can be concluded that, since it is all metal, it is self protecting and the best downconductors are the bodies of the tall structures, tanks and vessels themselves [6].

However there must be a voltage rise at the top of any structure that is struck by lightning and therefore any instruments mounted on that structure will rise in potential along with the structure. Unfortunately a potential difference will then exist between the instrument and the I/O of the SCADA or PLC equipment in the control room. Damage to both instrument and I/O is inevitable and no amount of structural lightning protection will solve this problema [6]. This will be discussed in the following Sections.

3 Equipotential Bonding Measures

Equipotential bonding is a very important measure in lightning protection. This measure allows a rapid flow of lightning current to ground and avoid dangerous potential differences over ground and upon the wiring electronic and electrical systems.

3.1 High Metallic Structures

Since most of the structures in refinery and chemical plant units are of steel, a high degree of protection is possible by the simple expedient of grounding the steel framework in several places.

3.2 Oil Storage Tanks or Vessels

IEC Standard [3] establishes that structures like tanks or vessels, should be equipotential bonded to ground once for 20 m diameter, or twice over 20 m. However, the author believes (and normally put in practice) that increasing the number of equipotential bonding to ground, the safety of the tank, vessel or structure is greatly increased. The higher the amount of equipotential bonding conductors to earth, the lesser the lightning current density in the bonding conductors and, consequently, the lesser the rise of potential. Figure 11 shows the concept of bonding to ground of tanks or vessels and Fig. 12 shows de concept of bonding to ground water pipes.

In the case of floating-roof tanks (FRT), the floating-roof shall be effectively bonded to the main tank shell. The design of the seals and shunts and their relative locations needs to be carefully considered so that the risk of any ignition of a possible explosive mixture by incendiary sparking is reduced to the lowest level practicable. On floating roof tanks, there are two natural points of contact between the floating roof and the tank shell, eventhough they do not guarantee a good and permanent contact. (a) when the floating roof reach the lowest level, resting on the base of the tank through the pontoons, see Fig. 3, (b) through the access metallic stairs from the top of the tank to the floating roof by the guide pole, see Fig. 13 [15].

Therefore, additional measures are required. (1) Multiple shunt connections shall be provided between the floating-roof and the tank shell at about 1.5–3.0 m intervals

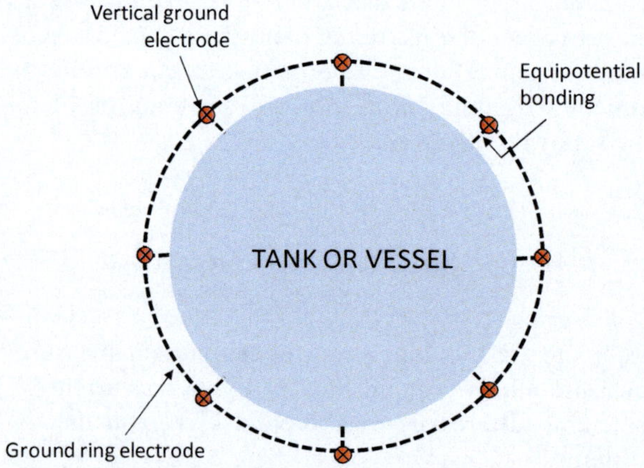

Fig. 11 Bonding to ground connections for tanks or vessels

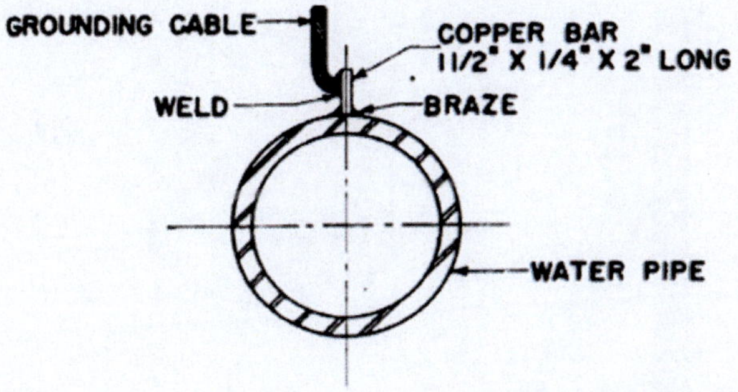

Fig. 12 Water pipe bonding to ground connection. Adpated form Benjamin and Cundelan [1]

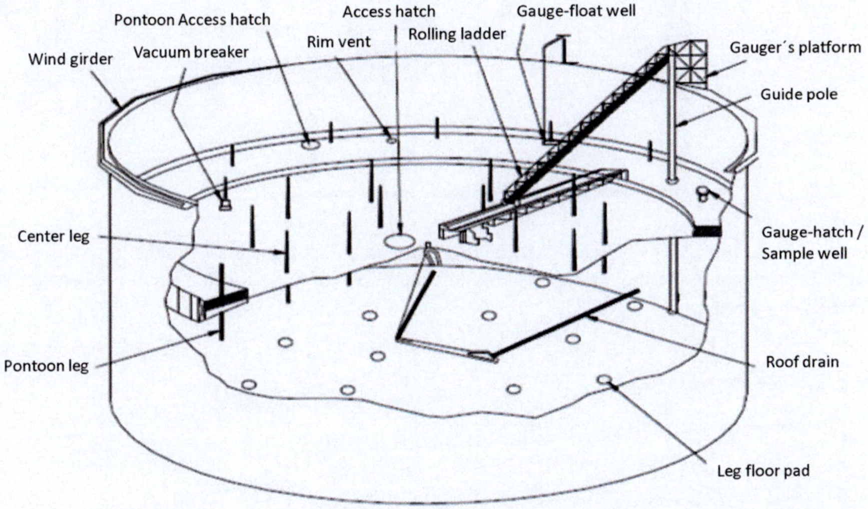

Fig. 13 Two natural points of contact between the floating roof and the tank shell (**a**) when the floating roof reach the lowest level, resting on the base of the tank through the pontoon legs, (**b**) through the access metallic stairs from the top of the tank to the floating roof by the guide pole. Adapted from API 2517 [16]

around the roof periphery. Material selection is decided by product and/or environmental requirements [1, 9], as shown in Fig. 14; (2) Alternative means of providing an adequate conductive connection between the floating roof and tank shell for impulse currents associated with lightning discharges are only allowed if proved by tests and if procedures are utilized to ensure the reliability of the connection [3]. For the last, retractil grounding conductors are able to be used in order to ensure the bonding connection between floating roof tank and the body of the tank [14], as shown in Fig. 15, where the aspect of conductor inductance should be taken into consideration.

Above-ground metal piping network inside a production facility but outside the process units should be bonded to ground every 30 m, or should be connected to

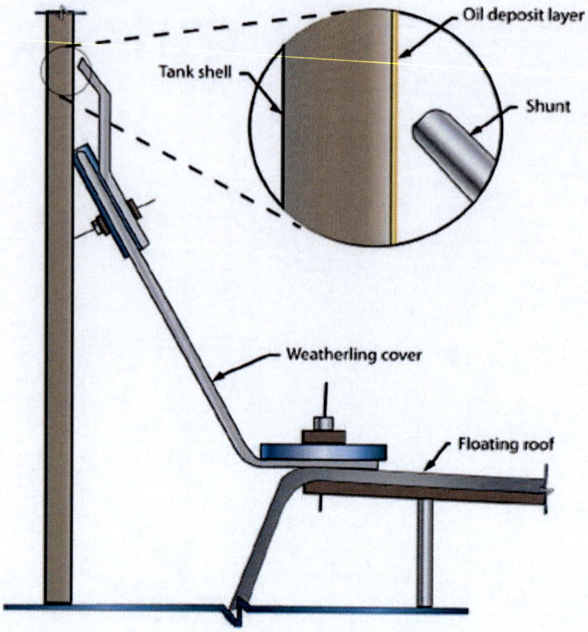

Fig. 14 Shunt connections between the floating-roof and the tank shell. Due to the loss of the tank circumference during the day (metal heating deformation), up to half of the shunt connections to the tank can be lost

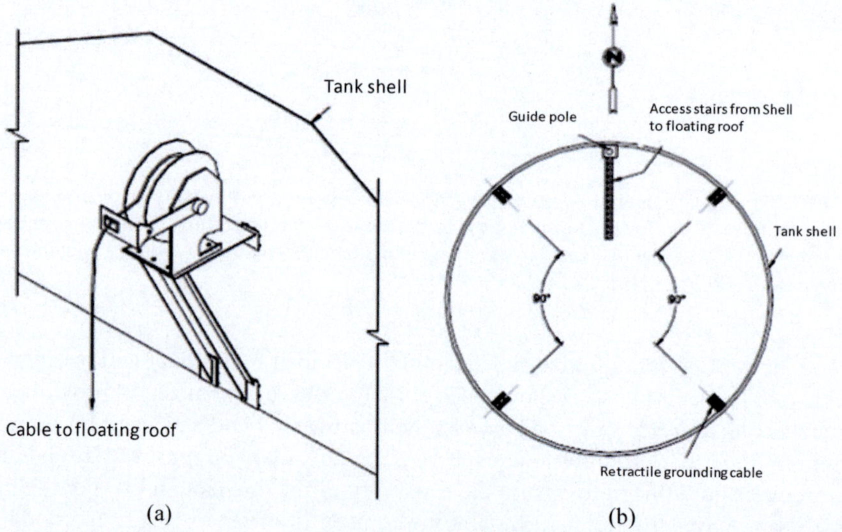

Fig. 15 **a** Retractile grounding cable device located at the top of the tank shell. One end is connected to the tank shell and the other one to the floating roof. **b** Recommended location of Retractile grounding cable devices along the perimeter of the FRT

ground by a surface earth electrode or an earth rod. Isolating supports of the piping should not be considered [3].

3.3 *Protection Against Accumulation of Static Charges*

Industrial plants such as refineries and chemical plants may have a potentially hazardous condition because of the accumulation of static charges due to a number of causes, among them, the movement of gases and liquids, the spraying of caustic solutions in tanks, acid bubbling, and even certain phases of the processes themselves. Liquids in a tank, if agitated, may accumulate static charges. It has been noted that liquids discharging from a pipe above the surface of the liquid in a storage tank will produce static charges in the liquid. Fortunately, most of the problems presented can now be dealt with. Static caused by the flow of gases or liquids in piping systems is probably carried off harmlessly by the piping itself. This is particularly true in refinery work because of the continuous electric path provided by present day pipe welding procedures. Potentials on piping may also be dissipated through pipe hangers and other supports electrically bonded to grounded structures. On occasion, however, it may be necessary to provide grounding for certain piping. This can usually be best determined after plant operation has begun. Each time that an oil tank is filled, a degree of turbulence exists which probably results in the formation of static charges in the oil and on its surface. Since petroleum is a dielectric, it has the capacity for holding electric charges. Grounding of the shell of the tank and of the piping attached thereto, while important, is not the entire solution since the electric charges are not readily given up by the petroleum product involved. When the potential gradient becomes steep enough an arc over long the surface of the oil to the tank shell may occur [1].

4 Equipment and Electrical System Bonding to Ground

Connections between fixed equipment to be grounded and the grounding system are usually made with stranded copper conductors. Sizes may vary as specified by the NEC [17], ranging from no. 8 to 3/0 American wire gauge depending upon the size of the service conductor. Exact grounding conductor size may also be determined by the maximum available fault current and the circuit protective device interrupting time. IEC Standard [18] recommends the cross-sectional area of protective bonding conductors for connection to the main earthing terminal need not exceed 25 mm^2 Cu or an equivalent cross-sectional area for other materials. The actual grounding connections between equipment and grounding electrode usually take the form of a loop to which the individual equipment may be joined by short cable connections. The loop is then connected at one or more points to grounding system. This main grounding loop or bus which will serve motors driving process equipment, lighting panels, motor control, and other electric equipment should be buried at least 18 inches below finished grade. All below grade connections to the ground bus should be made by copper brazing, fusion welding, or equivalent process capable of sustaining reasonable temperature. Bolted connections should be avoided because of the danger of high-resistance joints due to corrosión [1].

Connections to individual pieces of electric equipment take a variety of forms. Grounding of motors is best accomplished by drilling and tapping the motor frame after which the grounding conductor, equipped with a copper terminal, may be bolted directly to the frame. Satisfactory protection from physical damage may be afforded the grounding conductor, if a suitably formed length of 1/2-inch conduit has been cast in the pump foundation block. Figure 16 illustrates this clearly. Similar grounding connections should be provided at lighting panels, motor starters, switchboard frames, switchgear ground busses, steel instrument control boards, and all other metallic enclosures and supports of electric apparatus. Some equipment is supplied with grounding lugs by the manufacturer, others must be tapped—as are the motors—to receive a grounding connection. A point sometimes overlooked by the designer is the case of conduits terminating just above grade in motor control centers and cubicle-type switchgear. Since such conduits are not connected electrically to the equipment, they should be strapped to the enclosure or to the equipment ground bus if one has been provided. Grounding bushings are useful for this purpose [1].

Petrochemical plants are increasingly faced with the need to upgrade their control centers to support the installation of more advanced control systems and to better protect operations personnel in the event of catastrophe, in a hope of improving plant-wide communications, increasing operating efficiency, and reducing construction costs.

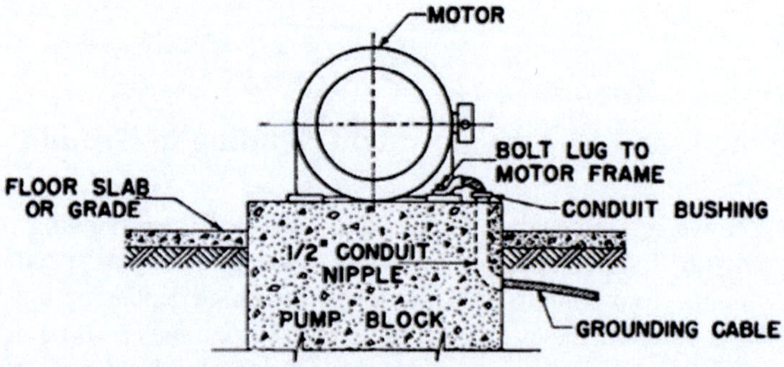

Fig. 16 Motor bonding to ground connection. Adapted form Benjamin and Cundelan [1]

5 Local and General Grounding Arrangements

5.1 Local Ground of Tall Structures

Since most of the structures in refinery and chemical plant units are of steel, a high degree of protection is possible by the simple expedient of grounding the steel framework in several places to the local grounding grid.

5.2 Local Ground of Oil Storage Tanks

Oil storage tanks and vessels should have a grounding electrode type B, according to IEC Standard [3], as shown in Fig. 17. Several aspects should be considered:

1. Keep the buried ground ring electrode electrically continuos for the entire path around the tank. Use welded connection to join final extremes of the ring.
2. Enhance the local ground ring electrode type B by three (at least 12 m long) buried cable as shown in Fig. 17 in order to reduce dramatically the lightning-generated rise of potential. The place of the installation should be at the points where the lightning masts are located or where the connection point to the ground ring from the tank is made.
3. Enhance the local ground ring electrode type B by vertical ground electrodes located every 6 m (or double the length of the vertical ground electrode). The

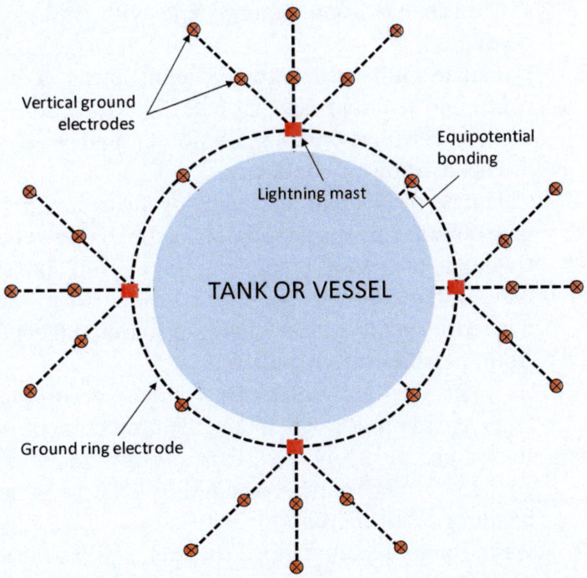

Fig. 17 Example of buried grounding ring type B for oil storage tanks, according to IEC standard [3] and additional protective arrangements. *Note* multi-path arrangements for the points where the lightning current flows to ground (lightning masts)

length of the ground vertical electrode will depend on the trend and value resistivity of the soil. Use welded connections to join the vertical electrods with the ground ring electrode.
4. It is strongly recommended to apply welded connections to the tank.

5.3 *Local Ground of Power Substations*

A satisfactory grounding system for substations may be achieved by installing a ground cable or ground bus underground, surrounding the substation area and connected at intervals to ground rods. A connection to metallic underground water piping is desirable. In larger substations, a grid may be formed by installing several underground cables across the substation area to connect opposite sides of the ground bus. All electric equipment installed in the substation should be connected to the ground bus. Substation fences should be similarly connected to the bus at frequent intervals and structural steel substation structures should also be grounded in at least two locations [1].

The objective of substation grounding as described above is to ensure that all parts of structures and equipment enclosures be at ground potential. In the event of a breakdown in insulation, flashover, or accidental contact with high voltage, the potential difference between the equipment and any point on the ground should not be great enough to constitute a hazard to personnel [1].

IEEE recommendation [19] may be used to design the grounding grid of the substation, considering the following aspects:

1. Design the grounding grid in such a way that the pieces of equipment and structures may connect to ground with suitable cables in a path as short as possible.
2. Calculate fault current for different levels of high voltages and identify the maximum fault current that will circulate in the grounding grid conductors in order to calculate the cross section of the buried conductor, taking into account the material of the conductor.
3. Calculate the ground resistance of the grounding grid by using soil resistivity data obtained by measurements (avoid using generic tables).
4. Calculate the rise of potential upon the soil considering the portion of the fault current that will circulate in the ground when the source that feed the fault is located at a remote place of the substation (take into account the current division from metallics return path).
5. Estimate the step and touch voltages according to the estimated allowable voltages, in order to guarantee the protection of personnel (staff).
6. In equipment of big size, it is strongly recommended to use more than one connection to ground, considering even to increase the cross section of the bonding cable to ground.
7. It is strongly recommended for each equipment to install vertical ground electrodes at the point of the bonding with the grounding grid, in order to guarantee

the connection to ground reach more stable moisture soil conditions. The length of the vertical ground electrodes will depend of the layers resistivity value of the soil.

8. Optimize the grounding grid by using suitable size of mesh and vertical ground electrodes upon the periphery of the grid.

5.4 *Local Ground of Distribution Lines*

The connection to ground of distribution lines obeys to the following two aspects:

1. Connect to ground the overhead grounding wire used to shield the line against direct ligthning stroke to live overhead conductors. Ii is recommended to ground the overhead grounding wire every pole of the line.
2. Connect to ground equipment and surge arrestes of equipment located in the line.

Both require low values of ground resistance, up to 25 Ω (when measured at low frequency). The arrangement of the grounding grid or electrode will depend of the soil resistivity.

5.5 *Local Ground of Structures and Buildings*

Eventough IEC recommendation [3] allows to use type A for grounding of structures and buildings, the author strongly recommend to apply arrangement type B (ground ring electrode), as shown in Fig. 18, for buildings and structures in petrochemical plants. Aspects mentioned in section 5.2 apply, with the additional considerations:

1. Ground resistance of the arrangement type B should be up to 10 Ω.
2. The ground ring electrode should preferably be buried at a depth of at least 0.5 m and at a distance of about 1 m away from the external walls, and be interconnected to concrete foundations of the building.
3. Ground electrodes shall be installed in such a way as to allow inspection during construction.
4. Interconnected reinforcing steel in concrete foundations, or other suitable underground metal structures, should preferably be used as a ground electrode (avoid using pre-stressed concrete). When the metallic reinforcement in concrete is used as a ground electrode, special care shall be exercised at the interconnections to prevent mechanical splitting of the concrete.

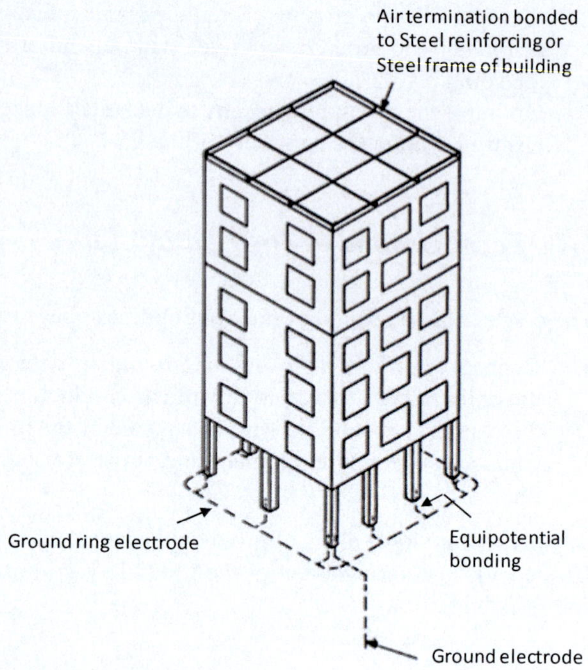

Fig. 18 Example of grounding arrangement type B for buildings, according to IEC recommendation [3]. Adapted from IEC 60364-5-54 [18]

5.6 Local Ground of Information Technology Systems

Electrical/electronic equipment damage from lightning may be placed into three major categories [20]:

1. Improper or insufficient grounding, which will result in the equipment being stressed and/or damaged (potential difference) from nearby equipment/objects.
2. Lack of protection from ground potential rise GPR, which will result in the equipment being stressed from its connection to remote earth at some distant location through communication wire-lines or power supply wiring and/or from intrabuilding GPR arising from the voltage drop between power and telecommunication ground references.
3. Lack of protection from lightning transients, namely surge protective devices SPDs. This will be discussed on next section.

Sites without telecommunication towers may experience Lightning Ground Potential Rise LGPR effects as much, if not more, than sites with towers as they are less likely to have extensive grounding infrastructure. Sites at most risk are areas with a higher occurrence of lightning and high soil resistivity.

IEEE recommendation [20] gives very useful measures to define the grounding arrangements. Following considerations should be taken into account [3, 20].

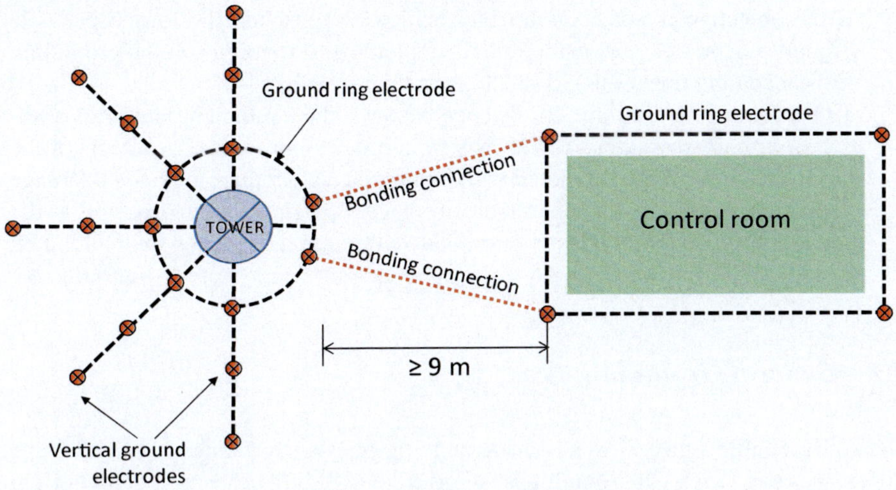

Fig. 19 Grounding arrangement recommended for telecommunication tower and control room

1. Both telecommunication tower and control room should have arrangement type B, according to IEC recommendation [3].
2. Enhance the local ground ring electrode type B by multi-path (at least 12 m long) buried cable in tower location as shown in Fig. 19 in order to reduce dramatically the lightning-generated rise of potential.
3. Enhance the local ground ring electrode type B by vertical ground electrodes located every 6 m (or double the length of the vertical ground electrode). The length of the ground vertical electrode will depend on the trend and value resistivity of the soil. Use welded connections to join the vertical electrods with the ground ring electrode.
4. Hardening against lightning ground potential rise (GPR) damage requires specially designed tower radial counterpoise grounding system with a grounding resistance not exceeding two (2) ohms. If the objective is not economically achievable, provide the lowest possible ground resistance value, using radial counterpoises, to minimize the grounding impedance (and thus GPR) as much as posible.
5. Hardening against lightning ground potential rise (GPR) damage requires an associated tower equipment building grounding system with a grounding resistance not exceeding two (2) ohms. If the objective is not economically achievable, provide the lowest possible ground resistance value, using radial counterpoises, to minimize the grounding impedance (and thus GPR) as much as possible
6. Both grounding arrangements should be joined together.
7. The total overall site ground resistance (tower and building) should not exceed one (1) ohm. This may require significant real estate space if the site soil resistivity is greater than 500 Ωm at the anticipated grounding electrode depth.

If the objective is not economically achievable, provide the lowest possible ground resistance value, using radial counterpoises, to minimize the grounding resistance (and thus GPR) as much as posible (see Fig. 19).

8. The recommended minimum distance between the equipment buildings associated with nearby antena towers is 9 m in order to minimize the effects of the electromagnetic field associated with lightning and to reduce the risk of damage to equipment circuits. In general, electromagnetic field strength drops off as the square of the distance. This is one of those rare exceptions in which a lengthy bond is an advantage in supporting a robust grounding system to lightning.

5.7 General Grounding Grid

General grounding grid aims to low grounding resistance, reduce hazard voltage differences and join local grounding systems (like lightning protection, fault current protection and signal reference subsystems to ground) [8]. There are two ways of implementing a general grounding grid: (a) Make a grounding grid meshed to interconnect local grounding grids or (b) Make a intermeshed local ground systems, as shown in Fig. 20.

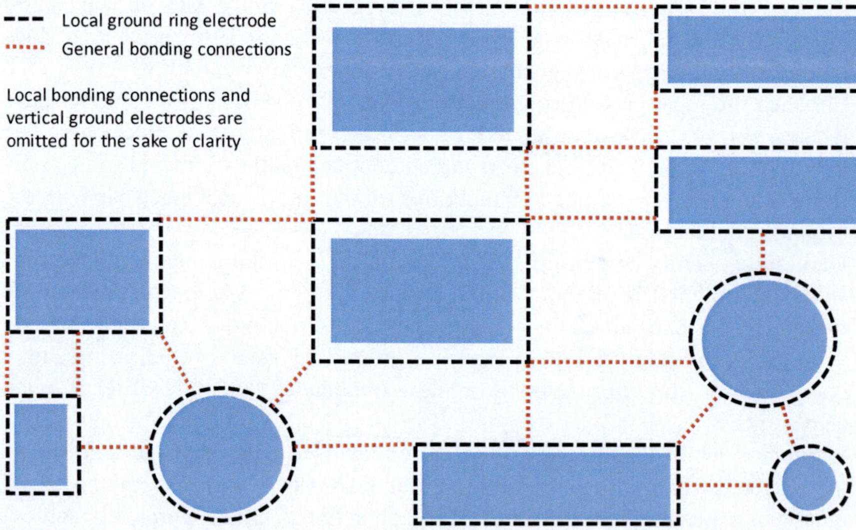

Fig. 20 Example on intermeshed local ground systems

6 Surge Protection Systems

6.1 Low Voltage Distribution System

The installation of a lightning and surge protection system for electrical installations represents the latest stage of the art and is an indispensable infrastructural condition for the trouble-free operation of complex electrical and electronic systems without consequential of damage [21]. The requirements on Surge Protective Device (SPD) are defined in IEC Standards as part of the lightning protection zone (LPZ) concept [22] and protection for power supply systems [23], as shown in Fig. 21.

According to IEC Standard [15], a low voltage distribution system in its entirety, from the power source to the las piece of equipment, is typically characterized by:

1. Earthing conditions of the power source (e.g. low voltage side of the distribution transformer) and,
2. Earthing conditions of the bodies of the equipment in the electrical consumer´s installation.

Due to the above earthing mentioned conditions (source-equipment), there are three basic types of distribution systems: **TN** (Terra-Neutral) system, **TT** (Terra-Terra) system and **IT** (Isolated-Terra) system. Depending on the type of the distribution system, the protective devices which can be installed in the various systems are:

TN system: (a) overcurrent protective device, (b) residual current protective device.

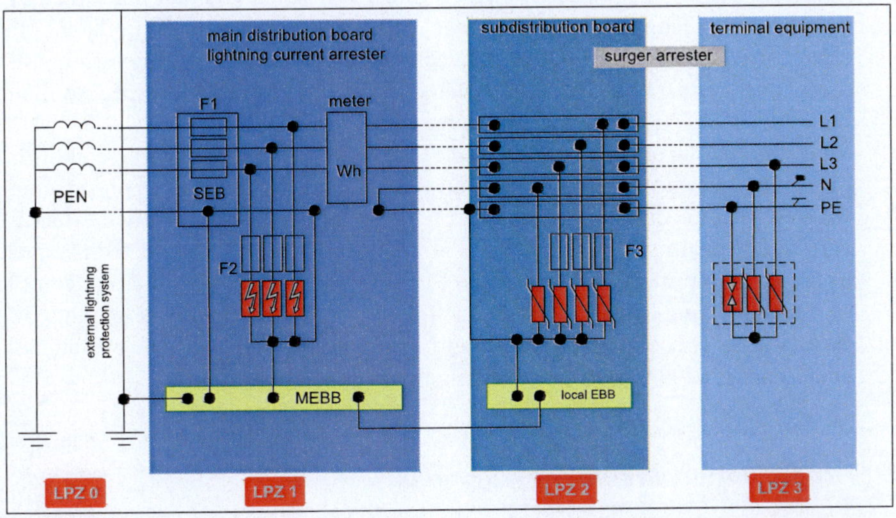

Fig. 21 Schematic diagram of use of arresters in power supply systems. Adapted from DEHN + SOHNE [21]

TT system: (a) overcurrent protective device, (b) residual current protective device, (c) fault-voltage-operated protective device (in special cases).

IT system: (a) overcurrent protective device, (b) residual current protective device, (c) insolation monitoring device.

According to IEC Standard [24], the designation class (types) of the SPD are:

Class (Type) 1. Lightning current arrester/combined arrester.

Class (Type) 2. Surge arrester for distribution boards, sub-distribution boards, fixed installations.

Class (Type) 3. Surge arrester for socket outlets/terminal devices.

The following considerations should be observed when using SPDs [21, 25]:

1. Ensure the energy coordination of the individual SPDs and make sure it takes into account all possible interferences such a switching overvoltages, partial lightning currents, etc.
2. SPD needs to withstand TOV (Temporary Overvoltage) related to power frequency surges.
3. For SPDs used in TN systems, both type 1 and 2 SPD should be used upstream of the residual current protective device in order to be effective for the protection against electric shock under fault conditions. Type 3 SPDs are installed downstream residual current protective device RCD.
4. For SPDs used in TT systems, both type 1 and 2 SPD must always be installed upstream of the RCD and must be arranged in such a way that the conditions for the use of overcurrent protective devices for the protection against electric shock under fault conditions are met. As in the case with TN system, type 3 SPDs are installed downstream residual current device RCD.
5. For SPDs used in IT systems, it is also advisable to install both type 1 and 2 SPD upstream of the RCD. Unlike TN and TT systems, the first fault in an IT system only creates an alarm. However, the voltage of the intact conductors to earth corresponds to the voltage between the phase conductors. Therefore, this stage must be taken into account when choosing the SPD with respect to their maximum continuous operating voltage.
6. Be aware that installation of SPDs in IT systems will depend on the condition of the incorporation or not of the neutral conductor in the distribution system.
7. Related to connecting cable lengths of SPDs, an optimum protective effect is achieved if the impulse voltage level at the installation to be protected is equal to the voltage protection level of the SPD. To accomplish this, the inductance of the connecting cable must keep as low as possible, and thus its length. IEC Standard [23] recommends that total cable length of SPD in cable branches should not exceed 0.5 m, and a maximum cable length of 1 m.

6.2 Information Technology Systems

Unlike power supply systems, the types of signals to be transmitted in automation and measuring and control systems have different parameter conditions, like [21]:

1. Voltage (e.g. 0–10 V).
2. Current (e.g. 0–20 mA, 4–20 mA).
3. Type of signal transmission (balanced, unbalanced).
4. Frequency (DC, LF, HF).
5. Type of signal (analog, digital).

The challenge in this case is to achieve that the useful signal must not be impermissibly influenced by lightning current and surge arresters in measuring and control systems. Therefore, the selection of arresters in order to protect downstream terminal devices (reducing also the risk of cable damage) depends, among other things, on the following criteria:

a. Lightning protection zones of the place of installation, if any.
b. Energies to be discharged.
c. Arrangement of the protective devices.
d. Immunity of the terminal device.
e. Differential mode and/or common mode protection.
f. System requirements, e.g. transmission parameters.
g. Compliance with product or application-specific standards, if required.
h. Adaptation of environmental/installation conditions.

The standard surge protective devices (SPD) in the telecommunications industry, for the termination of communication wire-line services is the gas discharge tube (GDT). GDTs are also called gas tubes. GDTs can be found on virtually every telephone pair terminated in homes, buildings, and similar locations. GDTs are designed to shunt most current to ground. If the magnitude shunted does not exceed a certain threshold the SPD will help protect equipment, and personnel, from harm [20].

Most shunting devices, however, do not fully protect network electronic equipment from a GPR or "outgoing current," whether induced from lightning or from a faulted power line. When shunting devices are connected to an elevated ground (outgoing current) during a GPR event, they merely offer an additional current path off the site to remote earth (the other end) [20].

When SPDs (GDTs, MOVs, ABDs, SCRs, SADs, SASs, etc.) are used as ground shunting devices, they will not protect equipment from GPR. These devices merely offer an additional path to remote earth through the communication pairs for any and all outgoing currents. When there is a GPR event the SPD provides a connection of the communication path in the reverse direction from which they were intended to operate and increases the possibility of equipment damage to telephone and power installations. The most susceptible locations are those where the equipment is located near, or under, towers and/or are located at a higher altitude than the surrounding área [20]. That's why the ground resistance should be kept as low as posible to protect electronic equipment from damage insulation.

Effective protection of sensitive equipment with SPD shunting devices is complex. A well-designed installation requires coordination of the protection for low-voltage power feeds (ac and dc) with the protection for telecommunications facilities in order to minimize the effect of intrabuilding GPR. The use of secondary SPD is recommended to supplement the primary SPD [26]. Surge resistibility and impedance of the terminal equipment must be compatible with the selected primary SPD. Even with a well-designed installation, part of the lightning current will reach the equipment and, in some cases, can affect service quality and/or cause equipment damage [20].

References

1. Benjamin EC, Cundelan JV (1955) Grounding, bonding, and lightning protection. Electr Eng 400–403
2. Galván DA, Gómez C (2013) Protection of oil storage tanks against direct lightning strikes: self protection scheme or standalone LPS? In: International symposium on lightning protection (XII SIPDA), Belo Horizonte, Brazil, 7–11 Oct 2013
3. IEC 62305-3 (2010) Protection against lightning—part 3: physical damage to structures and life hazards, 2010 edn
4. NMX-J-549-ANCE-2005 (2005) Protection against lightning—especification, materials and measuring systems. Mexican Standard (in Spanish)
5. Welker AJ (1998) Lightning—its effects and some simple safeguards in regards to oilfield operations. In: Proceeding of annual Southwestern petroleum short course, pp 329–346
6. Tompson P, Lightning protection of industrial plants, earthing, lightning & surge protection forum—IDC Technologies
7. NFPA 780 (2018) Standard for the installation of lightning protection systems, 2018 edn
8. MIL-HDBK 419A (1982) Grounding, bonding and shielding for electronic equipments and facilities
9. API Recommended Practice (2003) Protection against ignitions arising out of static, lightning and stray currents. American Petroleum Institute
10. API/EI Research Report (2009) Verification of lightning protection requirements for above ground hydrocarbon storage tanks, 1st edn. Energy Institute, London
11. Denov B, Zoro R (Nov 2017) Lightning protection system for oil and gas installation case on cilacap area at Central Java, Indonesia. In: Proceedings of the 2017 6th international conference on electrical engineering and informatics: sustainable society through digital innovation, ICEEI 2017, pp 1–7
12. IEEE STD 998-1996 (R2002) Guide for direct lightning stroke shielding of substations
13. IEEE STD 1410 (2010) Guide for improving lightning performance of electric power overhead distribution lines
14. Galván A (2006) Retractile grounding cable used for floating roof tanks: four-year experience. In: 28th international conference on lightning protection, Sep 2006, Kanazawa, Japan
15. IEC 60364-4-41 (2005) Low voltage power installations. Part 4-41: protection for safety. Protection against electric shock
16. API 2517 (1989) Evaporative loss from external floating roof tanks, 3rd edn
17. National Electrical Code NEC 2018
18. IEC 60364-5-54 (2011) Low voltage electrical installations. Part 5-54: selection and erection of electrical equipment. Arrangements and protective conductors
19. IEEE 80 (2000) Safety in AC substation grounding
20. IEEE 1692 (2011) IEEE guide for the protection of communication installations from lightning effects
21. DEHN + SOHNE (2014) Lightning protection guide, 3rd edn
22. IEC 62305-4 (2010) Protection against lightning. Part 4: electrical and electronic systems withing structures

23. IEC 60364-5-53 (2019) Low voltage electrical installations. Part 5-53. Selection and erection of electrical equipment. Devices for protection for safety, isolation, switching, control and monitoring
24. IEC 61643-11 (2011) Low voltage surge protective devices. Part 11: surge protective devices connected to low-voltage power systems. Requirements and test methods
25. Lightning protection book (2010) In: Cooray V (ed) Chap 17. The Institution of Engineering and Technology IET, ISBN 978-0-86341-744-3 (hardback)
26. ITU-T K.11. Protection against interference. Principles of protection against overvoltages and overcurrents

Protection of Selected Cases: PV Systems, Wind Turbines and Railway Systems

Michael Rock

Abstract The lightning protection and the surge protection of large ground-mounted photovoltaic power plants as well as of small roof-mounted photovoltaic systems is considered. Basics for external and internal lightning protection as well as special requirements, especially for surge protection, are presented. The measures result from experiences in the last years, are today recognized widely and are realized to a large extent. Lightning and surge protection of wind turbines has received increasing attention in recent decades due to considerable damage caused by direct lightning strikes. Today, onshore and offshore wind turbines are equipped with lightning protection systems according to the highest level of lightning protection and the declining damage shows the effectiveness of the measures. Special attention is paid to the protection of the rotor blades. But also surge protection for electrical energy and information technology systems including EMC measures is of great importance and is described. Many specialities have to be taken into account in the lightning and surge protection of wind turbines. For railway facilities and systems, national and international comprehensive concepts for lightning and surge protection have hardly been described. The considerations and measures presented here also do not claim to be complete, but are intended to address important aspects. These include above all measures for the protection of control and command technology systems as well as information on personal protection. In addition, lightning and surge protection of the power supply, measurement systems and EMC measures are addressed.

Keywords Photovoltaic · Solar power plant · Wind turbine · Rotor blade · Railway facilities and systems · Lightning and surge protection · Air termination system · Earthing system · Equipotential bonding · Surge protective device (SPD)

M. Rock (✉)
Group for Lightning and Surge Protection, Technische Universität Ilmenau, Ilmenau, Germany
e-mail: michael.rock@tu.ilmenau.de

C. Gomes (ed.), *Lightning*, Lecture Notes in Electrical Engineering 780,
https://doi.org/10.1007/978-981-16-3440-6_7

a. **Protection of PV Systems**

1 Limitation of PV Systems to be Considered

In the following, large ground-mounted photovoltaic (PV) systems or solar power plants and roof-mounted photovoltaic systems or small PV systems will be considered [1].

The photovoltaic system includes not only the solar modules for the direct conversion of solar radiation into electrical power, but also other components such as the d.c. connecting cables between modules and inverter(s), usually fuses, earthing, lightning protection, disconnecting devices, meters and undervoltage or overvoltage monitoring relays. Some of these components are mandatory for legal reasons, while others, e.g. lightning protection, are often only recommended depending on the exposure of the system [1] (IEC 62093 2005).

Solar power generation are part of today's electrical engineering. Their professional design includes lightning and surge protection, which contributes to the longest possible trouble-free use of this regenerative electrical power source [2].

2 Lightning and Surge Protection for Ground-Mounted Photovoltaic Power Plants

i. Importance of lightning protection for solar power plants

Every year, photovoltaic ground-mounted power plants with an electrical output of several gigawatts are newly installed. Together with the high number of small rooftop photovoltaic systems, photovoltaics is developing into a relevant part of modern and, in particular, regenerative power supply in many countries [3].

Large power plants with 100 MW and more are now being built, which often feed directly into the medium or high-voltage level. Photovoltaics has thus become an integral part of the supply system and must therefore meet the requirements for stable grid operation. Consequently, supply interruptions and loss of yield must be avoided or minimized. Furthermore, the high investment volume and the required 20-year minimum service life make it necessary to assess the risk of damage from lightning strikes and to implement protective measures [2].

a. Lightning risk for solar power plants

Photovoltaic systems are exposed to lightning due to slightly elevated systems with a large area in the open field (lightning impact area roughly equal to the base area of the PV system, some ha to km^2) [4, 5].

There is a correlation between solar radiation, humidity and frequency of lightning strikes. Regions with high sun intensity and high humidity in which PV power plants are built are exposed to an immediately higher risk of lightning. The regional lightning frequency (ground strikes per square kilometer and year) and the location and size of the PV power plant are the basis for calculating the probability of lightning strikes into the system.

In general, PV power plants are exposed to the local weather influence of thunderstorms for decades.

Studies of 25 PV solar parks in Germany in the period from 2005 to 2015, i.e. with construction around the year 2010 and the consideration of localized ground flashes +/–5 years before and after construction have shown:

- negative flashes: significant shift towards larger amplitudes, increase in amplitudes and no change in numbers after construction of PV parks
- positive flashes: slight shift towards larger amplitudes, no change in amplitudes and no change in numbers after PV parks have been constructed
- exceeding probability distribution: displacement of the part with higher negative lightning current amplitudes in the direction of CIGRÉ characteristic of the negative first stroke and of the mean amplitude part in the direction of CIGRÉ negative subsequent stroke characteristic after construction of the PV parks.

The installation of a solar power plant results in an increased lightning risk to it.

b. Necessity of lightning and surge protection for solar power plants

Damage in PV systems is caused both by the destructive effect of direct lightning strikes and by inductive or capacitive coupled voltages from the electromagnetic lightning field. Voltage peaks from switching operations in the upstream a.c. or d.c. grid can also cause damage. Defects can occur in PV modules, inverters and their monitoring and communication systems. In addition to the replacement and repair costs, the economic damage is also reflected in the loss of yield and culminates in the costs for calling up reserve power plant capacity.

The aim is to protect both the operating building and the PV system from fire damage, i.e. direct lightning, and the electrical and electronic systems, such as inverters, generator main lines and remote diagnosis systems, from the effects of lightning electromagnetic pulses (LEMP).

Lightning pulse impacts also lead to premature aging of bypass diodes, power semiconductors and the input and output circuits of the data systems, which in turn means increased repair work for the subsequent period. In addition, grid operators make demands on the availability of the generated power, which are determined by grid codes. Increasingly, these points are also being considered by financing and insurance companies, and lightning protection measures are also used in the so-called Due Diligence examinations for financing purposes.

c. Measures to protect solar power plants against the effects of lightning

A lightning protection system (LPS), which is often designed for LPL III according to [6, 7], usually meets the requirements for PV power supply systems. The risk of damage due to lightning strikes must be determined on the basis of [8] and the results taken into account during planning.

Effective protection requires a LPS whose elements are coordinated with each other. Starting with the air termination system, earthing system, lightning equipotential bonding up to surge protective devices (SPD) for the power and data side.

Surge protective devices (SPD) can be used according to [6] or [7] In addition to the minimum discharge capacity of SPD, information is also given on the design of the earthing system for ground-mounted photovoltaic power plants.

ii. External lightning protection for solar power plants

a. Air termination system and down conductors

In order to protect against direct lightning strikes into the electrical systems of a PV power plant, it is necessary to arrange these in the protective area of air termination system. The number of air-termination rods can be determined using the rolling sphere method and, as a rule, protection class III (Fig. 1) [9]. They form a protected space across module tables, operating rooms and wiring. With regard to the inductive coupling of faults, it is recommended that generator junction boxes and decentralized inverters mounted on module tables are mounted as far away as possible from air termination devices. All down conductors must be connected to the terminal lugs of the earthing system. Due to the risk of corrosion at the outlet point of the terminal lugs from the ground or concrete, these must be corrosion-resistant (stainless steel or shrink tubing). For mechanical fastening, the air termination devices can often be connected to the module tables.

In addition to air-termination rods to protect the solar modules, small air termination tips at the highest points of modules (Fig. 2) with frames capable of carrying lightning currents can also be used. In general, it makes sense to mount the modules on earthed metal tables/frames.

b. Earthing system

The earthing system (Figs. 2 and 3) is the basis for the effective implementation of lightning and surge protection measures in PV power plants. An earthing resistance of less than 10 Ω is recommended for the earthing system [9]. With flat strip 30 mm ×

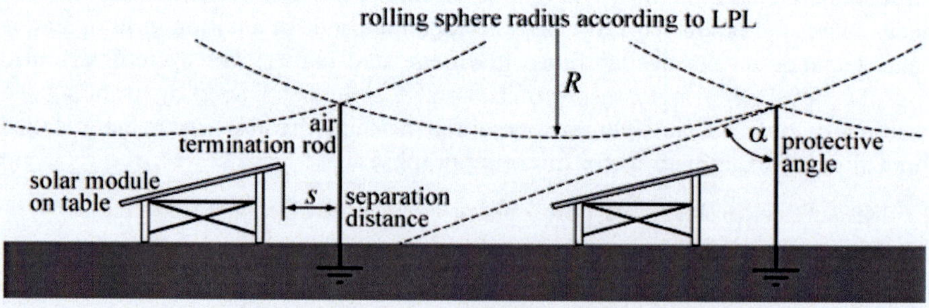

Fig. 1 Determination of the protective space by means of rolling sphere method or protective angle method [2]

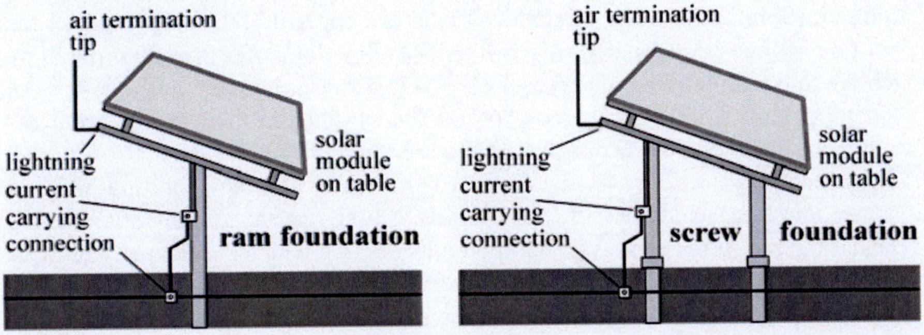

Fig. 2 Ram and screw foundation with lightning current carrying connection of air termination and earthing system [7]

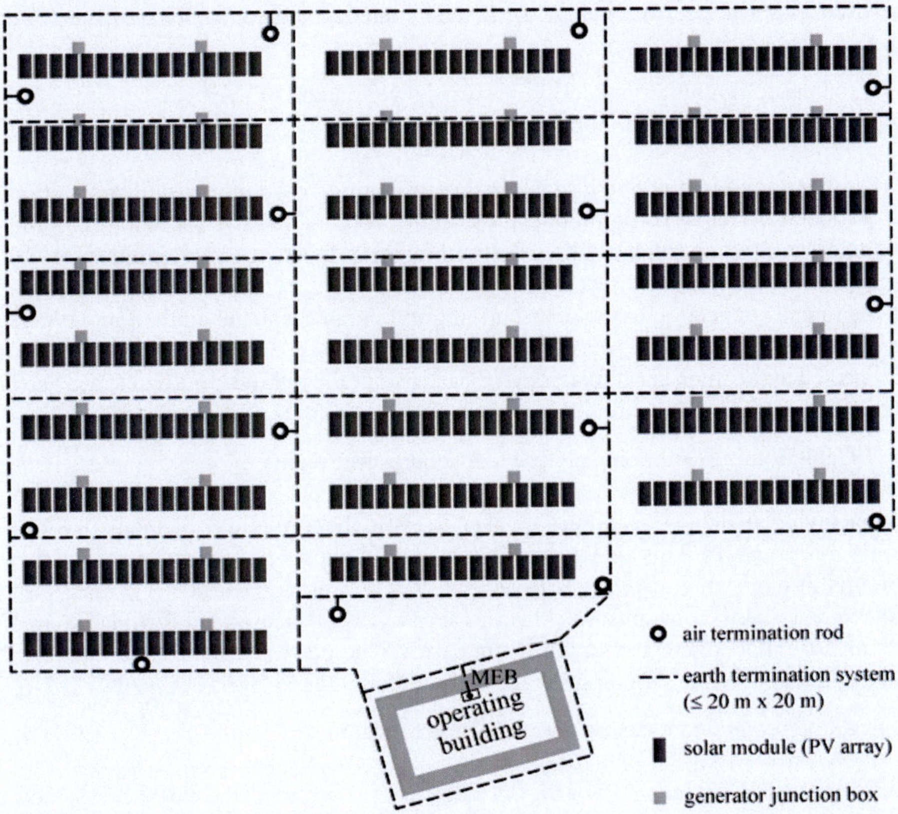

Fig. 3 Schematic of the earthing system of a large outdoor solar power plant (PV generator) including operating building [2, 7]

3.5 mm or 10 mm wire made of stainless steel or copper or galvanized steel in the form of closed meshes (meshed ring earth electrode), not larger than 20 m × 20 m, which are laid below freezing depth, there is a long-term resistant earthing system. The metal module tables can be used as part of the mesh if they meet the requirements for natural components according to [9] and are interconnected.

The individual earthing systems of the PV generator and the operating building must be connected, which reduces the overall earthing resistance. The intermeshing of the earthing systems creates an equipotential surface which significantly reduces the voltage stress on the electrical connecting cables when lightning strikes between the PV module field and the operating building. The meshes are to be connected with lightning current-tested connecting components. The metal support frames on which the PV modules are mounted must be connected to each other and to the earthing system. Frame constructions using pile driving or screw foundation technology can be used as earth electrodes (Fig. 3), if their material and wall thickness meet the requirements of [9].

iii. Internal lightning protection for solar power plants
a. Lightning equipotential bonding

Lightning equipotential bonding is the direct lightning current carrying connection of all metal systems. If the modules, the entire cabling and the operating building together with the weather station are within the protective space of the external lightning protection, no direct lightning currents are to be expected on the cables. If the mains connection to the distribution grid operator is made on the low-voltage level, SPD type 1 lightning current arresters must be used at this transition point and connected to the main earthing bar (MEB), as part of the lightning current flows here. The same applies to incoming telecommunication cables.

b. Cable routing within installations in solar power plants

When laying all cables, care must be taken to avoid the formation of conductor loops over large areas [1]. This applies within the single-pole serial connections of the d.c. circuits in one string, as well as to several strings among each other. It must also be avoided that data or sensor cables run across several strings and form large-area conductor loops together with the string cables. The power cables for d.c. and a.c., data and equipotential bonding should be routed together as far as possible.

c. Surge protection measures for solar power plants

Surge protective devices (SPD) (Fig. 4) must be used to protect the electrical systems within PV power plants [10]. When a strike occurs in the external lightning protection of an open field system, on the one hand high voltage pulses are coupled into all electrical conductors, on the other hand partial lightning currents occur within the open field wiring (d.c., a.c. and data cables), the peak value of which is influenced by the design of the earthing system, the specific earthing resistance on site and the design of the wiring.

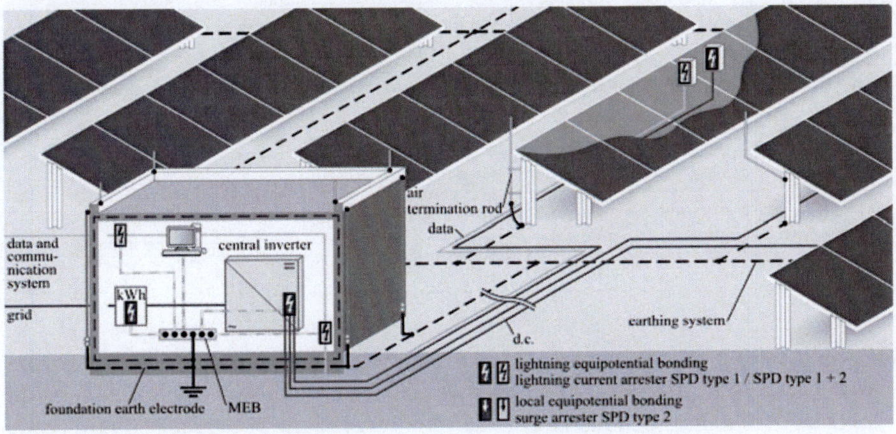

Fig. 4 Lightning protection concept for PV power plant with central inverter

Plant concepts with central inverter technology (Fig. 4) involve extensive d.c. cabling in the field.

Use voltage limiting type 1 d.c. SPD with a minimum discharge capacity of 10 kA (10/350 μs). SPD with high short-circuit strength must be used, also because of possible reverse currents.

DIN EN 62305-3, 5 [6] or IEC TR63227 [7] also contains an estimate of the lightning current distribution. To calculate the lightning current distribution, the down conductors of the lightning protection system, the possible earthing connections of the module array and the d.c. cables must be taken into account. The peak value of the partial lightning currents which are fed into the d.c. lines via the SPDs depends not only on the number of down conductors but also on the impedance of the SPD. This SPD impedance in turn depends on the rated voltage of the SPD, the SPD topology and the SPD type, voltage-switching (crowbar-type) or voltage-limiting (clamping-type). Characteristic for partial lightning currents through SPD on the d.c. side of the PV system is a shortening of the pulse shape. When selecting suitable SPD, both the maximum surge current and the pulse charge must be taken into account. In order to simplify the SPD selection for the user, the necessary lightning impulse current carrying capacity of type 1 SPD can be selected depending on the SPD type, voltage-limiting varistor arrester or voltage-switching spark gap arrester according to Table 1. The maximum occurring surge currents are considered as well as the partial lightning currents of the waveform 10/350 μs, so that the SPD can discharge the pulse charge of the lightning currents.

d. Solar power plants with decentralized string inverters

If PV power plants are designed with decentralized string inverters, a large proportion of the power cabling is shifted from the d.c. side to the a.c. side. The inverters are mounted in the field under the module tables of the respective solar generators. Due to its proximity to the modules, the inverter also performs typical functions of generator

Table 1 Minimum discharge capacity of voltage-limiting or combined SPD type 1 and voltage-switching SPD type 1 for a solar power plant at lightning protection level (LPL) III [6, 7]

Highest lightning current peak value I_{imp}	Clamping and combined (series connected) SPD Type 1				Crowbar and combined (parallel connected) SPD Type 1	
	$I_{10/350}$		$I_{8/20}$		$I_{10/350}$	
	Per protected path	I_{total}	Per protected path	I_{total}	Per protected path	I_{total}
100 kA	5 kA	10 kA	15 kA	30 kA	10 kA	20 kA

junction boxes. The lightning current distribution is influenced by the power wiring of string or central inverters. In string inverters, too, the power cabling acts as an equipotential bonding conductor between the local earth potential of the module field in which the lightning strikes and the remote potential of the mains transformer. The difference to the system with a central inverter is that in systems with string inverters, the partial lightning currents flow on the a.c. lines. Accordingly, type 1 SPD must be installed on the a.c. side of the string inverters and on the low-voltage side of the mains transformer.

The minimum discharge capacity of Type 1 SPD is listed in Table 1, depending on SPD technology. Type 2 SPD are sufficient on the d.c. side of the string inverters. The string inverters and the associated module field form a local equipotential surface when the earthing system is suitably designed, so that no lightning currents are to be expected in the d.c. wiring, but the surge protective devices essentially limit induced interference pulses and thus also protect the modules. Several a.c. outputs of outdoor inverters are combined and fused in so-called a.c. collective distributors. If type 1 SPD are used there, they protect all inverter outputs at a cable distance of up to 10 m. The additional a.c. field cabling is brought together in the operating building; SPD type 1 or combined SPD type 1 + type 2 must also be used at this grid transition point. Other operating equipment such as grid and system protection, alarm center or web server, which are less than 10 m cable length from this SPD, are also protected with regard to their network supply.

e. Surge protection measures for information systems

In operating buildings, the data information from the field is combined for remote maintenance by the plant operator and for power measurement and control by the grid operator. Reliable data transfer must be ensured at all times so that service personnel can determine the causes of faults by remote diagnosis and remedy them on site. String and inverter monitoring, weather data acquisition, theft protection and external communication are based on a wide variety of physical interfaces. The sensors, electrical interfaces and data lines must be protected with information technology surge protective devices.

Large-area induction loops are formed in the interaction of power cables, metal module table rows and data cables (Fig. 5). This is an ideal environment for transient surges due to lightning discharges that can be coupled into these lines. Such voltage

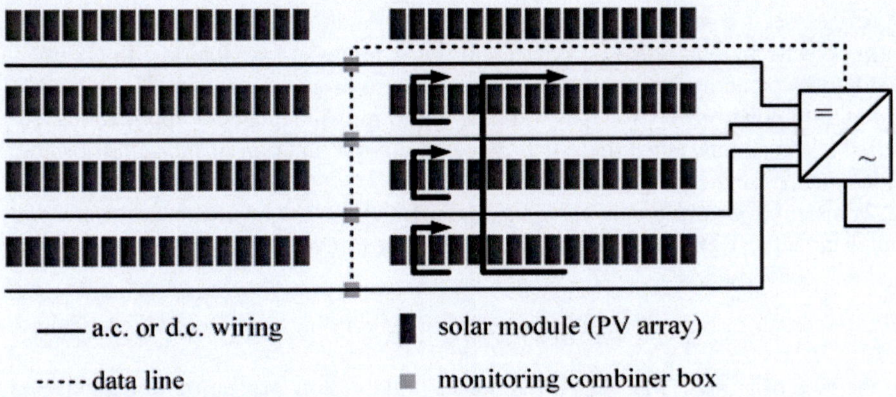

Fig. 5 Schematic diagram of induction loops in PV power plants

peaks are capable of exceeding the insulation/impulse strength of these systems. Overvoltage damage is the result. SPD must therefore also be used for data transmission in monitoring generator connection boxes or in decentralized string inverters. Cable shields must be properly connected at all connection points. In order to prevent malfunctions such as ripples and vagabonding currents, this can also be done with indirect shield earthing via arresters.

3 Lightning and Surge Protection for Roof-Mounted Photovoltaic Systems

i. Importance of lightning protection for roof-mounted PV systems

Roof-mounted photovoltaic systems are also exposed to lightning because they usually represent the highest installation on a building roof (lightning impact area equal to occupied roof area plus three times building height; area of the PV system several 10 m^2) [4, 5].

Many millions of small PV systems are currently installed worldwide. The worthwhile consumption of power and the pursuit of a degree of independence in power supply will make PV systems an integral part of electrical installations in the future. The PV systems are exposed to the effects of weather and must withstand these for decades. The high and rising number of small rooftop photovoltaic systems is also already contributing to renewable power supply in some countries. Small roof PV systems are connected at the low-voltage level. Usually, PV cabling is introduced into the building and there are often long cable runs to the grid connection point.

Lightning discharges cause field and conducted electrical interference. This effect increases with the length of the line or the size of the conductor loop. Damage caused by surges occurs not only on the connected PV modules, inverters and their monitoring electronics, but also on devices in other domestic installations. In commercial buildings, additional damage can be caused to production systems, resulting in production downtimes. If surges are coupled to grid-independent applications, so-called PV island systems, these can then cause a malfunction of the solar-powered systems, e.g. medical equipment, water supply.

Transient overvoltages from switching operations of the upstream a.c. or d.c. grid cause damage. Defects can occur in PV modules, inverters, charge controllers or d.c./d.c. converters.

a. Necessity of lightning protection system on buildings with roof PV systems

In the case of a direct lightning strike into a building, the protection of persons and fire is the first priority. The released energy of a lightning discharge is one of the most frequent causes of fire. With PV systems, a distinction must be made between installations on buildings with and without lightning protection. Even when planning a PV system, it is usually clear whether the building is already equipped with lightning protection. For public buildings (e.g. meeting places, schools and hospitals), for example, building codes require lightning protection systems. In the case of commercial or private buildings, the necessity of lightning protection is differentiated according to location, type of construction and use. It must be determined how easily a lightning strike can occur or whether a lightning strike can lead to severe consequences. Systems in need of protection must therefore be equipped with permanently effective lightning protection systems. In privately used buildings, lightning protection is often not installed. This is partly for financial reasons, but also due to lack of sensitivity to the subject.

According to the current state of the art, the installation of PV modules on buildings does not increase the risk of lightning strikes, so that the need for lightning protection cannot be directly deduced from the mere existence of a PV system. However, a lightning strike can increase the risk for the building's electrical systems. This is due to the fact that the cabling of the PV lines inside the building in existing risers and cable trays can cause strong conducted and radiated interference from lightning currents. The risk of damage from lightning strikes must therefore be determined in accordance with [8]. The results must be implemented during the construction of the roof PV system.

DIN EN 62305-3, 5 [6] and IEC TR63227 [7] describe that a LPS designed for lightning protection level III (LPL III) meets the normal requirements for PV systems (>10 kWpeak) on buildings. In principle, photovoltaic systems on buildings must not impair with existing lightning protection measures.

b. Necessity of surge protection within roof PV installations

In case of lightning discharges, surges are coupled into electrical conductors. Surge protective devices (SPD) have proven their worth in protecting electrical systems against these destructive voltage peaks. In many cases, this surge protection is already required in insurance conditions for photovoltaic systems. In particular, the inverter must be protected by SPD. In PV installations on the a.c. and d.c. side as well as in existing signal and communication circuits, surge protection measures must therefore be provided.

c. Cable routing in installations of roof-mounted PV systems

When laying the cables, make sure that no large conductor loops are formed [1]. This applies to the wiring of the d.c. circuits to the string and also to several strings among each other (Fig. 6). Furthermore, it must be avoided that data or sensor cables run across several strings and, in combination with the string cables, form large-area conductor loops. This must also be taken into account when connecting the inverter to the grid connection. It is important that the power lines (d.c. and a.c.) are laid together with the equipotential bonding throughout.

ii. External lightning protection for roof-mounted PV systems

a. Earthing of roof PV systems

PV modules are mainly mounted on metal mounting systems. The active d.c.-side PV components have double or reinforced insulation. The combination of a variety of technologies on the module side and on the inverter side, e.g. with or without galvanic isolation, results in different grounding requirements. Functional earthing of the metal substructure is carried out if the system is within the protective area of interception devices and the separation distance is maintained. A conductor cross-section of at least 6 mm^2 copper or equivalent is required for this functional earthing

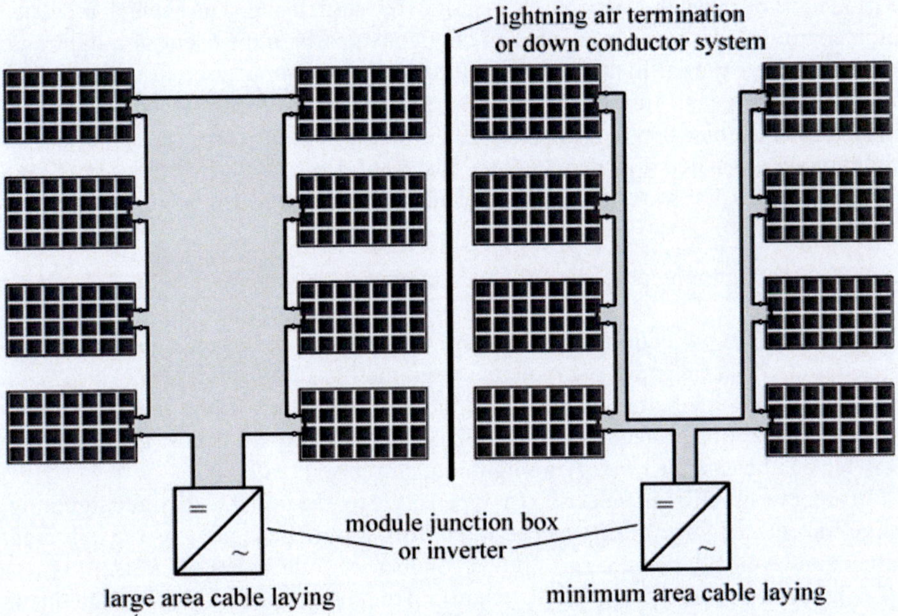

Fig. 6 Low-induction laying of d.c. cables. Avoid large conductor loops [6, 7]

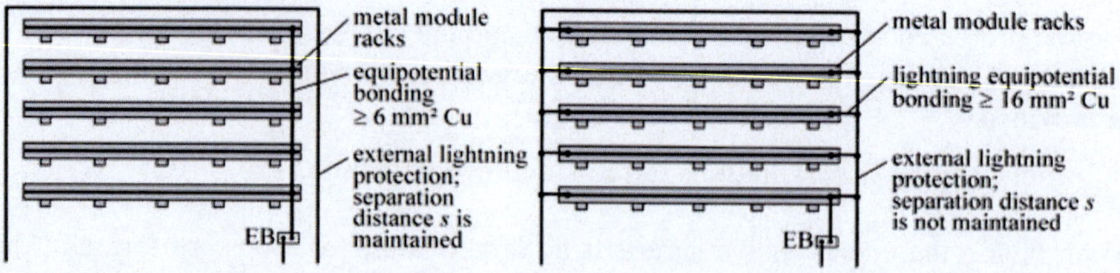

Fig. 7 Functional earthing of module racks without external lightning protection or if separation distance is maintained (left)—Lightning equipotential bonding on module racks if separation distance is not maintained (right) [7]

(Fig. 7). With conductors of this cross-section, the modular rack rails must also be permanently connected to each other. If the mounting system is directly connected to the external lightning protection because the separation distance cannot be maintained, these cables become part of the lightning equipotential bonding. A lightning current carrying capacity of these elements is therefore a basic requirement. The minimum requirement for a lightning protection system according to LPL III is 16 mm^2 copper or equivalent. Here, too, the modular rack rails must be permanently connected to each other. The requirements for natural components according to [9] (Fig. 7) apply.

b. Separation distance

The separation distance between the lightning protection system and the photovoltaic system must be taken into account. It describes the sufficient distance which prevents an uncontrolled flashover into adjacent conductive parts in the event of a lightning strike into the external lightning protection [9]. In the worst case, an uncontrolled flashover can trigger a building fire. But this also results in damage to the PV system.

A further technical possibility to realize the separation distance is the use of high-voltage-insulated down conductors. Insulated down conductors can come into contact with the PV system directly after the voltage-controlled termination range (sealing unit).

iii. Internal lightning protection for roof-mounted PV systems

In general, surge protection on the d.c. and a.c. sides of the inverter or inverters is necessary to protect against overvoltages from the PV system and the a.c. grid. The shielded routing of the d.c. cables from the module strings to the inverter is best suited as supporting protective measure or as compensating measures. The shielding may have to be capable of carrying lightning currents [11, 12].

In order to simplify the selection for type 1 SPD for the user, the required lightning surge current carrying capacity can be taken from Table 2 depending on the lightning protection level and the number of down conductors of the external LPS [2, 11].

If there is no lightning protection system or if the separation distance is maintained, equipotential bonding must be carried out with at least 6 mm^2 copper (Table 3). If the

Table 2 Selection of the minimum discharge capacity of voltage-limiting SPD type 1 (varistor) or combined SPD type 1 (series connection of varistor and spark gap); according to [7]

Lightning protection level LPL	Highest lightning current peak value I_{imp} (kA)	Number of down conductors of external lightning protection			
		<4		≥4	
		Clamping and combined (series connected) SPD type 1			
		Per path $I_{8/20}/I_{10/350}$ (kA)	I_{total} $I_{8/20}/I_{10/350}$ (kA)	Per path $I_{8/20}/I_{10/350}$ (kA)	I_{total} $I_{8/20}/I_{10/350}$ (kA)
I	200	17/10	34/20	10/5	20/10
II	150	12.5/7.5	25/15	7.5/3.75	15/7.5
III, IV	100	8.5/5	17/10	5/2.5	10/5

Table 3 Summary of separation distance, equipotential bonding and surge protection

Initial situation	Measure	Separation distance maintained	Equipotential bonding	Surge protection
External lightning protection	Adjust lightning protection system	Yes	At least 6 mm^2 copper	d.c.: Type 2 a.c.: Type 1
		No	At least 16 mm^2 copper	d.c.: Type 1 a.c.: Type 1
No external lightning protection; ground wire terminal	Analysis of requirement of building regulations or risk management	–	At least 6 mm^2 copper	d.c.: Type 2 a.c.: Type 2

cable length to the inverter is more than 10 m, an additional equipotential bonding with protective devices must be installed near the inverter. The connection to the equipotential bonding must be made with at least 6 mm^2 copper parallel to the d.c. lines. If partial lightning currents occur, grounding must be carried out with at least 16 mm^2 copper (Table 3). Metal cable routing and cable support systems reduce magnetic coupling to a minimum. The cables of the PV system must always be routed tightly and parallel. The reduction in the area between the lines reduces the inductively coupled overvoltage.

a. Selection of SPD according to the protection level

The d.c. side of photovoltaic systems can have different operating voltages, depending on the system. Values up to 1500 V d.c. are currently possible. Accordingly, the terminal devices also have different dielectric strengths. To ensure effective protection of the system, the protection level U_P of the SPD must be lower than the dielectric strength of the system to be protected. At least 20% safety distance should be maintained between the dielectric strength of the system and the U_P. The energy coordination between SPD type 1 or SPD type 2 and the device input must be observed. If arresters are already integrated in the terminal device, then the coordination between

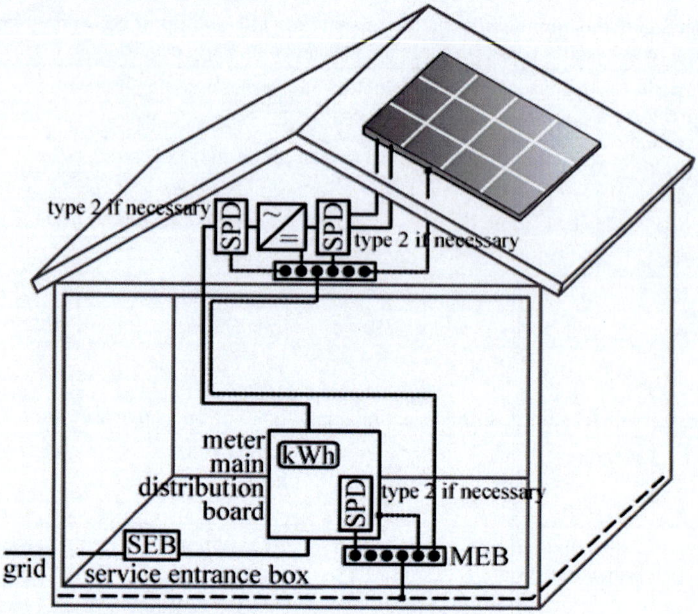

Fig. 8 PV system on building without external lightning protection (situation A) [7, 13]

SPD type 2 and the input circuitry of the terminal device is usually already taken into account by the manufacturer.

The surge protection concept for a PV system on a building without external lightning protection (situation A) shows Figs. 8 and 9. Here, dangerous overvoltages are inductively coupled into the PV system by close strikes of lightning or act on the consumer system from the supply grid via the house connection.

Each d.c. input (Maximum Power Point Tracker MPPT) of the inverter must be equipped with a type 2 surge protective device. IEC 61643-32 [11] provides for an additional type 2 d.c. SPD on the module side for distances of more than 10 m between inverter input and PV generator. If PV inverters and other electronic components, such as a.c.-coupled battery storage systems, are not more than 10 m away from the installation location of the SPD at the grid connection point (low voltage feed-in), they are adequately protected. For longer cable lengths, an additional type 2 SPD must be used. The use of an SPD type 1 + type 2 (combined lightning current and surge arrester) is recommended at the mains connection point (low-voltage main distributor).

If inverters with data and sensor lines for yield monitoring have wired communication interfaces, suitable IT surge protective devices are required.

Figure 10 shows the surge protection concept for a PV system with external lightning protection and with sufficient separation distance between the PV system and the external lightning protection (situation B). The primary protection objective is to avoid personal injury and damage to property, in particular building fires, caused

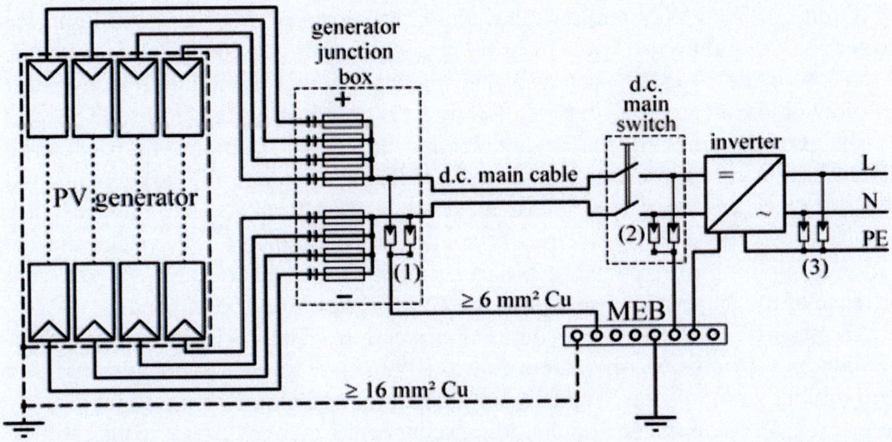

Fig. 9 Surge protection for a PV system without LPS (example with TN system). The surge arresters (2) in front of the inverter are recommended for long d.c. main lines (10 m). Surge arresters (2) and/or (3) are often already integrated in the inverter. The surge arresters (1) in the generator junction box can be dispensed in smaller PV systems with input-side varistors inside the inverter

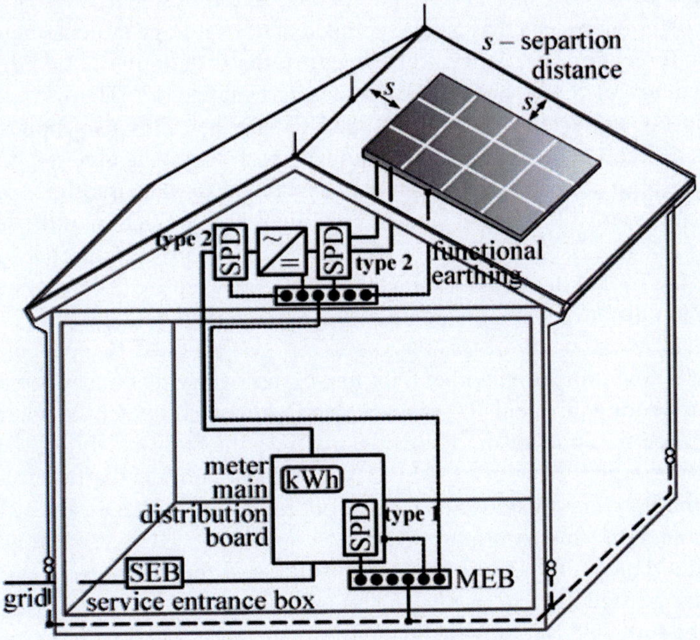

Fig. 10 PV system on building with external lightning protection while maintaining the separation distance (situation B) [6, 7]

by lightning. The PV system must not impair the function of the external lightning protection. In addition, it itself must be protected against a direct lightning strike; it must be installed in the protected space of the external lightning protection. Air termination devices, e.g. air-termination rods, form this protected space and prevent direct lightning strikes into the PV modules and the wiring. The protected space can be determined using the protective angle method (Fig. 14) or the rolling sphere method (Fig. 14) in accordance with [9]. It must be ensured that the separation distance is maintained between all electrically conductive parts of the PV system and the external lightning protection. At the same time, a core shadow, e.g. by sufficient distance of the air-termination rods to the PV modules, must be avoided.

An essential component of a lightning protection system is lightning equipotential bonding. It must be carried out for all conductive systems and cables fed into the building from outside. Lightning equipotential bonding is achieved by directly connecting all metal systems and indirectly connecting all live systems to the earthing system via SPD type 1 lightning arresters. The lightning equipotential bonding should take place as close as possible to the building entrance in order to prevent partial lightning currents from entering the building. The mains connection point must be equipped with a multi-pole SPD type 1 (combined lightning current and surge arrester). For cable lengths less than 10 m between the SPD, the inverter and other electronic components, such as an a.c.-coupled battery storage system, there is sufficient surge protection. For longer cable lengths, the installation of additional SPD type 2 upstream of the devices to be protected is mandatory. The d.c. side of the inverter must be protected with an SPD type 2 PV arrester. This also applies to transformerless devices. If the inverters are equipped with wired data lines, e.g. for yield monitoring, surge protective devices must be installed for data transmission.

In buildings with a PV system and external lightning protection, if the separation distance is not maintained (situation C), the following protection concept can be applied. If the roof skin is made of metal or if it is formed by the PV system itself, the separation distance cannot be maintained from an installation point of view.

The metal components of the PV mounting system must be connected to the external lightning protection with a lightning current carrying connection of at least 16 mm^2 copper or equivalent. This means that lightning equipotential bonding must now also be carried out for the PV cables fed into the building (Fig. 11). The d.c. cables must be equipped with a PV-SPD type 1 or PV combined arrester. Lightning equipotential bonding must also be carried out in the low-voltage supply. If the PV inverter and e.g. the battery storage system are more than 10 m away from the SPD type 1 required there for the grid connection point, a further SPD type 1 or combined arrester must be used. If yield monitoring is planned, suitable surge protection devices must also be provided for the wired data lines.

If the separation distance cannot be maintained, DC cables can be shielded and routed down to ground level outside the building (see Fig. 12). In this case, the cable shield has to be connected to the air-termination system at the high-point and to the earthing system at the base-point immediately before entering the structure. The cable shield must be designed to carry lightning currents and must be integrated into the lightning equipotential bonding as described in [14].

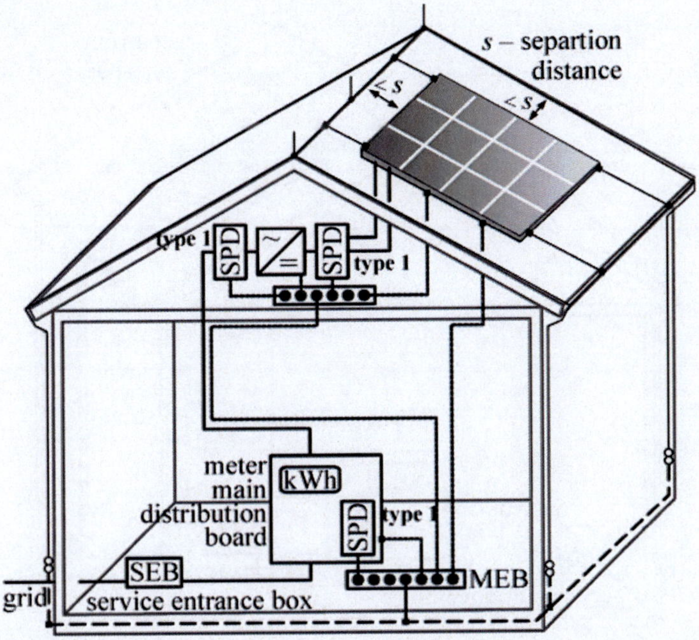

Fig. 11 PV system on building with external lightning protection without maintaining the separation distance (situation C) [6, 7]

It should be noted that the separation distance *s* between the PV components (e.g. modules) and metal parts carrying lightning current such as the external lightning protection, rain gutters, roof windows or antenna systems has to be maintained (Figs. 13 and 14).

b. Roof-mounted PV systems with module inverters

Module inverters (micro inverters) require a different surge protection concept. The d.c. cable of a module or a module pair is connected directly to the small inverter. The module d.c. cables must be laid avoiding unnecessary conductor loops, but direct inductive couplings into such small d.c. structures usually have only a low energetic destruction potential. In systems with module inverters, the extensive wiring of the PV system is carried out via the a.c. side (Fig. 15). If the module inverter is located directly on the module, then the wiring with surge protective devices can only be carried out on the a.c. side.

- Buildings without external lightning protection require SPD type 2 for alternating or three-phase current in the immediate vicinity of the module inverters and at the low voltage feed-in.
- Buildings with external lightning protection and maintained separation distance require SPD type 2 in the immediate vicinity of the module inverter and SPD type 1 with lightning current carrying capacity at the low voltage supply.

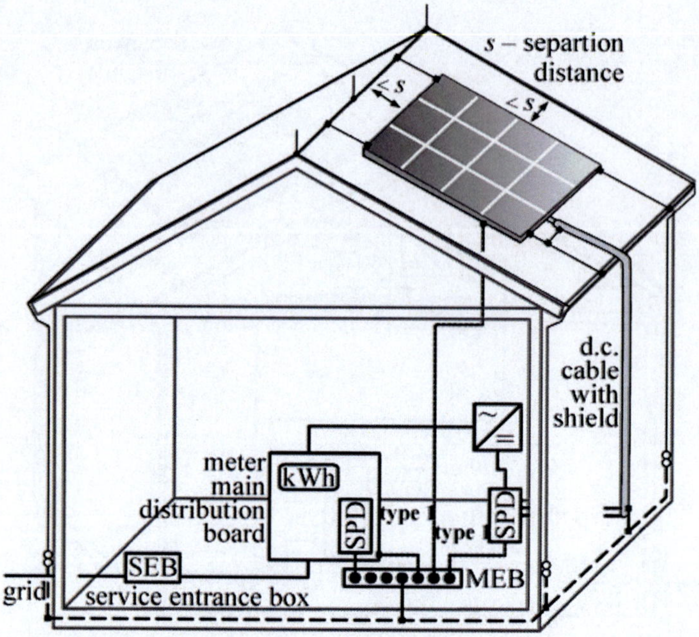

Fig. 12 PV system on building with external lightning protection without maintaining the separation distance (situation C) and using d.c. cable with shield capable for carrying lightning current [7, 13]

Fig. 13 Determination of protected space by protective angle method and PV modules in protected space of air-termination rods taking into account the separation distance [6, 7]

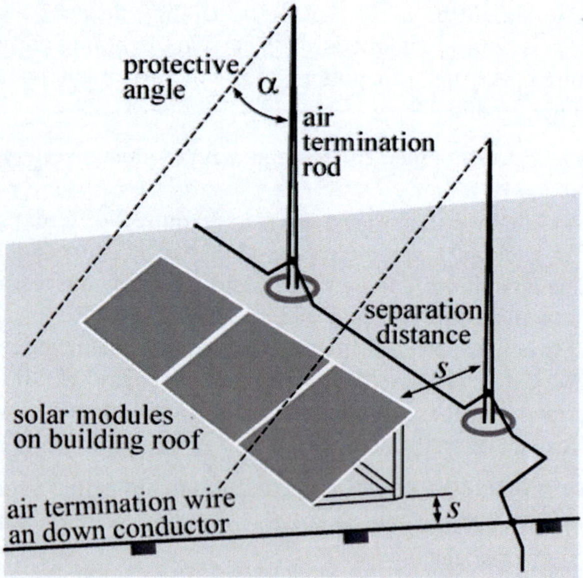

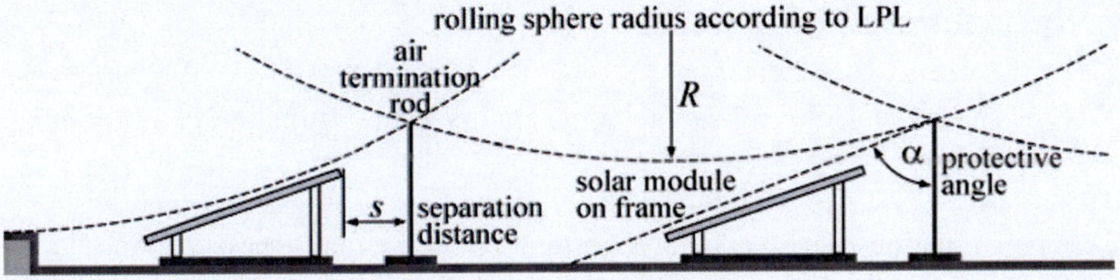

Fig. 14 Protected space by means of rolling sphere method or protective angle method [2, 6, 7]

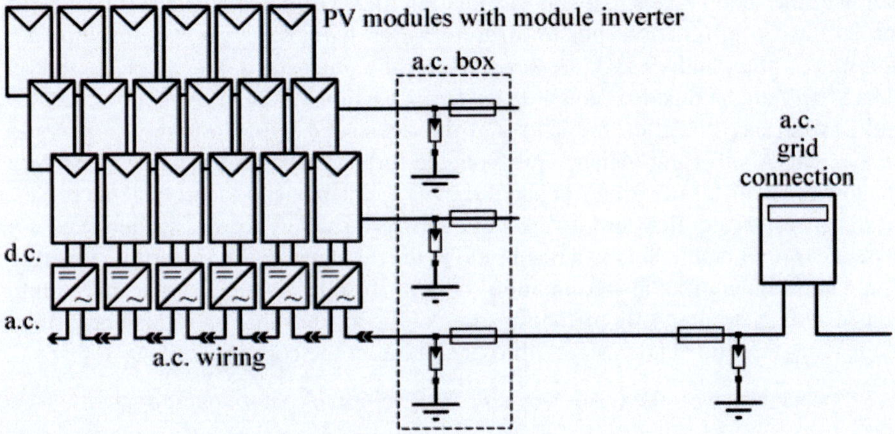

Fig. 15 Building without external lightning and surge protection for module inverters in the connection housing of the cabling provided by the customer

- Buildings with external lightning protection and without maintaining the separation distance require SPD type 1 in the immediate vicinity of the module inverters and at the low voltage supply.

Module inverters from all manufacturers have equipment monitoring systems. If the data is modulated onto the a.c. lines via the module inverters, surge protection must be provided on the separate receiver units (output, data processing). The same applies to wired interface connections and their power supply to downstream bus systems, e.g. Ethernet, ISDN [2].

4 Specialities of PV Systems

i. Core shadow on solar cells of solar generators and construction of air terminations

The air terminations of the external lightning protection are necessary. An uncontrolled lightning strike into the PV system would result in the flow of lightning currents in the electrical system and lead to serious damage within the system. When installing the external lightning protection, care must be taken to avoid relevant shadowing of the solar cells, e.g. by air-termination rods. Diffuse shadows, such as those formed by air-termination rods or air termination wires far away, are insignificant in terms of plant and yield. Core shadows, on the other hand, are characterized by clearly defined darkened contours on the surface behind them, causing power losses and unnecessary stress on the cells and the associated bypass diodes, and should be avoided. A sufficient distance prevents the formation of core shadows. The core shadow of an air-termination rod is reduced as the distance to the module increases. At distances greater than in Fig. 16, only a diffuse shadow remains. The necessary distance is predictable; it is in a fixed ratio to the diameter of the air-termination rod. The core shadow of an air-termination rod with 10 mm diameter has scattered into a diffuse shadow after 1.08 m. The shadow of the air-termination rod is not visible. [6] in Annex A and [7] deals with the calculation of the core shadow.

ii. Special surge protective devices for the d.c. side of photovoltaic systems

In PV systems with central inverters, fuses are used to protect against reverse currents, whereby the maximum current occurring depends on the current irradiation. Under certain operating conditions, reverse current fuses only respond after a few minutes (Fig. 17). Surge protective devices in generator junction boxes must therefore be designed for the possible total current, consisting of operating current and return current, and disconnect automatically in the event of overload without forming arcs [11, 13].

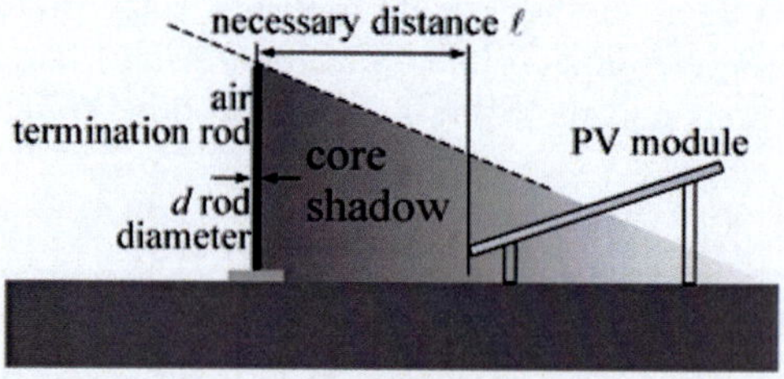

$\ell = 108 \times d$

rod diameter d	necessary distance ℓ
10 mm	1.08 m
16 mm	1.73 m
20 mm	2.16 m

Fig. 16 Distance from air-termination rod to module free of core shadows

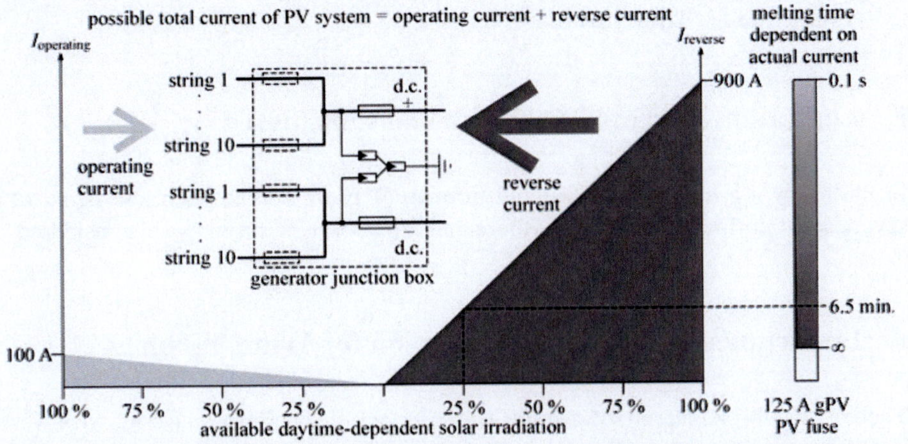

Fig. 17 PV system with d.c. maximum current of 1000 A and daytime-dependent prospective short-circuit current at the PV surge protective device [2]

The U/I characteristics of photovoltaic power sources differ significantly from those of conventional direct current sources. They have a non-linear characteristic with constant current behaviour (Fig. 18) and are the cause of the long maintenance of ignited arcs. This peculiarity not only results in a larger design of PV switches and PV fuses, it also requires suitable surge protective devices with a specially adapted disconnecting device. These PV-SPD must be capable of extinguishing PV-d.c. follow currents. Safe operation, even in the event of overload of SPD on the d.c. side, is required in [6, 7, 11]. On the d.c. side of PV systems, special SPD suitable for PV must always be used, which can handle PV currents with PV characteristics and high d.c. system voltages and thus avoid an increased fire risk [2].

The maximum continuous operating voltage of the d.c.-SPD must be selected in such a way that it is higher than the open circuit voltage of the PV generator, which is to be expected at maximum solar radiation on a cold day in winter.

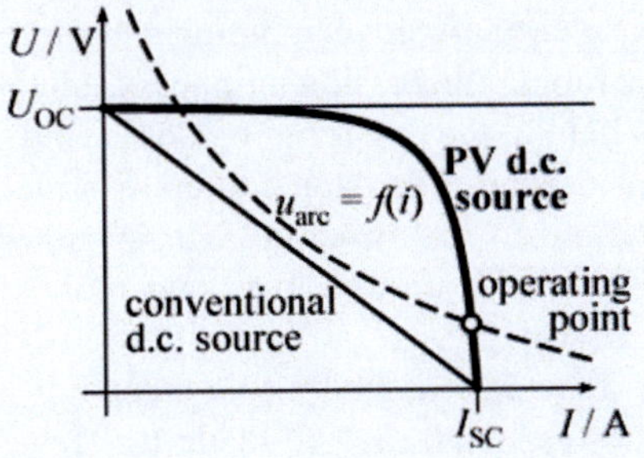

Fig. 18 Characteristic curves of conventional d.c. source and PV generator as well as U/I characteristic of electric arc

b. **Protection of Wind Turbines**

5 Limitation of Wind Turbines to be Considered

In the following, the individual wind turbine (WT) or a wind farm consisting of several wind turbines as well as onshore and offshore wind turbines are considered.

6 Importance of Lightning Protection for Wind Turbines

There is an unbroken trend towards the use of renewable energies such as wind power [15]. This results in an enormous market potential not only for the power industry, but also for the suppliers of the power industry and the electrical trade, worldwide. The wind power industry continues its impressive growth. In 2018, more than 20,000 new wind turbines with a capacity of more than 50 GW were installed worldwide [16]. The importance of wind turbines is obvious. The reliable availability of power is an important factor in the growth rates of this electricity market [3].

Wind turbines (WT) are exposed to lightning strikes as they are free-standing, slender and, in particular, tall structures. Wind turbines are often exposed to direct lightning due to their exposed position and height. Several studies have shown that wind turbines of the multi-megawatt class are exposed to at least 10 direct lightning strikes per year. The high investment costs must be amortized in a few years due to the feed-in tariff, i.e. downtimes due to lightning and overvoltage damage and the associated repair costs must be avoided. Comprehensive lightning and surge protection is therefore necessary.

With the increasing size of wind turbines and thus the increasing length of rotor blades, lightning protection measures are becoming more and more important. Therefore, the rotor blades installed today are equipped with lightning protection consisting of several receptors along the rotor blade, which are guided via a metallic wire (down conductor) to the blade root, from where they are connected to the earthing system [17, 18]. The great importance of effective lightning and surge protection is, however, not only derived from the rapidly increasing number of wind turbines, but, due to the rapid increase in overall height, the danger to modern wind turbines from lightning strikes has increased considerably. The danger of being struck by lightning increases quadratically with the height of the structure. Megawatt wind turbines with their blades reach an overall height of up to 150 m and more today (Fig. 19) and are therefore particularly at risk [4, 5].

Lightning strikes are unavoidable with all larger wind turbines. Most lightning strikes hit the rotor blades especially in the blade tip area. The result is considerable damage. The initial view that rotor blades made of non-conductive glass fiber composite material could do without a lightning protection did not prove to be true in practical operation. For this reason, the demands for effective lightning protection

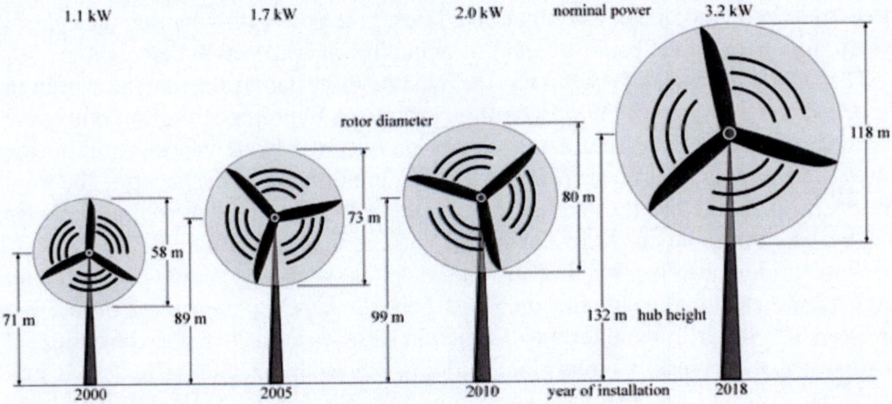

Fig. 19 Increase in overall height of onshore wind turbines in recent years

became louder and louder as wind turbines became more widespread and larger, especially among insurers. Today, a lightning protection system (LPS) is a matter of course for all new rotor blades [19, 20]. The carbon fiber is characterized both by the longest tearing length and by a high modulus of elasticity. The stiffness of carbon fiber components is comparable to that of steel constructions. The fatigue strength properties are good. Only the still high price of carbon fiber speaks against it. Carbon fiber is therefore often used in combination with glass fiber material for particularly stressed areas. Carbon fiber has practically no corrosion problems, but when used for rotor blades it requires special precautions for lightning protection, because the electrical conductivity is not high enough. Down conductors in the rotor blades with transitions to the shaft and to the tower, specially designed for current conduction, as well as an effective (low-impedance) foundation earth electrode allow damage to be limited. For this purpose, e.g. metal caps are attached to the blade tips and large-surface copper mesh is placed under the blade surface in order to conduct lightning currents without major damage [19].

In measuring and control circuits, on the generator as well as on supply facilities, etc., the system is protected against overvoltage damage by powerful coarse and fine protection devices, which can be caused by voltage increases on the generator or by direct or indirect lightning strikes. Direct lightning strikes usually result in major damage. The main components of modern wind measuring systems are wind sensors [18], wind measuring mast and measuring computer, which allow fully automatic and maintenance-free operation. Prerequisites for this are their weatherproof design, internal lightning protection and an efficient power supply [17].

The electrical safety devices include the aviation obstacle lighting with day and night marking and the fire alarm system. Depending on the location, a warning device for the danger of icing or even an electric resistance heating for the defrosting of the rotor blades may be required. The safety devices must be protected by lightning and surge protection so that their function is also ensured in the event of lightning strike

[17]. The central module of a wind turbine houses the power distribution, the control system for charging the batteries and the switching and protection devices.

The rolling bearings of a wind turbine must be protected against the flow of high electrical currents, in particular lightning currents. A lightning strike can otherwise cause very expensive bearing damage, especially to the blade bearings and in the tower head bearing. This measure is of particular importance for the generator bearing arrangement. In the event of a short circuit, there is a risk that the rolling bearings will become unusable due to fusion craters and corrugation formation.

Wind turbines are not particularly fire-prone plants such as vehicles or aircraft with large quantities of flammable fuel on board. Nevertheless, experience has shown that fires can also occur in wind turbines, which in some cases have led to the complete destruction of turbines. Combustible building materials, insulating materials and operating materials are also present in wind turbines. For example, the large-area cladding of the nacelle made of glass fiber composite material is flammable. Fires are primarily triggered by lightning strikes (Fig. 20) and electrical short-circuits; extinguishing the fire in these tall structures is often hopeless. Lightning protection, which is first and foremost fire protection, is therefore of great importance today.

i. Lightning environment of wind turbines

At 50 onshore wind farms in Germany, each with 1–21 wind turbines from 120 to 200 m in height, observations were made on localized earth flashes. A period of 10 years around 2010 with 5 years before and 5 years after the installation of the wind farms was investigated [22].

Fig. 20 Burning wind turbine as result of a lightning strike [21] (NDR)

- negative flashes: significant increase in number (+64% on average), no noticeable change in amplitude and slight shift towards larger amplitudes after construction of wind farm
- positive flashes: increase in number (+29% on average), noticeable change in amplitude and significant shift in lightning current amplitudes towards higher values after construction of wind farm

In addition, 2 offshore wind farms in the German North Sea were considered.

- negative flashes: significant increase in number compared to onshore wind farms and small increase in amplitude after construction
- positive flashes: increase in number and noticeable increase in amplitude after construction

Overall, it can be seen in the exceedance probability distributions that the area with higher positive lightning current amplitudes shifts towards the CIGRÉ characteristic for positive first strokes and the middle area of negative flashes shifts towards the CIGRÉ subsequent stroke characteristic after construction of the wind farms.

For the dimensioning of lightning protection measures, it must be taken into account that for objects with a height greater than 60 m and lightning-exposed position, in addition to cloud-to-earth flashes, earth-to-cloud flashes, so-called upward lightning, as well as side flashes must also be taken into consideration. The upward flashes occur above all in winter with charges even higher than 300 °C. It therefore makes sense to set higher requirements for air termination and down conducting systems. The reason for this is that the charge is the cause of fusion on turbine components and thus has a decisive influence on the maintenance of down conducting systems, e.g. spark gaps. As an example, the lightning charge in Japan reaches values of 600 °C due to upward flashes during winter thunderstorms [23].

Studies in Japan, North America but also in Central Europe have shown that wind turbines are exposed to a large number of upward flashes due to the strong increase in height. Operating experience shows that especially in areas where winter thunderstorms are to be expected, the normative maximum value described in [24] for the charge of the flash $Q_{flash} = 300$ °C does not reflect the actual possible load. For this reason, the charge value was doubled to $Q_{flash} = 600$ °C for such systems. This value then also forms the basis for the tests of the melting out on rotor blades and rotating components. The increase of the current parameters for the first positive surge current to 20 MJ/Ω, which roughly corresponds to a lightning current of 280 kA 10/350 μs, is also being discussed. This increased pulse current is anchored in an additional national annex to the Japanese standard as a normative threat variable. This current parameter is often used as a test parameter by international wind power manufacturers, especially for offshore applications. IEC 61400-24 [25] states that there is currently no evidence that the lightning current parameters for offshore turbines deviate significantly from the parameters for onshore turbines. Therefore, the standard requires that these threat values also be used for offshore wind turbines [23].

Table 4 Relationship between percentage of upward lightning and winter lightning activity [23]

Winter lightning activity	Percentage of upward lightning (%)
High	80–99
Medium	40–90
Low	20–50
None	10–40

The risk analysis in [25] is not used to assess whether a lightning protection system (LPS) is necessary and according to which lightning protection level (LPL) it is to be designed. It is determined that wind turbines must be protected against lightning and that the lightning protection must be dimensioned according to LPL I as standard. The risk analysis according to [25] is mainly used to estimate the frequency of dangerous events and especially the number of direct impacts into the wind turbine. The effect of flashes with several impact points on the earth is taken into account by doubling the values of the earth flash density according to [26].

It should be emphasized that winter thunderstorm activities must be taken into account when estimating lightning frequencies. Field experience shows that lightning damage to wind turbines occurs especially in areas with increased winter thunderstorm activity. Table 4 shows that, depending on the degree of winter thunderstorm activity, such winter lightning flashes can represent the absolutely dominant damage scenario. It should also be noted that the percentage of upward flashes may be even higher for wind farms constructed in mountainous terrain or at high altitudes above sea level.

7 Lightning and Surge Protection for Wind Turbines

The design of the protection concept is based on [25], the standards of the [8, 9, 14, 24] and the guidelines of (Germanischer Lloyd) [25] and [27] recommend that all subcomponents of the lightning protection system (LPS) of a wind turbine (WT) be protected according to the lightning protection level LPL I, unless a risk analysis can prove that a lower LPL is sufficient [8]. A risk assessment can also show that different sub-components are assigned different protection levels. IEC 61400-24 [25] recommends a complete lightning protection concept as the basis for lightning protection.

The LPS of a WT consists of the external lightning protection and the surge protection measure (SPM) to protect the electrical and electronic equipment. When planning the protective measures, it is advantageous to divide the WT into lightning protection zones (LPZ). The lightning protection of wind turbines includes the protection of two subsystems which are only found in WT: the rotor blades and the mechanical drive train. The complex problems of the protection of rotor blades and rotating parts or bearings require detailed investigation and are manufacturer and type specific.

i. Lightning protection zones concept

The lightning protection zones concept is a structuring measure to create a defined electromagnetic compatibility (EMC) climate within an object. The defined EMC climate is specified by the interference immunity of the electrical equipment used. The lightning protection zones concept therefore contains the protective measure of reducing the conducted and field-bound interferences at interfaces to agreed values. For this reason, the object to be protected is divided into protection zones [28] (Fig. 21).

The determination of the zones LPZ 0_A, i.e. the system parts which may be exposed to direct lightning strikes, and LPZ 0_B, which is assigned to those system parts which are protected against direct strikes by external air terminations or air-termination devices integrated in system parts, such as in the rotor blade, is carried out by the rolling sphere method. According to IEC 61400-24 [25], the rolling sphere method is not applicable to the rotor blades themselves (moving parts). Figure 23 shows the principle application of the rolling sphere method and Figs. 22 and 28 the possible division of a WT into different lightning protection zones. The division into lightning protection zones depends on the structure of the WT. They should take their structure into account. However, it is decisive that the lightning parameters acting from outside

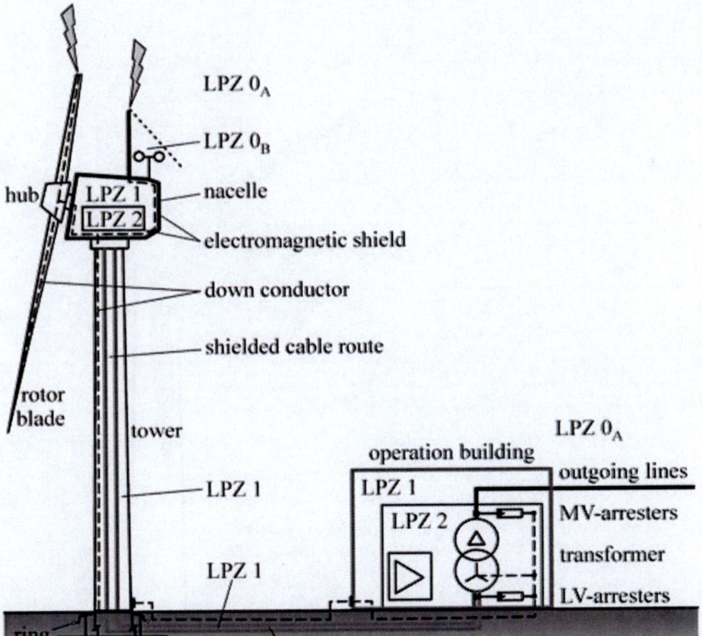

Fig. 21 Lightning protection zones concept for wind turbine—lightning protection and earthing system (Germanischer Llyod)

in the lightning protection zone LPZ 0_A are reduced at all zone boundaries by suitable shielding measures and the installation of surge protective devices to such an extent that the electrical and electronic devices and systems located within the WT can be operated without interference.

The lightning protection system (Fig. 22) must protect the mechanical components against damage and protect the electrical and electronic components against destruction and overvoltage.

a. Electromagnetic shielding measures in wind turbines

The nacelle should be constructed as a self-contained metal shield. Within the nacelle, a volume with an electromagnetic field considerably weakened on the outside is achieved. A tubular steel tower, as it is often used in large wind turbines, can be regarded as an almost perfect Faraday cage for electromagnetic shielding.

In concrete hybrid towers, the function of the galvanic cage must be ensured by means of reinforcement steel as well as earthing and through-hole bonding of the individual components. The switch and control cabinets in the nacelle and, if available, in the operating building should also be made of metal. The connecting cables should be provided with an outer screen capable of carrying lightning currents or be routed in a closed metal cable duct. In terms of interference protection, shielded cables are only effective against EMC couplings if the shields are connected to the

Fig. 22 Rolling sphere method at wind turbine

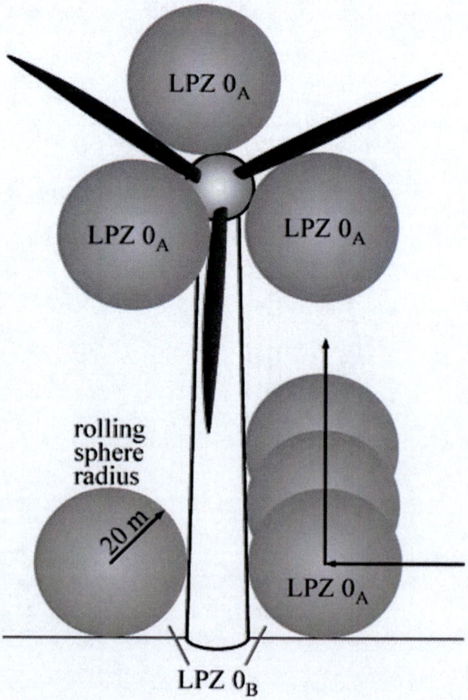

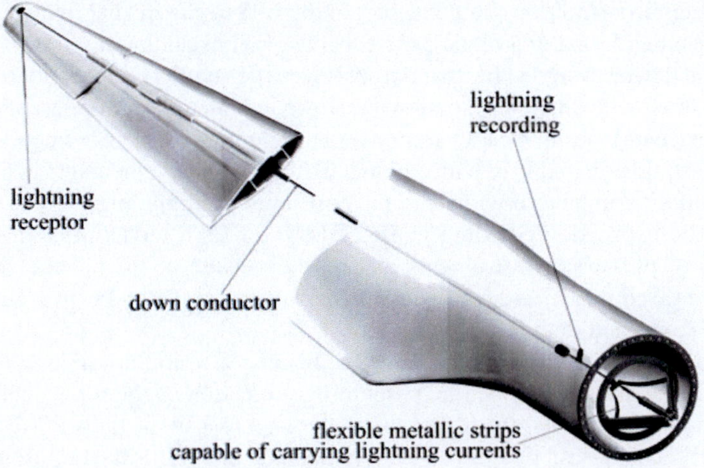

Fig. 23 Example for the lightning protection of a rotor blade [19]

equipotential bonding on both sides. The shields must be contacted with all-round connection terminals.

Magnetic shielding and cable routing should be carried out in accordance with [14]. Shielding measures include, for example, metal braiding on nacelles with glass fiber reinforced plastic (GRP) coating, a metal tower, metal switch cabinets, metal control cabinets, shielded connecting cables with lightning current carrying capacity (metal cable duct, shielded pipe or similar) and cable shielding.

ii External lightning protection for wind turbines

The external lightning protection has the task of intercepting direct lightning strikes, including strikes on the WT tower, and diverting the lightning current from the strike point to earth. Furthermore, it serves to distribute the lightning current in the ground without causing thermal or mechanical damage or dangerous sparking which could cause a fire and endanger persons. The potential points of strike into a wind turbine can be determined using the rolling sphere method, except for the rotor blades (Fig. 23). Lightning protection level (LPL) I is recommended for WT. A rolling sphere with a radius $R = 20$ m is therefore rolled over the WT to determine the strike points. Wherever the sphere touches the WT, potential lightning strike points and thus air terminations are required.

The external lightning protection measures include air-termination and down-conductor devices in the rotor blades, air-termination devices to protect the nacelle superstructures, the nacelle and the hub, use of the tower as an air termination (for side flashes) and down conductor as well as a foundation earth electrode in combination with a ring earth electrode as an earthing system.

The nacelle construction should be part of the LPS to ensure that lightning strikes the nacelle either hit natural metal parts capable of withstanding the stress or an air-termination device designed for that purpose. Nacelles with a GRP sheath or the like should be fitted with an air termination and down conductors forming a cage around the nacelle (metal mesh). The air termination, including the exposed conductors in this cage, should be capable of withstanding lightning. Other conductors in Faraday's cage should be designed to withstand the portion of the lightning current to which they may be exposed. According to IEC 61400-24 [25], air terminations for the protection of measuring instruments etc. on the outside of the nacelle should be designed in accordance with the general regulations in [9] and down conductors should be connected to the cage described.

Natural components made of conductive materials, which always remain in or on the WT and are not changed, e.g. lightning protection of the rotor blades, bearings, machine frames, hybrid towers, may be used as part of the LPS. If WT are made of a metal construction, it can be assumed that this fulfils the requirements for external lightning protection of LPL I. The prerequisite is that the lightning is safely intercepted by the lightning protection of the rotor blades and can be diverted to the earthing system via the natural components such as bearings, machine carriers, tower and/or bypass systems, e.g. open spark gaps or carbon brushes.

a. Air termination and down conductor on wind turbines especially on rotor blades

As can be seen from Figs. 23 and 28, the rotor blades, the nacelle with superstructures, the rotor hub and the WT tower can be struck by lightning. If these are all able to safely catch the maximum expected lightning surge current of 200 kA and conduct it to the earthing system, they can be used as natural components of the external lightning protection of the WT.

If the rotor blades or their spars are made of steel, they form an ideal lightning air-termination and down-conductor device and do not require any further lightning protection devices. In the past, rotor blades made of glass fiber material were manufactured without special lightning protection measures. With the increasing spread of wind turbines, however, the number of damages caused by lightning strikes rose sharply. In the insurance statistics, these cases of damage were at times at the top of the list, so that economic pressure arose to limit this type of damage. Rotor blades today therefore without exception have special lightning protection devices. The lightning protection system of the entire wind turbine consists of metal connections, copper brushes and elastic copper strips at the critical transition points which divert the lightning current into the earthing system of the foundation. (Germanischer Llyod)

For the lightning protection of the rotor blades, metallic receptors are often inserted into the GRP blade tip and sometimes also into the sides of the rotor blades as shown in Figs. 21 and 22, which represent defined strike points for the lightning. In the simplest case, this is a screwed-in and therefore easily replaceable metal part. From the receptor, thick metallic wire or rope is routed to the blade root as down conductor inside the rotor blade. There this conductor is connected to the rotor hub with flexible metallic strips. When lightning strikes, it can be assumed that the lightning strikes the tip of the blade or the receptor and then takes its way via the lightning conductor inside the blade to the earthing system via the nacelle and the tower [19].

With regard to the lightning protection measures to be foreseen, [29] provides some general information, particularly with regard to rotor blades. For wind turbines erected in areas with high winter thunderstorm activity, such as the west coast of Japan, continuing lightning currents have been recorded which clearly exceed the total charge of 300 As required for LPL I (up to more than 1000 °C). Multiple lightning strikes during the lifetime at the same rotor blade end (receptor) are a realistic threat scenario. In this case, it must be taken into account that the charge of the lightning current that is decisive for material melting and ablation at the arc root point has an accumulating effect [30] (Fig. 24).

b. Earthing system for wind turbines

The earthing system of a wind turbine must combine several functions, such as personal protection, lightning protection and the improvement of EMC. Information on material (Fig. 23) is absolutely necessary for the distribution of lightning currents and to prevent destruction of the WT. The earthing system must also protect people from electric shock.

In the event of lightning strikes, the earthing system must discharge any high lightning currents into the ground and distribute them there without causing dangerous thermal and electrodynamic effects. It is generally important that an earthing system is set up for a WT, which is used both for lightning protection and for earthing the power supply grid.

Ring earth electrode and the reinforcement in the foundation must be connected to the tower construction. The reinforcement of the tower foundation should always be used for earthing of WT. The grounding of the base of the tower and the service building should be connected by a grounding mesh in order to obtain the largest possible grounding system. In order to avoid high step voltages in the event of a lightning strike, potential-controlling, corrosion-resistant ring earth electrodes should be laid in the ground around the base of the tower for the purpose of personal protection (Fig. 25).

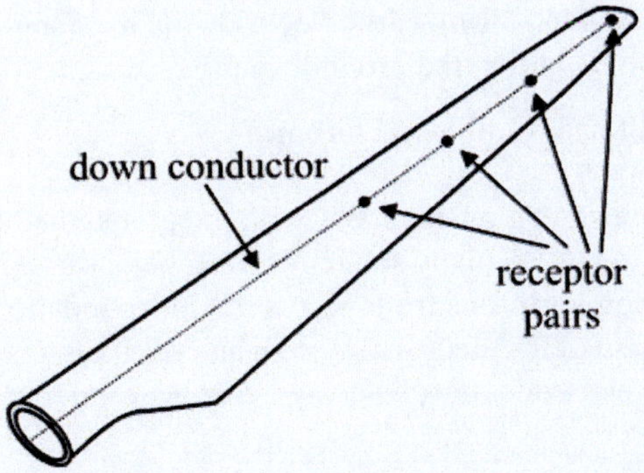

Fig. 24 Metal receptors on the surface of rotor blade and connected to down conductor inside the blade [4]

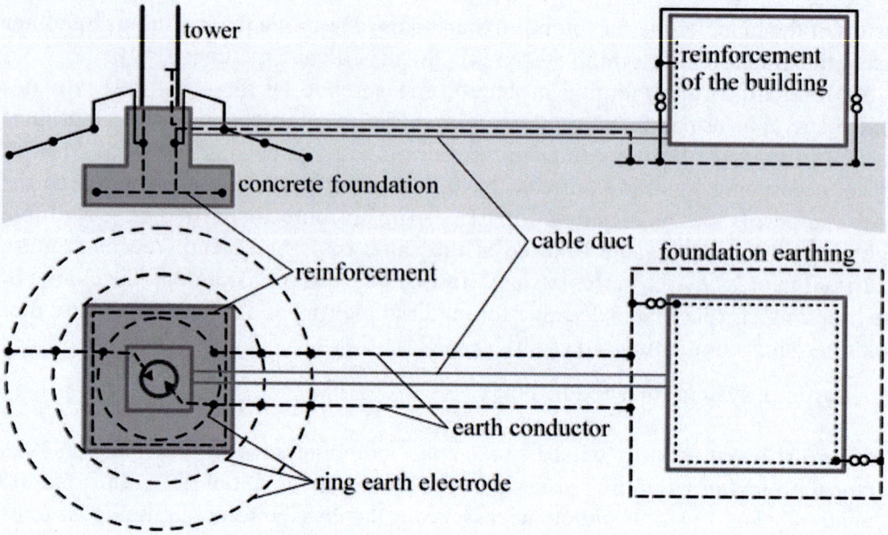

Fig. 25 Meshed network of earthing systems of a wind turbine [2]

A foundation grounding system is advantageous from both a technical and an economic point of view and is required in the technical connection conditions of the distribution grid operators. The foundation earth electrode is considered a component of the electrical system and fulfils essential safety functions. Metals for earth electrodes must comply with the materials listed in [9]. The behavior of the metals with regard to corrosion in the ground must always be observed. Round or strip steel, which can be either galvanized or non-galvanized, must be used as the material for the foundation earth electrode. Round steel must have a diameter of at least 10 mm, for strip steel the dimensions have to be at least 30 mm × 3.5 mm. It must be taken into account that this material is covered with at least 50 mm of concrete to protect it from corrosion. In addition, a connection must be made between foundation earth electrode and main earthing bar (MEB) in the WT. Corrosion-resistant connections must be made via fixed earthing points or connection lugs made of high-alloy stainless steel (Material No. 1.4571). A ring earth electrode made of stainless steel (Material No. 1.4571) must also be laid in the ground.

c. Earthing system for onshore wind turbines

For the earthing systems of an onshore WT with integrated medium-voltage system, which has to fulfil many tasks, many standards must be observed during installation. As a rule, a foundation earth electrode is used. The foundation earth electrode is considered a component of the electrical system and fulfils essential safety functions. The foundation earth electrode must be designed as a closed ring and arranged in the foundation of the tower.

The lightning protection earthing has the task of safely taking over the lightning current from the down conductors and diverting it into the ground. From the point of view of lightning protection, a single, common earthing system of WT is advantageous for all purposes (e.g. medium voltage system, low voltage supply, lightning protection, electromagnetic compatibility, telecommunications and control systems). For this purpose, the foundation of the WT made of reinforced concrete should be used primarily as foundation earth electrodes. They result in the most effective earth electrode with a low earthing resistance and represent an excellent basis for equipotential bonding.

The design of earthing systems must meet the following requirements:

- mechanical strength and corrosion resistance
- control the highest fault current (usually calculated) from a thermal point of view
- ensure safety of persons with regard to voltages at the earthing system occurring during the highest earth fault current or lightning surge currents (touch and step voltages).
- ensure function for complete lifetime of WT.

Consequently, the following parameters are important for the design of the earthing system:

- properties of the surrounding soil
- type of neutral-point connection and resulting short-circuit currents in the event of a fault.

In an installation where different nominal voltages are used, the requirements for each voltage level must be met. The medium-voltage earthing system should also be used for lightning protection. The earthing resistance recommended in the lightning protection standard [9] is less than or equal to 10 Ω.

A connection between foundation earth electrode and main earthing bar (MEB) is to be established in the WT via a connection part. In the case of reinforced foundations, as used at WT, the round or strip steel is placed on the lower reinforcement layer. It must be safely connected to the reinforcement at intervals of 2 m in an electrically conductive manner.

Figure 26 show the example of an earthing system of a WT where the foundation is designed as a circular ring. In the foundation a foundation earth electrode is installed as a ring and 2 further inner rings. The radial connecting lines between the 3 rings are continued to the center of the circle on a cross clamp. The rings and the connecting lines are connected to the reinforcement by means of clamp connections. From the internal inner ring, connection lugs are led to one earthing fixed point each, which is used to connect the earthing system with the equipotential bonding bar in the tower. There are also connection lugs for connection to the tower down conductors. Outside the foundation, a ring earth electrode is installed at a typical distance of 1 m from the outer edge of the foundation. Connecting leads run from the ring earth electrode to the outer inner ring and the connection of depth earth electrodes is optionally possible at the ring earth electrode.

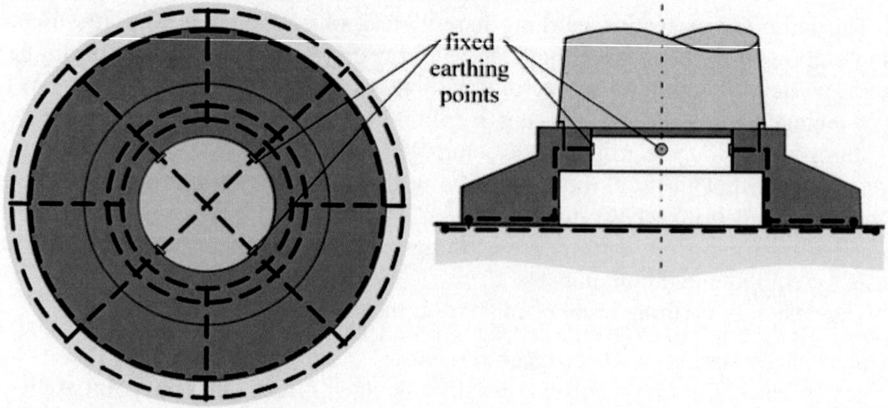

Fig. 26 Plan view of foundation (left) and section of foundation with earthing system (right) [2]

Due to the growing hub heights of wind turbines, concrete towers as well as hybrid towers consisting of a lower concrete tower and an upper steel tube tower are increasingly being erected. Since these towers also contain the earthing system, they must also be considered as part of the electrical system.

For concrete towers with reinforcing steel, the reinforcement can be used for the lightning down conductor if it is ensured that 2–4 parallel vertical connections with sufficient cross-section are available, which are connected horizontally to each other at the top and bottom as well as at intervals of 20 m. Connected in this way, the reinforcing steel offers an effective weakening of the magnetic field and a reduction of the lightning current within the tower. The tower is to be regarded as a primary protective earth conductor (PE) and as an equipotential connection. Due to the tower height direct lightning strikes into the tower construction are to be expected, and this circumstance is to be considered with the construction of the tower. External lightning protection equipment may be considered for use on concrete towers, but should always be connected to the tower's reinforcement steel. For the connection between the individual tower elements, fixed earthing points can be used in conjunction with bridging ropes. It must be ensured that all components used can carry the lightning current.

With tubular steel towers, all requirements, in particular the lightning current carrying capacity according to LPL I with lightning currents up to 200 kA, are fulfilled due to the existing cross-section and the completely metallic design. All metallic components used in the tower must be integrated into the equipotential bonding system. Conductor systems have to be connected to equipotential bonding at each end, at intervals of 20 m and on each platform. Components such as tensile ropes, hoisting ropes and rail systems must be connected at both ends to an equipotential bonding system.

iii. Internal lightning protection for wind turbines

The internal lightning protection must be designed correctly, as the system voltages of the WT are increasing more and more. The trend is towards 690 and 1000 V for TN systems in order to keep the cable cross-section small in large plants [28].

In addition to earthing and equipotential bonding, internal lightning protection measures include spatial shielding and separation distance, cable routing and cable shielding as well as the installation of coordinated surge protective devices (SPD).

a. Protection of cables at transition from lightning protection zones

For the safe operation of electrical and electronic devices, protection against conducted interferences at the interfaces of the lightning protection zones (LPZ) must be implemented in addition to shielding against field-bound interferences (Figs. 27 and 28). At the transition LPZ 0_A to LPZ 1 (classically also referred to as lightning equipotential bonding) protective devices must be used which are able to discharge considerable partial lightning currents non-destructively. These protective devices are referred to as SPD type 1 lightning current arresters and are tested with surge currents of the waveform 10/350 μs. At the transition LPZ 0_B to LPZ 1 and higher, low energy surge current pulses as a result of voltages induced by external processes or overvoltages generated in the system itself, e.g. switching overvoltages, must be controlled. These protective devices are referred to as SPD type 2 surge arresters and are tested with surge currents of waveform 8/20 μs.

According to the lightning protection zone concept, at the interface between LPZ 0_A and LPZ 1 or between LPZ 0_A and LPZ 2, all external cables and lines with SPD type 1 must be included in the lightning equipotential bonding without exception. This applies to both power and communication cables. For each additional zone interface within the volume to be protected, an additional local equipotential bonding must be set up, in which all cables and lines that penetrate this interface

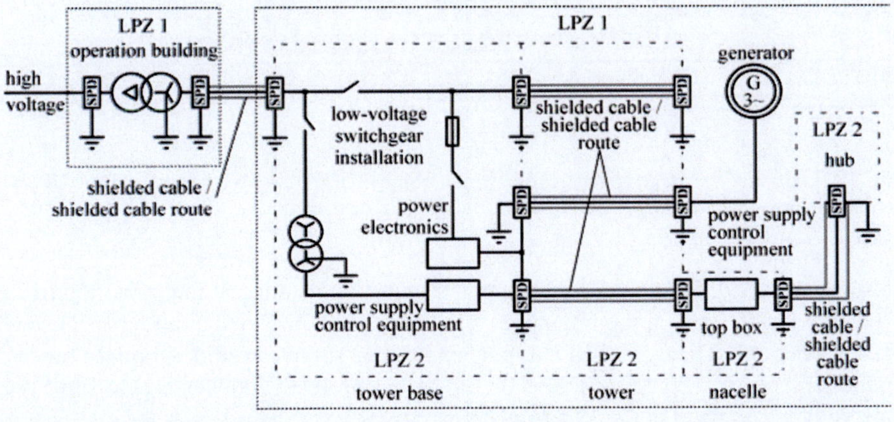

Fig. 27 Use of SPD at the LPZ boundaries of wind turbine with an operating building [2]

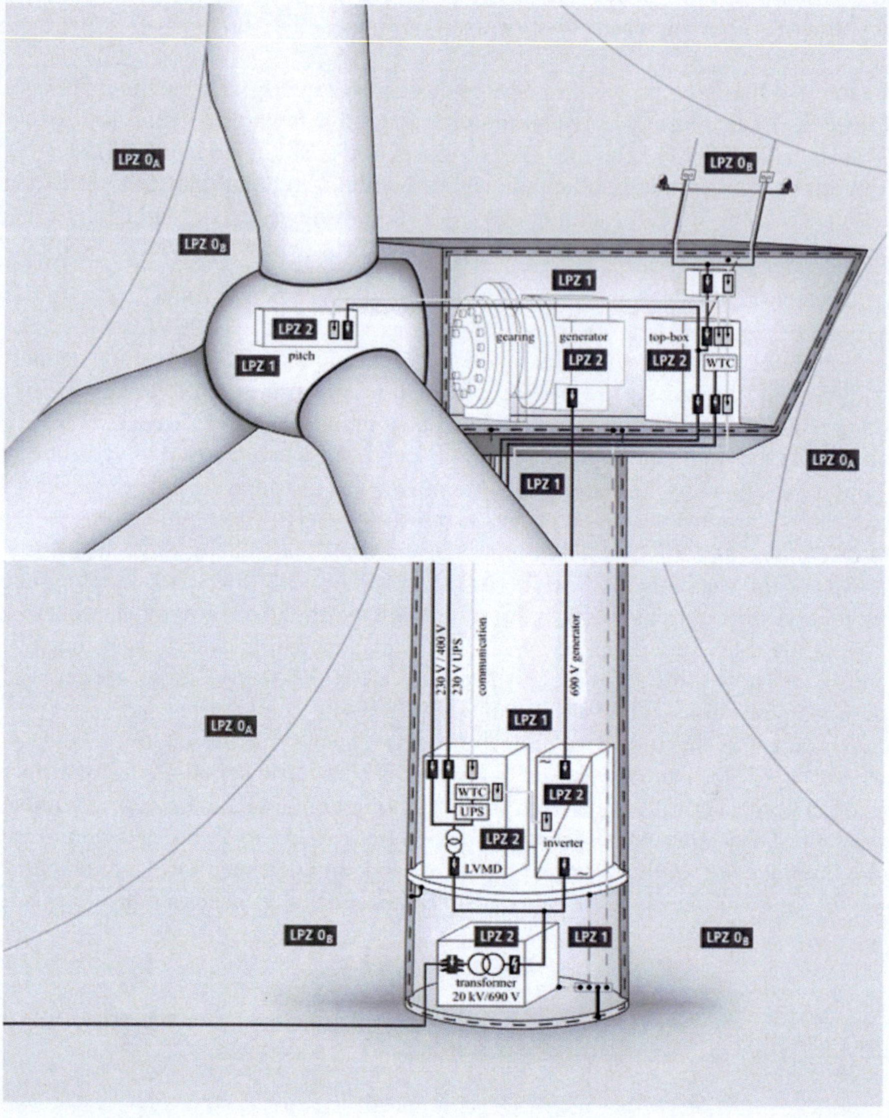

Fig. 28 Lightning and surge protection of a wind turbine [2] (WTC—wind turbine control, UPS—uninterruptible power supply, LVMD—low-voltage main distribution board)

must be included. SPD type 2 and/or type 3 must be installed at the transition from LPZ 0_B to LPZ 1 and at the transition from LPZ 1 to LPZ 2 and at all other internal zone transitions. The task of SPD type 2 and type 3 surge arresters is both to further reduce the residual interference of the upstream protective stages and to limit the overvoltages induced in the WT or generated there.

b. Protection of electrical power systems

The transformer of the WT can be accommodated in different places, in the separate switching house, in the tower base, in the tower or in the nacelle. In very large WT, for example, the unshielded 20 kV cable is fed into the tower base to the medium-voltage switchgear, consisting of a vacuum circuit breaker, a mechanically interlocked busbar disconnector, an outgoing earthing switch and a protective relay. The medium-voltage cables then run from the medium-voltage switchgear in the WT tower to the transformer, which can be located at the base of the tower or in the nacelle (Fig. 27). The transformer supplies the control cabinet in the tower base, the control cabinet in the nacelle and the pitch system in the hub with a TN-C system (L1, L2, L3, PEN conductor). The electrical equipment in the nacelle is supplied with a.c. low voltage from the control cabinet in the nacelle. All electrical equipment installed in the WT have a rated impulse withstand voltage corresponding to the rated voltage of the system. This means that the SPD to be installed must have at least the specified protection level, again in accordance with the rated voltage of the system. The SPD used to protect the 400/690 V supply, for example, must have a protection level of at least $U_P \leq 2.5$ kV and the SPD used to protect the 230/400 V supply, for example, must have a protection level of $U_P \leq 1.5$ kV to protect sensitive electrical and electronic equipment.

The protected area includes power and information technology equipment such as:

- power supply hub; pitch system in hub; signal cables nacelle—hub
- protection of the aviation obstacle lighting on the sensor mast in LPZ 0_B; protection at the respective zone transitions (LPZ 0_B to 1, LPZ 1 to 2)
- signal cable weather station and control cabinet in the nacelle; nacelle superstructures
- control cabinet in the nacelle; voltage supply
- protection of the generator, the transformer, the mains filter, and of the measuring equipment; the MV transformer supply is protected by medium-voltage arresters; these must be adapted to the medium-voltage grid to its grid configuration and voltage
- power supply of the control cabinet at the base of the tower (TN-C system)
- main supply TN system (SPD type 1 with high follow current limitation required)
- protection of the inverter on both sides of the inverter (mains and machine side) as well as on the generator; if double-fed asynchronous generators are used, an arrester combination with increased dielectric strength (continuous voltage up to 1000 V and surge currents up to 40 kA 8/20 μs) must be used on the rotor side
- signal cables in the control cabinet at the base of the tower.

c. Surge protection for information technology installations

Surge arresters for protecting electronic equipment in telecommunications and signal processing networks against indirect and direct effects of lightning strikes and other transient surges are described in accordance with [31] and installed at the zone boundaries according to the lightning protection zone concept [32]. In the case of arresters consisting of several stages, it must be ensured that the various protection stages are coordinated with each other. Information technology cables are often fed into the WT and the control cabinets are connected from the base of the tower to the nacelle via fiber optic cables. The cabling of the actuators and sensors from the control cabinets, on the other hand, is carried out using shielded copper cables. The fiber optic cables do not need to be equipped with SPD, as interference from an electromagnetic environment cannot occur, unless the fiber optic cable has a metal sheath, which must then be included in the equipotential bonding directly or via SPD.

In general, the following shielded signal lines must be connected:

- signal lines (4–20 mA interfaces), which lead from the sensors of the weather station into the control cabinet, come from the LPZ 0_B, run into the LPZ 2 and are best protected with combined arresters; shield grounding is carried out by means of shield connection terminals for permanent and low-impedance shield contacting of the protected and unprotected side of the SPD; if wind measuring devices (anemometers) are equipped with a heater, this must be protected with a suitable energy-coordinated combined arrester
- signal lines running between the nacelle and the pitch system in the hub; SPD suitable for high frequencies must be used
- signal lines for the pitch system, wiring of the signal lines depends on the sensors used, which may have different parameters depending on the manufacturer; wiring should be on both sides, in the pitch system and in the controller
- signal lines to the inverter; signal lines to the fire extinguishing system

d. Condition Monitoring

The subject of plant availability at WT is becoming increasingly important, especially for offshore wind farms. Condition monitoring of the lightning current and surge arresters for preload is important. Through the targeted use of condition monitoring, service calls can be planned and costs saved. In addition, SPD function monitoring with a potential-free contact is usually possible. The signal to replace the SPD in the next service interval is transmitted to the turbine controller via the telecommunication contact.

8 Specialities of Wind Turbines

i. Direct lightning strike in wind turbine

Since wind turbines are very high, they can trigger lightning when exposed to a thunderstorm electric field. In fact, a wind turbine could be a better trigger for lightning than a stationary tower of similar height [4]. In the presence of an electric background field, all pointed structures at earth level form corona, even at high towers corona discharges are generated (Fig. 29a). Corona discharges create a space charge area around the top, and this space charge can screen the top of the tower from an electric field. As the electric field generated by the storm cell slowly increases, the corona discharge increases and continues to screen the tip from the increasing strength of the electric field. So, to create a connecting upward leader from the tip, the field must grow very rapidly to overcome the screening effect of the corona. In a slow-growing field, the corona generated at the tip prevents the formation of a connecting upward discharge. In wind turbines, however, these space charges cannot accumulate at the tips because they are displaced by the rotation of the blades. When a rotor blade is parallel to the ground, the electric field at its tip is rather small and increases rapidly as the blade turns upwards towards the cloud (Fig. 29b). The electric field is greatest when the blade is perpendicular to the ground and away from the ground (Fig. 29c). The electric field at the tip of the rotor blade oscillates while the blade rotates in the background field. As the rotor blade rotates, the electric field at its tip increases rapidly without forming a significant amount of corona space charge. Without screening the tip from the electric field, the conditions for starting upward leaders in the electric background field are achieved that no stationary tower of similar height would have produced [4].

Wind turbines operating with a rotating rotor thus offer more favorable conditions for lightning strikes.

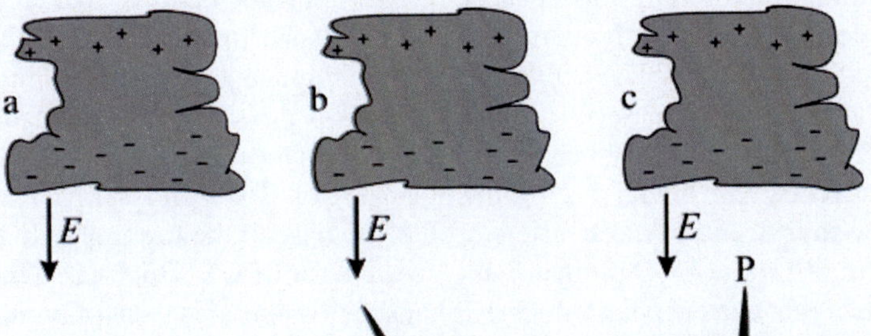

Fig. 29 Due to the constant electric background field, charges are accumulated at the upper head of the wind turbine (**a**). The electric field intensity at the tip of the horizontal rotor blade (P in **b**) increases during rotation and reaches its maximum when the blade is vertical (P in **c**) [4]

ii. Temporary overvoltages

In the event of faults in the power generation system, e.g. short circuits, temporarily high mains voltages can occur due to the continued presence of the propulsion, the rotation of the hub. The surge arresters in the area of the generator and the inverter must have a correspondingly high continuous voltage strength in order not to be damaged. This applies in particular to SPD type 2 on varistor basis without serial spark gaps which could be overloaded.

iii. Lightning caused cross overvoltages at tower top

Metal towers, towers with reinforcement or extensive coaxial metal elements of the tower structure have a very good magnetic shielding effect for electrical cable systems inside the tower. Nevertheless, high lightning surge voltages occur at the upper end of the tower between cables and the tower when lightning impulse currents are diverted via the tower. Therefore, surge protection measures at the top of the tower are absolutely necessary. Measures for lightning equipotential bonding and in particular SPD must therefore be used at the top and bottom of the tower to safely prevent dangerous sparking.

iv. Lightning current peak values at tower base

Due to large tower heights and long rotor blades, travelling wave effects occur with steep lightning impulse currents because the wind turbine behaves like an electrical line with comparatively high characteristic impedance (tower 150–250 Ω). At the base of the tower, this line is terminated with low impedance by the earthing system. This can lead to an increase in the current peak values at the base of the tower. However, this travelling wave behavior is only pronounced with very steep lightning impulse currents, with negative subsequent strokes or also with negative first strokes with a very short front. Surge arresters used at the base of the tower should therefore be powerful SPD type 1 lightning current arresters for LPL I, which can dynamically control the higher current peak values. Higher specific energies than for typically used type 1 SPD are not to be expected due to this effect, since no peak value increase occurs with slowly rising high-energy positive first lightning current.

v. Strike frequency and side flashes on wind turbines

According to the latest evaluations, there are usually more strikes on wind turbines than would be expected according to the height. At least 8–9 strikes per year per wind turbine instead of 2–3 strikes per year are to be expected. This refers to a single wind turbine, even if it is located in a smaller or larger wind farm.

With wind turbines as high structures, side flashes are not only possible, but more likely. Similar to the lightning protection of normal high buildings, it is also necessary to protect the areas above 80% of the hub height against side flashes at WT, i.e. to integrate air-termination devices and the connection to down conductors or to ensure that natural components are resistant to strikes. This applies in particular to the nacelle including existing weather measurement equipment, communication technology, etc. Strikes below 80% of the hub height occur with a significantly lower probability and with significantly smaller lightning current amplitudes, so that protective measures can usually be omitted there.

vi. Wind turbines with guyings

For very high wind turbines, with nacelle heights greater than approx. 200 m, 3 or 4 guy ropes (made of steel) are used for mechanical stabilization. Due to the rope cross-sections required to achieve mechanical strength, all partial lightning currents can be safely controlled. However, direct strikes into these guy ropes are also possible, although lower lightning current parameters occur. The effects at the root points of strike, in particular melting with severing of partial wires, are only permissible to the extent that no impermissible weakening or even tearing of the ropes is possible.

vii. Personal safety

It should be noted that the erection of WT may take several days and that persons are particularly exposed to the risks of lightning strikes during this time. In the documentation of the wind turbine, safe areas have to be indicated which must be visited in the event of a thunderstorm. In this context, the term "personal safety distance" was introduced. This minimum distance can be calculated according to the calculation of the separation distance according to [9]. If this minimum distance is maintained, uncontrolled flashovers to persons staying in the designated safety areas are to be prevented.

viii. Lightning current measuring devices in wind turbines

There are recommendations for the use of monitoring systems that provide information on lightning strikes into the WT. This enables the preventive maintenance of systems to be optimized and downtimes to be minimized. Such lightning current measuring devices are particularly important for lightning protection of rotor blades. Undetected lightning strikes in rotor blades and possibly associated serious consequential damage can thus be avoided. Therefore, such monitoring systems are not only capable of recording the relevant lightning current parameters, but also indicate which rotor blades have been struck directly by lightning. Continuing lightning currents have a special influence on the loading of the lightning protection of rotor blades and rotating components. Measuring systems should also be able to detect continuing currents without superimposed or subsequent pulse currents, so-called ICC-only lightning currents, since such lightning flashes of such current waveforms cannot be detected by lightning location systems.

c. **Protection of Railway systems**

9 Limitation of Railway Systems to be Considered

With regard to lightning protection and surge protection, the following section considers stations with railway installations and buildings, railway facilities, especially for passenger transport, electrical railway control technology and the rail network with track installations, overhead lines and signaling systems (railway infrastructure). In particular, electrified (electrically operated/powered) railways are considered [33, 47].

The lightning and surge protection discussed below primarily considers passenger traffic, including freight traffic.

10 Importance of Lightning Protection for Railway Facilities and Railway Systems

Railway station buildings, overhead lines, moving and stationary trains are often the highest facilities in the immediate vicinity and are therefore endangered by direct lightning strikes. In addition, overhead contact lines, rails and electrical signal lines represent long, large-scale loops into which high pulse voltages and pulse currents can be coupled by electromagnetic induction at close lightning strikes.

Lightning protection systems are required for immobile railway installations, as these are structures for which the necessity exists by law. This is because they are structural installations, such as railway stations, for a larger number of people. Last but not least, crowds of people are a particular problem with lightning strikes.

Railway networks are considered to be exposed to the effects of lightning, since many km to many 100 km of lines and thus many strikes occur each year (often routes through different regions with different lightning densities).

Rail-bound transport ensures mobility and has an important function as a sustainable means of transport for passengers and freight. The development of safe and highly available transport routes has given the railways a constantly growing significance. As a result, the railway infrastructure will be massively renewed and expanded over the next few decades. As the railway network stretches over large distances, its size and exposed location make it an ideal target for the effects of atmospheric discharges, e.g. in the form of direct lightning strikes. Thus, buildings, installations and electronic equipment of the railway are considerably endangered by lightning strikes and the resulting electromagnetic disturbances.

Damage is caused by direct lightning strikes in overhead contact lines, rails, masts and buildings (Fig. 28). This results in the following protection goals:

- personal protection
- fire protection

- protection against mechanical damage
- protection of power supply, radio systems and control and command technology (CCT)
- protection of electronics and digital interconnection systems
- ensuring plant availability.

A further threat are railway-specific overvoltages, e.g. caused by switching operations or continuous influencing voltages in adjacent line cables.

Railway stations are not just places where countless people pass by every day. They also offer attractive areas for shopping, eating and enjoying other services. Within these complex infrastructures, solutions ensure a trouble-free flow of electricity and data. These solutions meet the highest safety standards with tested systems for fire protection, lightning protection and surge protection. In buildings where many people spend a lot of time, three protection objectives must be achieved: Limiting the speed of fire, securing escape and rescue routes, maintaining the function of important safety and electrical systems.

No railway operator can afford a lightning-related technical device or system failure. Effective protection is guaranteed by correctly installed lightning protection systems. Damage is also frequently caused by electrical surges, which can be triggered by a nearby lightning strike or by switching in large electrical systems.

Thus lightning and surge protection should include protection for stations and railway buildings as well as for the railway network and signaling equipment.

i. Railway environment

The railway environment is dominated by the overhead structure, which forms a huge lightning antenna. In rural areas this overhead structure is a main target for lightning strikes (Fig. 30). An earthing cable on the masts ensures that the entire structure is at the same potential. Every third to fifth mast is connected to the traction return rail, the other rail is used for signaling. In d.c. traction areas the masts are isolated from the ground to avoid electrolysis, while in a.c. traction areas the masts are in contact with the ground. Sophisticated signaling and measurement systems are mounted on

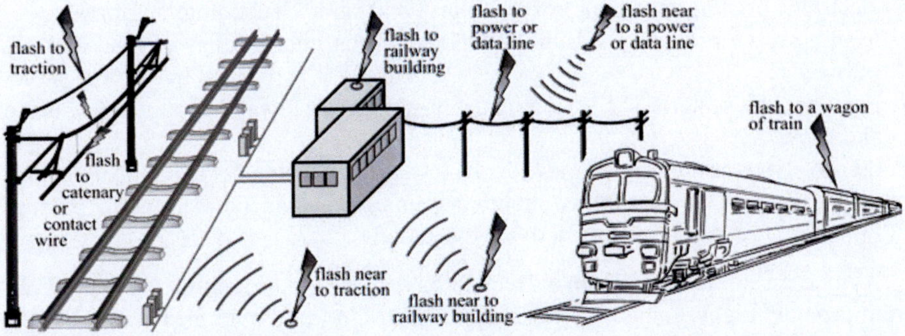

Fig. 30 Points of lightning strikes as sources of damage of railway systems [34]

or near the rail. These devices are exposed to lightning strikes in the rail which are absorbed by the overhead construction. Sensors on the rail are connected via cables to track-side measuring systems which are connected to earth. This explains why rail-mounted devices are not only exposed to induced overvoltages, but also to conducted direct lightning surges [36].

ii. Minimize downtime and operational disturbances with lightning protection

For railway technology to run reliably, many highly sensitive electrical and electronic systems must function reliably. However, this continuous availability of the systems is at risk. Lightning strikes or electromagnetic interferences damage or destroy cables, control components, assemblies or computer systems and in most cases lead to operational faults and time-consuming troubleshooting. This means delays in rail traffic and high follow-up costs.

Causes of damage, losses and rail operation downtimes are:

- Direct lightning strikes: Lightning strikes the overhead contact line, the rail or a mast. Malfunctions or rail operation downtimes are usually the result.
- Indirect lightning strikes: Lightning strikes an adjacent building or the ground. Now lightning surges spread via the cables or are inductively coupled in, damaging or destroying unprotected electronic components.
- Electromagnetic interferences: Overvoltages can occur if different systems, e.g. a railway traffic light bridge, the high-voltage road and the railway overhead line, influence each other due to their spatial proximity.
- Causes within the railway system: A further danger is switching operations or triggering fuses. They can also generate overvoltages and cause damage.

With a consistent, optimized concept for lightning and surge protection, cost-intensive malfunctions can be reduced and system downtimes minimized.

iii. Specific railway facilities in the lightning and surge protection concept

The following railway facilities are to be protected by an integrated lightning protection zone concept:

- electronic and digital signal boxes; control and safety technology systems
- level crossing protection systems ensure safety at intersections of road and rail traffic
- point heating systems should guarantee trouble-free railway operation even in ice and snow
- d.c. rail systems are used worldwide in metropolitan area in local public transport. d.c. rail systems such as trams, suburban trains and underground trains are gaining in importance and are being actively expanded.

As modern control and command technology (CCT) is increasingly digitalized and equipped with highly sensitive electronics, it is now more susceptible to faults than in the past. The consequences of system failures due to lightning strikes or overvoltages can lead to delays in rail traffic, often associated with high costs. Availability, even during thunderstorms, can be increased with a carefully planned lightning and surge protection concept.

iv. Measuring systems and signaling elements for railway

Various rail-bound measuring systems are used to monitor the condition of the wagon fleet and the undesirable loads in the rail structure. Some of these systems are hot bearing detectors, hot brake detectors, wheel profile measurement system, weighing or wheel impact measurement, inclined bogie detector, track measurement, vehicle identification system, weighbridge, etc. The following measuring elements are important and must be available for an effective signaling system, track circuits, axle counters, turnout detection and associated power supplies.

The environment for railway signals consists of a variety of different types of devices with complex connections distributed along a railway line and located in exposed locations. The railway line and the structures around it tend to receive lightning strikes [35].

In the vicinity of railway signaling, lightning and/or surges occur, enter or are passed into the signaling system by one or more of the following devices and processes:

- power supply points, e.g. 240 V a.c. or 120 V a.c., derived from overhead line systems, e.g. 33 kV/11 kV/2.2 kV transmission/distribution systems
- e.g. 240 V a.c. or 120 V a.c. power supply air cables (in rural areas)
- e.g. 240 V a.c. or 120 V a.c. power supply cable
- overhead line 1500 V d.c. contact wire structure in the electrified area
- rail and track connections to signaling sites
- cables for signaling control and display circuits (connected to field devices)
- communication lines
- communication equipment on high masts
- consequence of earth potential rise (EPR)
- induction into the power supply, communication, control wiring, etc.

Typically, the wiring standards and arrangements allow the transmission of very high overvoltages and currents through the signaling system.

In the environment of railway signaling there are various earthing systems. These earthing systems are: signal ground, electrical 240 V earth, electrical high-voltage ground, public/municipal 240 V multiple earth neutral (MEN) ground and communication ground. (TN 030 2018).

11 Lightning and Surge Protection for Railway Facilities and Railway Systems

The progressive automation of operational processes has already covered entire railway systems. There are high demands on the availability and reliability of the railway systems used. System failures, e.g. at an electronic interlocking or parts thereof, affect the railways not only locally, but also nationally. Since the operation of the railways must also be guaranteed during thunderstorms, both the operator and the industry of the rail suppliers are asked to deal with transient surges caused by switching operations and lightning discharges [37].

- Railway buildings: Lightning protection level (LPL) III generally applies to buildings frequented by the public and LPL II to buildings with extensive information technology equipment. When selecting a lower lightning protection level or dispensing with a lightning protection system, risk assessment must be carried out and documented.
- CCT systems: The highest lightning protection level I generally applies. Surge protection in railway technology ensures a safe and trouble-free infrastructure.

i. External lightning protection for railway systems

External lightning protection systems with air terminations, down conductors and earthing system shall be planned and installed. Separation distances must be calculated, documented and observed individually. Preferably foundation earth electrodes shall be erected. If no foundations are laid, earth electrodes shall be designed as rod earth electrodes or ring earth electrodes. In existing installations without foundation earth electrodes with CCT systems installed, ring earth electrodes shall be retrofitted or, if this is not possible, rod earth electrodes. Earthing systems of neighboring buildings of the railway should be included in the earthing concept. For building distances of up to 3 m, the earthing systems must be connected to each other (Fig. 31). At distances of up to 20 m, influences must be demonstrably considered. The total earthing resistance should not exceed 10 Ω. Railway tracks are not suitable as lightning protection earthing systems. However, the tracks must be integrated into the equipotential bonding system in an appropriate manner.

For new installations, the reinforcement of the entire building must be included in the equipotential bonding. For buildings in reinforced concrete construction, the reinforcement should preferably be used as down conductor system. The prerequisite for this is that the reinforcement, including its joints and connections, is capable of carrying lightning currents. Concrete module buildings without external lightning protection must have lightning current carrying reinforcements included in the equipotential bonding. The lightning current carrying capacity of foundations, building walls and ceilings must be confirmed.

External lightning protection for components of the outdoor railway systems is generally not required. If possible, the protected space of overhead catenary or contact wires or bridges should be used (Fig. 32).

In the outdoor area, the aim is to lay the signal cables and install signal systems in the protected space of the overhead lines or the traction and/or conductor rails. Lightning strikes in traction circuits or building (Fig. 30) lead to voltage craters, which can cause insulation damage to cables laid in their vicinity. Minimum distances between signal cables and earthing systems (mast foundations, building earth electrodes, etc.) must therefore be observed. Signaling equipment such as interlocking buildings,

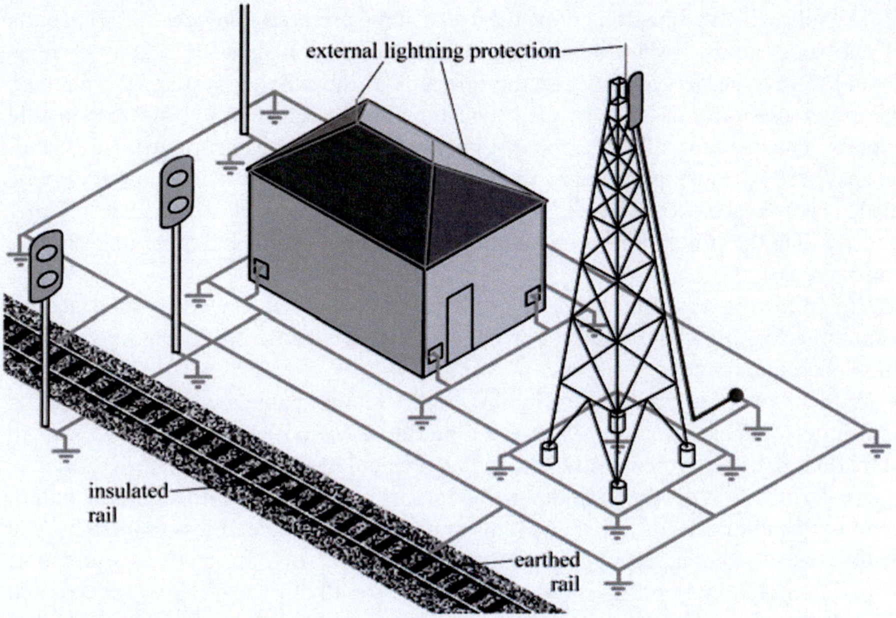

Fig. 31 Railway earthing and external lightning protection

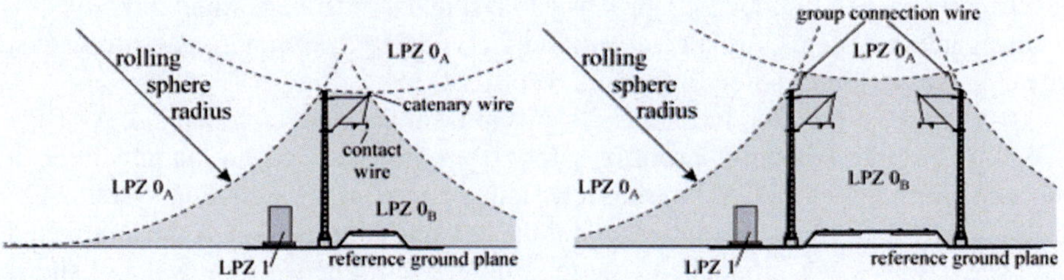

Fig. 32 Lightning protection zones of traction systems [34]

switchgear houses, signals or cable systems must always be installed at less exposed locations.

ii. Internal lightning protection for railway systems

A comprehensive protection of buildings and electrical or electronic systems against the effects of lightning electromagnetic pulses (LEMP) can be achieved by LEMP protection measures system (LPMS). It consists of the individual combination of

- earthing and equipotential bonding measures
- lightning protection measures for magnetic and spatial shielding
- cable routing and shielding measures as well as
- measures for energetically coordinated protection by SPD [38].

The following applications show the use of surge protective devices (SPD) in parts of railway systems. It should be noted that protection concepts for railway systems should always be agreed between the operator of the railway system, the planner, the system supplier as well as the client (general contractor) and the responsible experts. The question of the scope of lightning and surge protection measures shall be answered by a risk analysis in accordance with [8]. The decision cannot only be made according to economic aspects. The protection of persons always has a higher priority than the protection of installations or material assets and must therefore be considered.

The following descriptions mainly refer to future signaling systems, such as electronic interlockings (EI), railroad crossings (RRX), electrically located points and all telecommunications systems.

With a view to improving availability and reducing damage caused by lightning and surge voltages, signaling systems will in future always be designed in LPL I. An individual risk assessment and the associated calculations are no longer required.

Air-termination devices and down conductors on buildings with signaling equipment with a floor area of less than 10 m^2 or an enclosed space of less than 25 m^3 can be dispensed with, but not with other measures such as surge protection by means of SPD. Equipotential bonding should be provided for all electrical railway systems. It is always necessary to set up a lightning equipotential bonding system covering all trades. All active electrical conductors should preferably be included in the equipotential bonding at the entrance to the building via SPD indirect with low impedance. Electrically passive conductors, such as cable shields, must be earthed directly.

In signaling systems, the special requirements arising from functional safety must be taken into account (safety integrity level SIL 4).

Overvoltage protection in indoor systems must always be implemented. All SPD may only be used if there is proof of safety from the CCT system manufacturer. It must be ensured that the SPD has no retroactive effect on the signaling system. The safety verification must take into account the behavior of this SPD in the unaffected case as well as in the influenced case (e.g. inductive influence by travel and short-circuit currents in the overhead line and the traction return current paths or lightning effects) with CCT systems free of earth faults or earth-faulted systems and also with failures or faults of the SPD. SPD must be installed in circuits galvanically connected to outdoor components, e.g. signals, points or axle counters.

The implementation of EMC measures shall be taken into account. In practice, for shielding measures, existing metal facades and reinforcements of walls, floors and ceilings on or in the building are often used to form shielding cages, which are ideally combined to form shielding cages, e.g. in the case of control and command technology concrete buildings. Cable shields must be earthed on both sides. For floor openings of shielded cables, their shields must be connected to the meshed equipotential bonding. Current-carrying cable shields in outdoor installations must be earthed on both sides. Otherwise, it must be noted that continuous influencing voltages of up to 250 V and short-term influences of up to 1500 V must be expected. To reduce induction loops, cabling in interlocking buildings should be separated into cable ducts according to cable categories. All CCT external cabling must be introduced into the building at a central point. Other trades may have their own central entrances.

Main equipotential bonding and earthing bars, e.g. MEB, shall be located close to lightning protection zone boundaries. Equipotential bonding rails shall be accessible and shall be located in close proximity to sub-distribution boards or cable termination racks. Suitable earthing fixed points shall be available. The track system shall only be connected to MEB at one single position. For new installations, TN-C-S systems with insulated PEN conductors and central earth connection, TT systems or IT systems shall be installed. TN-C systems are not suitable.

a. Standard and backup power supply

For standard 50 Hz power supply systems, such as electronic interlockings, lightning protection is provided by combined lightning current and surge arresters SPD Type 1. This SPD is capable of discharging lightning currents up to 100 kA (10/350 μs) several times. Even at high short-circuit currents of up to 50 kArms, the occurring mains follow currents are significantly reduced. Selectivity to upstream overcurrent protection devices (fuse) up to 20 A gL/gG is achieved. This enables maximum system availability, which ensures the functionality of the electrical power supply system for the operator of the railway system.

In various railway system applications, the traction power supply system is increasingly being used as a supply system for other systems, such as signaling systems or even as a backup power supply. The overhead line voltage up to 25 kVrms 50 Hz is converted into low voltage of e.g. 2 × 230 V by means of transformers. The short-circuit currents vary between 3 and 20 kArms depending on the location. Since there are different loads and thus different powers of the lightning current arresters used, especially in railway systems with operating frequencies of 16.7 Hz, the applicability of protective devices should be documented in the test report of an independent test laboratory.

b. Turnout control for local public transport railways

The electronic turnout control is used in local traffic railways. It safely controls turnouts for various drive technologies in signaling systems and turnout areas. These turnout controls can be flexibly adapted to the railway operator's requirements and comply with safety integrity level SIL 3 according to [39]. In addition to the safety level with signal control, turnout control and monitoring system, the evaluation unit of the track switching device and, if necessary, an isolating transformer unit for the power supply are installed in an external cabinet. The surge protection devices are SPD of category D1 according to [15]. They are capable of safely discharging lightning currents up to 5 kA 10/350 μs. In addition, the SPD selected for this application offer energetically coordinated surge protection for the parts of the system using combined lightning current and surge arresters. Both lightning currents and currents inductively coupled into the periphery, e.g. in turnout operating devices, signal transmitters, key switches, are safely controlled.

c. Railway power supply systems—application of surge arresters for medium-voltage systems

Electrified alternating current railways form a dense network that leads to a large equivalent collection area for direct and indirect lightning strikes. Surge arresters for medium-voltage systems are used to minimize the effects of such sources of damage on the traction current network and the associated systems.

In order to select the correct MV surge arrester for the application, the energy absorption capacity, the maximum continuous voltage in the rail network and the ambient conditions on site must be taken into account [40]. A class 3 surge arrester used on a supply transformer in a 25 kVrms/50 Hz traction power supply system and also used every 10–20 km, for example on high-speed lines, provides surge protection. Classification in line discharge class 3 offers sufficient energy absorption capacity. This allows the surge arresters to withstand local loads without damage over a long period of time. When further designing the MV surge arresters, it must be taken into account that the maximum non-permanent voltage U_{max2} occurring in the railway system is defined, which must only occur for a limited period of time and not longer than five minutes. The continuous voltage U_c of the surge arrester must be selected in accordance with $U_c \geq U_{max2}$.

Arresters for d.c. rail systems are also based on the above selection criteria. MV surge arresters for d.c. rail networks can effectively protect them against the effects of lightning strikes [41, 42]. An example for 3 kV d.c. traction is shown in Fig. 33.

Tested surge arresters are to be used in the low and high voltage range.

d. Track circuits and open earthing of railways

In alternating current (a.c.) circuits, track circuits are used for signaling occupied or free track sections. The insulation of one or both rails is required (Figs. 33 and 34). In the event of a fault, the insulation of the tracks of a track section creates a hazard

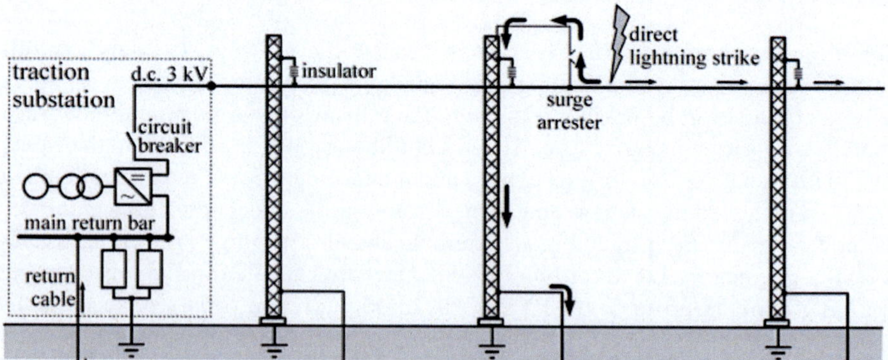

Fig. 33 Lightning current distribution in the case of connection to one rail [34]

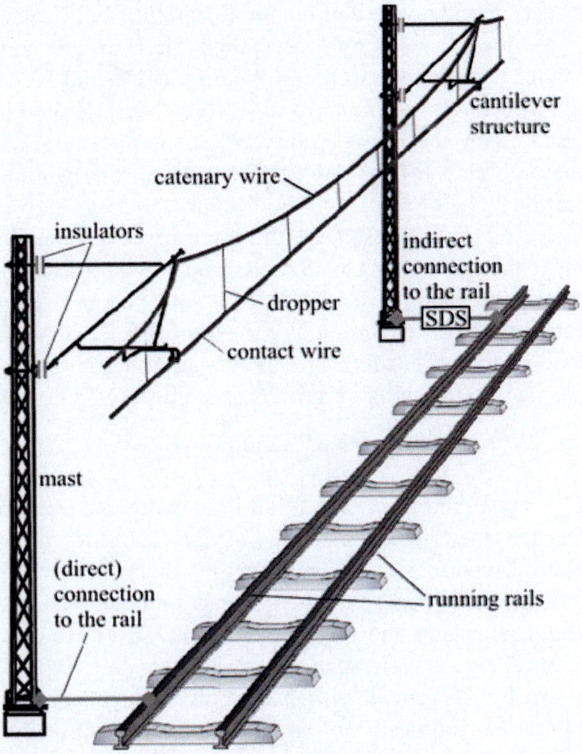

Fig. 34 Lightning protection zones of traction systems [34, 37]

potential for persons through indirect contact. According to [43, 44], the busbars with alternating current circuits are generally earthed. For railway earthing systems, [43, 44] must be observed. It defines the requirements for protective measures for electrical safety in fixed installations connected to a.c. and d.c. railways. However, [43, 44] also contains the requirements for all installations which are threatened by power supply systems of electric railways. It includes the requirements for the protection of persons and installations.

The protection against accidental contact must be ensured by rail earthing (Fig. 34). The track earthing implies the connection between conductive parts and the track earthing. The rails represent the earth. They are used as reverse circuits and are deliberately connected to earth in alternating current circuits. The orbital earth contains all conductive parts connected to it. A distinction is made between direct railway earthing, direct connection between conductive parts and railway earthing, open railway earthing and indirect connection of conductive parts with railway earthing by means of voltage limiters or short-circuiters, e.g. track circuits insulated on one side, direct current circuits with insulated tracks.

The protective devices used must ensure the protection of persons and property both in the event of an interruption of the overhead contact line and in the event of a lightning strike. Voltage limiting devices (SDS) are used for this purpose (Fig. 34).

These provide safe equipotential bonding both under lightning current load and under loads caused by a permanent short-circuit connection in the case of short-circuit currents due to the high-current-resistant welding of the SDS electrodes. If the SDS is short-circuited and its electrodes are therefore welded, this SDS must be replaced by the railway operator's maintenance personnel. Experience has shown that lightning strikes, unlike malfunctions, are the most common cause of electrode welding.

Based on this result, lightning current resistant SDS are required, which avoid a permanent welding of the electrodes when discharging lightning currents. With lightning current resistant SDS, a safe potential equalization is only temporarily established for the duration of the lightning flash. If, however, a short-circuit current load occurs, the safety of persons and material assets, as with an SDS that is not resistant to lightning current, is guaranteed by safe welding of the electrodes [37].

e. EMV of railways and lightning protection

The signal and train protection systems of modern railways are highly automated. Digital electronics can realize a high degree of intelligence and flexibility in the control functions and is increasingly used in railway systems. However, due to the low destruction energies associated with modern electronics, there are more concerns about reliability and safety. Before the advent of modern electronics, key components in signal and control networks were electromechanical and largely insensitive to electromagnetic disturbances such as transient events caused by lightning strikes. Therefore, the design of signal and control networks at that time was not always carried out according to the strict rules of electromagnetic compatibility. Later, individual subsystems in the network were replaced by units with modern electronics and new functions such as automatic train stop (ATS), automatic train control (ATC) etc. were added. These developments made the signal and control system more susceptible to lightning transients.

During thunderstorms, the railway systems are exposed to electromagnetic pulses (LEMP), either by direct lightning strikes on any part of the overhead contact line or by the induced voltages generated in the overhead contact line by nearby lightning strikes. A part of the lightning energy reaches the electronics and causes disturbances and destruction.

The main objective is to provide effective lightning protection to avoid or minimize traffic congestion and thunderstorm delays. The system to be protected should include the rails and overhead contact line system above the conductive earth, track circuits, amplifier and auto transformers, buried communication cables near the tracks, etc. The system to be protected should also include the tracks and overhead contact line system above the conductive earth.

The following system components must be included in the lightning equipotential bonding indirectly via isolating spark gaps:

- systems with cathodic corrosion protection and stray current protection measures
- earthing systems of high voltage installations above 1 kV, if impermissibly high earthing voltages can be carried off

- railway earth for alternating current and direct current railways (railway tracks may usually only be connected after approval).

f. Low voltage SPD for a.c. and d.c.

When it comes to dimensioning common a.c. systems, there is a wide range of SPD on offer. These can be TN or TT systems, fed networks from the overhead railway line, emergency standby power system with mobile diesel aggregates or special IT systems, e.g. for transmission lines with different voltage levels. As a rule, they are easy to plan on the basis of the manufacturer's documentation. Basic aspects must be observed when dimensioning SPD in a.c. systems. In the railway sector, further aspects must be taken into account. Above all, the railway earthing affected by reverse current and the resulting influences should be included in the planning when selecting SPD. A long-term influence of 250 V a.c. and a short-term influence of 1500 V a.c. should be assumed. Thus in railway networks a strict separation of the system against the railway earth should be established, either by spark gaps or by SPD with gas arresters connected to earth. The use of such SPD also ensures that no leakage currents occur between the active conductors.

In the railway sector there have always been direct current (d.c.) systems with typical voltage levels of 48/60 V d.c. for telecommunications and CCT and 36/48 V d.c. for railroad crossing protection systems. However, new railway projects require new system architectures. In future, the voltages used will also be implemented as d.c. systems. Here, CCT 400 V d.c. are the most important. Power buses, which are led from the track field concentrator into the field to the field element connection box, or also the CCT system 48 V d.c. level (own requirements). When planning d.c. systems, existing a.c. SPD are often used. This is also possible in many areas. In certain cases, however, the framework conditions should be examined more closely. Particularly due to the photovoltaic boom of recent years and the recognition by many manufacturers that special technologies or circuit designs are necessary for d.c. systems, it is advisable to discuss the applications in detail with experts or companies.

g. SPD for telecommunications and control and command technology

In addition to safeguarding the power supply side, the areas of telecommunications and control and command technology (CCT) must also be taken into account. In general, the selection of information technology SPD depends on considerations similar to those of power technology SPD:

- lightning protection zones at the installation site and their arrangement
- compliance with product- or application-specific standards
- adaptation to ambient conditions, installation conditions
- mounting type and environment.

In contrast to the selection of SPD in power systems, which usually have uniform conditions with regard to voltage and frequency, there are different types of signals to be transmitted in control and command technology systems with regard to

- voltage (e.g. 36, 48, 60 V)
- current (e.g. 0–20, 4–20 mA)
- signal reference (balanced, unbalanced)
- frequency (d.c., Low Frequency, High Frequency)
- signal type (analog, digital).

Each of these electrical quantities of the useful signal to be transmitted can contain the information actually to be transmitted. For this reason, the useful signal must not be inadmissibly influenced by the use of SPD, e.g. in control and command technology systems. Therefore, some aspects have to be considered for the selection, which are roughly described in the following [45].

CCT systems are supplied from the local supply or from the internal network of the railway operator or from stationary or mobile emergency power supply systems. Lightning and overvoltage protection must be effective for all operating states. Even in systems without external lightning protection, protection against partial lightning currents must be implemented in the main power supply for reasons of availability. When feeding in with or without isolating transformer, SPD type 1 must be used directly at the building entrance. According to the lightning protection zone concept, sub-distributions must always be equipped with SPD. The lightning protection zone boundary between indoor and outdoor installations is formed by the cable termination frame. Dimensioning of the SPD must take into account, for example, the system voltage, additional influencing voltages or short-term influencing voltages from the railway system. SPD in CCT systems must always be built up from series connection of varistor (MOV) with gas discharge tube (GDT) between each signal wire and earth (Fig. 34), in particular to ensure absence from feedback. Reliability against feedback and protection of the system must be proven (Fig. 35).

These SPD must have a rated discharge surge current of 3 kA 8/20 μs in accordance with category C2. The SPD protection level (<1.5 kV) must be well below the insulation strength of the equipment and the insulation strength between the signal wires and earth.

h. Protection of mobile radio systems for railways and SPD for antenna systems

SPDs for antenna cables differ in particular according to their suitability for coaxial, symmetrical or waveguide systems, depending on the physical design of the antenna

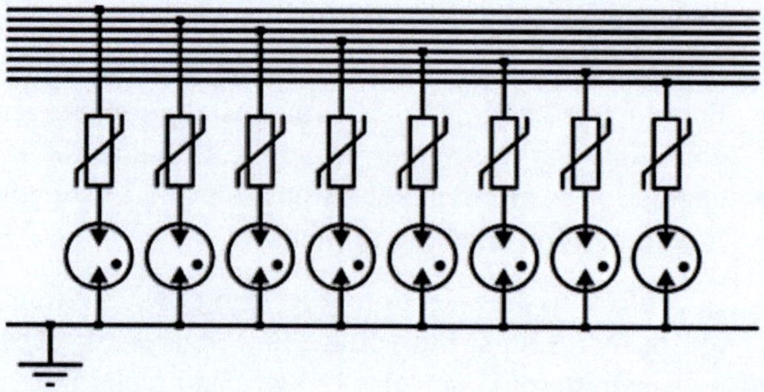

Fig. 35 Each signal core with series connection of MOV and GDT wired

cable. In coaxial and waveguide systems, the outer conductor can usually be connected directly to the equipotential bonding.

Remote radio head (RRH) technology is not only used for commercially used mobile radio applications. This technology is also used in digital authority radio (BOS) systems, e.g. in police and rescue services or in railway communications, where high reliability and system availability are top priorities.

i. Protection for data communication systems

Data communication systems are available for various railway infrastructures, whether for high-speed trains or local public transport. They usually consist of an antenna installed on a mast, a power source and a modem. These devices are essential to ensure the safety of rail transport. It is therefore necessary to make every effort to avoid malfunctions of these systems. The risk of direct and indirect lightning damage to installations must be taken into account. Lightning can strike antennas directly, but can also affect other external elements, an overhead line, a traction substation, etc., causing surges and induced partial lightning currents.

To protect against direct lightning effects, the installation of lightning air-termination rods together with a metallic lightning down conductor, a rod earth electrode as a simplified earthing system and an equipotential bonding bar is proposed. Under certain conditions defined in the international standard [9], the metallic mast can serve as a natural down conductor. It will also be necessary to establish equipotential bonding with the metal parts of the system (Fig. 31).

To protect against indirect lightning strikes, the installation of coaxial SPD to protect coaxial cables and the installation of a SPD type 1 + 2 to protect power lines is recommended. Grounding kits are also used for the coaxial cables.

j. Influence on electrical railway systems in the event of lightning strike

The railway systems are usually struck by lightning on the overhead line (Fig. 30). Statistical studies have shown that the number of lightning strikes on a substation is very low compared to overhead contact lines. Normally, a lightning strike on a contact line leads to very local damage. And usually, for economic reasons, only important safety devices and connection terminals to substations are protected. In view of this, the choice of surge arrester (SA), technology, design, installation and location are important to minimize incoming surges in the substation. But this surge arrester may not be sufficient to protect the entire substation, and more SA tuned to the first one are needed. Surge protective devices (SPD) on power lines and communication/control lines are also required. Response shows that telecommunications and signaling systems are the most frequently destroyed.

k. Earthing the computers

A common problem exists with all measurement or information systems that use computers to perform data analysis and other functions. Typically, computer enclosures are grounded via the power cord and computers' 0 V (reference line/PE) is also grounded. This situation usually violates the principle of making the measuring/information system float to protect it from external lightning strikes. The only way to overcome this dilemma is to feed the computer via an isolating transformer and separate the computer frame from the system cabinet in which it is mounted. Electrical connections to other devices will again lead to an earthing problem, for which a fiber optic connection is proposed as the solution. The key word is the consideration of the overall system and the search for a holistic solution.

12 Specialities of Railway Facilities and Railway Systems

i. Lightning protection of persons

During thunderstorms, platforms, stops and stations offer protection for commuters and travelers. Earthing of roofed systems is particularly important. If the lightning finds a way into the ground in this area, it can lead to increased touch and step voltages and thus to fatal injuries. Other faults, such as a broken overhead contact line, can also result in inadmissibly high touch voltages and thus endanger people.

In context with external lightning protection, lightning-induced touch and step voltages in the station area must be minimized by suitable measures such as standing surface insulation or potential control.

It must be taken into account that panic in crowds is possible with direct impact in station buildings.

a. Protection of persons near overhead wires

In the event of a lightning strike into an overhead line, the lightning current flows through the various masts closest to the point of strike into the ground. The risks at these masts are the same as with lightning down conductors, touch voltages and step voltages. It is generally accepted that people within a radius of 3 m of the masts are in a dangerous situation without protective measures. Although the protection of technical personnel can be considered through technical or organizational measures (thunderstorm warning and lightning information systems), it is difficult to consider evacuating/clearing a platform during thunderstorms. Studies have helped to define the risks, their probability of occurrence and the personal protection measures to be taken for masts. These are usually insulation of the line connecting the overhead contact line and possibly its traction SPD, and equipotential bonding. It can also be an insulation of the earthing system [46].

Rails and overhead lines can introduce partial lightning currents into station areas.

b. Passengers in railway wagons

Persons in railway wagons, especially passenger wagons, are well protected against lightning (Fig. 27).

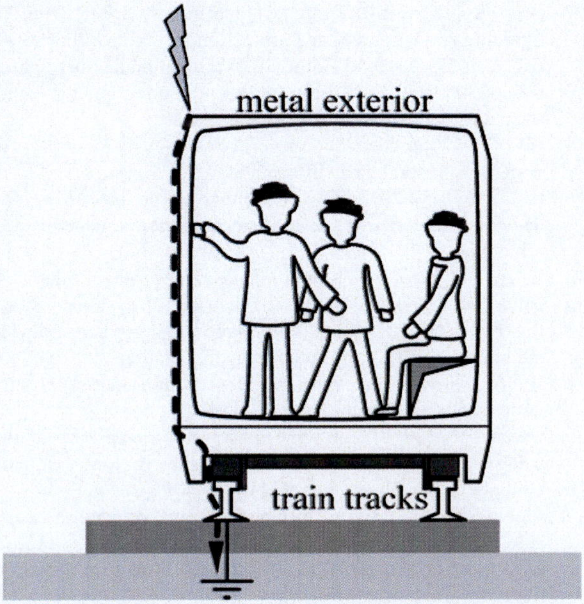

Fig. 36 Railway passenger wagon during lightning strike

The metal shell of a train forms a so-called "Faraday cage" which blocks electric fields and lightning currents in this shell and protects passengers from lightning strikes, currents and high potential differences. When lightning strikes, this cage conducts the electrical current through the outer shell of the train, not through the cabin, and lightning current flows through the steel wheels to the steel rail. The train is earthed through the track (Fig. 36).

When lightning strikes a train, the electrical charge in the metal surface of the wagon flows to the tracks and thus to the ground. Since the outside conducts the lightning current outside the enclosed space and no current flows through the interior, the passengers inside are safe.

References

1. Häberlin H (2012) Photovoltaics—system design and practice, Wiley, Ltd, Chichester
2. Dehn+Soehne—Lightning Protection Guide (2007) 2nd updated edition, DEHN+SÖHNE GmbH + Co.KG., Neumarkt, Germany
3. International Energy Agency, Electricity information (2018) IEA statistics, Paris, OECD Publishing, 2018; World Energy Outlook 2018, November 2018, www.iea.org/weo
4. Cooray V (2015) An introduction to lightning. Springer, Dordrecht, Netherlands
5. Cooray V (ed) (2010) Lightning protection, IET power and energy series 58, Stevenage, United Kingdom
6. DIN EN 62305-3 VDE 0185-305-3 Beiblatt 5:2014-02: Blitzschutz – Teil 3: Schutz von baulichen Anlagen und Personen – Beiblatt 5: Blitz- und Überspannungsschutz für PV-Stromversorgungssysteme

7. IEC TR63227:2018: Draft technical report (DTR) from IEC TC 82:37A/334/INF:2019-01: lightning and surge voltage protection for photovoltaic (PV) power supply systems
8. IEC 62305-2:2010-12 Edition 2.0: protection against lightning—part 2: risk management
9. IEC 62305-3:2010-12 Edition 2.0: protection against lightning—part 3: physical damage to structures and life hazard
10. IEC TR 62066:2002-06: surge overvoltages and surge protection in low-voltage a.c. power systems—general basic information
11. IEC 61643-32:2017-09: low-voltage surge protective devices—part 32: Surge protective devices connected to the d.c. side of photovoltaic installations—selection and application principles
12. IEC 61643-12:2008-11: low-voltage surge protective devices—part 12: surge protective devices connected to low-voltage power distribution systems—selection and application principles
13. IEC 61643-31:2018-01: Low-voltage surge protective devices—part 31: requirements and test methods for SPDs for photovoltaic installations
14. IEC 62305-4:2010-12 Edition 2.0: protection against lightning—part 4: electrical and electronic systems within structures
15. Owens BN (2019) The wind power story, a century of innovation that reshaped the global energy landscape, IEEE Press. John Wiley & Sons Inc., Hoboken, New Jersey
16. Global Energy Wind Council, http://www.gwec.net
17. McNiff B (2002) Wind turbine lightning protection project: 1999–2001, National Renewable Energy Laboratory (NREL), NREL/SR-500-31115, Cole Boulevard, Golden Colorado USA.https://doi.org/10.2172/15000382, https://www.osti.gov/biblio/15000382-wind-turbine-lightning-protection-project
18. Earnest J, Rachel S (2019) Wind power technology, 3rd edn. PHI Learning Private Limited, Delhi, Delhi
19. Hau E (2016) Windkraftanlagen – Grundlagen, Technik, Einsatz, Wirtschaftlichkeit, 6. Auflage, Springer Vieweg, Berlin, Deutschland
20. Breeze P (2019) Power generation technologies, 3rd edn. Newnes, Elsevier, Oxford, United Kingdom, Cambridge, MA, United States
21. Klinger F (2012) State of the art and new technologies of direct drive wind turbines, IRENEC, Istanbul
22. Birkl J, Diendorfer G, Thern St, Shulzhenko E, Kolb J, Rock M (2016) Initial investigation of influence of wind farms to lightning events. In: 33rd international conference on lightning protection, 25–30 September, Estoril
23. Birkl J, Shulzhenko E, Heidler F, Diendorfer G (2017) Measuring lightning currents on wind turbines. In: 4th international symposium on winter lightning, ISWL 2017, Joetsu, Niigata-ken
24. IEC 62305-1:2010-12 Edition 2.0: protection against lightning—part 1: general principles
25. IEC 61400-24:2019-07 Edition 2.0: Wind energy generation systems—part 24: lightning protection
26. IEC 62858:2019-0 Edition 2.0: lightning density based on lightning location systems (LLS)—general principles
27. Germanischer Llyod (1993) Vorschriften und Richtlinien IV, Teil I Windenergie, Richtlinien für die Zertifizierung von Windkraftanlagen; Germanischer Llyod 2010, Vorschriften und Richtlinien IV – Teil 1: Richtlinie für die Zertifizierung von Windenergieanlagen
28. Stiebler M (2008) Wind energy systems for electric power generation, springer series in green energy and technology, Springer, Berlin, Heidelberg
29. IEC 61400-1:2019-02: Edition 4.0: Wind energy generation systems—part 1: design requirements
30. Carriveau R (2011) Fundamental and advanced topics in wind power, InTech, Rijeka, Croatia, Chap 4: Madsen, S. F., Verification of Lightning Protection Measures, pp 65–88
31. IEC 61643-21:2000-09: low voltage surge protective devices—part 21: surge protective devices connected to telecommunications and signalling networks—performance requirements and testing methods

32. IEC 61643-22:2015-06: low-voltage surge protective devices—part 22: surge protective devices connected to telecommunications and signalling networks—selection and application principles
33. Code of Practice for Railway Protection (2004) October 2004 Edition, Development and building control department, land transport authority
34. Zielenkiewicz M, Maksimowicz T, Burak-Romanowski R The protection of DC railway traction power supply systems against direct lightning strike. In: 34th international conference on lightning protection (ICLP 2018), Rzeszow, Poland, September 2–7, 2018
35. NSW Transport RailCorp: SPG 0712, Lightning and surge protection requirements, engineering specification—signals—construction specification, Issued 23 August 2012, Technical Note—TN 030: Update to SPG 0712 Lightning and Surge Protection Requirements, Version 1.4—partly superseded, November 2018, Section 5.8, Appendix C, D, and F, https://www.transport.nsw.gov.au/system/files/media/asa_standards/2018/spg-0712.pdf
36. Theethayi N, Mazloom Z, Thottappillil R, Lindeberg P-A, Schütte T (2007) Lightning interaction with the Swedish railway network, railway engineering conference
37. Pusch H (2015) Lightning and surge protection for railway systems, e&i elektrotechnik und informationstechnik, praxis+wissen. http://media.klinkmann.lt/catalogue/content/data_en/Dehn/Dehn_Lightning_and_Surge_Protection_for_Railway_Systems_en_0515.pdf
38. IEC 61643-11:2011-03: Low-voltage surge protective devices—part 11: surge protective devices connected to low-voltage power systems—requirements and test methods
39. IEC 61508-2:2001-07: functional safety of electrical/electronic/programmable electronic safety-related systems—part 2: requirements for electrical/electronic/programmable electronic safety-related systems
40. IEC 62497-2:2010-02 Edition 1.0: railway applications—insulation coordination—part 2: overvoltages and related protection
41. IEC 62848-1:2016-06 Edition 1.0: railway applications—DC surge arresters and voltage limiting devices—part 1: metal-oxide surge arresters without gaps
42. IEC 62848-2:2019-06 Edition 1.0: railway applications—DC surge arresters and voltage limiting devices—part 2: voltage limiting devices
43. IEC 62128-1:2013-09: railway applications—fixed installations—Electrical safety, earthing and the return circuit—part 1: protective provisions against electric shock
44. IEC 62128-3:2013-09: railway applications—fixed installations—electrical safety, earthing and the return circuit—part 3: mutual interaction of a.c. and d.c. traction systems
45. Zoro R, Pakki RR, Komar R Lightning protection for electric railway in Indonesia telecommunication and signalling system. In: 2017 international conference on high voltage engineering and power system, Bali, Indonesia, October 2–5, 2017, pp 476–478
46. Lightning Protection—Railway Facilities—FR–V1, SEFTIM, France, June 2015, https://seftim.com/wp-content/uploads/2015/09/Railway-facilities-protection-EN-V1-032015.pdf
47. Brenna M, Foiadelli F, Zaninelli D (2018) Electrical railway transportation systems, the institute of electrical and electronics engineers Inc, IEEE Press, Piscataway, NJ. Wiley Inc., Hoboken, New Jersey

Lightning Injury: Occurrence and Medical Treatment

Ronald L. Holle, Mary Ann Cooper, and Norberto Navarrete-Aldana

Abstract Lightning deaths and injuries have greatly been reduced in both number and population-weighted rate in more developed countries in the last few decades. However, this reduction has not taken place in many developing nations. The most important factor affecting lightning casualties is not an excessive occurrence of lightning, although this is often a contributor, but the vulnerability of people in developing nations. In these locations, people continue to rely on subsistence agriculture and have no lightning-safe buildings or vehicles nearby, a poor understanding of lightning, weak medical systems, and no access to lightning data in real time. Although direct strike is often considered the most common mechanism of injury, it is quite rare. Instead, ground current, side flash, upward leader, and direct contact are more common. Injuries are commonly related to cardiac issues and neurologic impacts rather than burns which are usually less consequential. Medical treatment at the time of a mass casualty event should concentrate on those who appear to be dead, and CPR can be lifesaving. Long-term sequelae are often permanent and difficult to manage without substantial intervention which is usually not available in developing areas.

Keywords Isokeraunic level · Kerauna medicine · Injury mechanisms · Barotrauma · Step potential · Ventricular fibrillation

---

R. L. Holle (✉)
Holle Meteorology and Photography, Oro Valley, AZ, USA
e-mail: rholle@earthlink.net

M. A. Cooper
University of Illinois, Chicago, IL, USA
e-mail: macooper@uic.edu

N. Navarrete-Aldana
Hospital Simón Bolívar, Bogota, Colombia

C. Gomes (ed.), *Lightning*, Lecture Notes in Electrical Engineering 780,
https://doi.org/10.1007/978-981-16-3440-6_8

1 Lightning Injury and Occurrence

A dramatic reduction in lightning casualties has taken place in more developed regions of the world. In contrast, there has not been perceptible progress in many developing countries. Figure 1 shows the latest map of fatality rates by country based on multiple-year publications describing national fatality studies since 1990. Most notable is the concentration of large fatality rates in southern Africa. Note that fatality data are missing in many of the nations in Africa where lightning is frequent (Fig. 2), the countries are densely populated, safe locations from lightning are often not available, and lightning fatality rates are therefore likely to be very large.

Lightning safety can be achieved in well-developed nations due to nearly universal access to well-constructed buildings and fully enclosed metal-topped vehicles that can be reached in very short time periods when lightning becomes a threat. In less developed countries, these refuges from lightning are often not available at all and seldom reachable in a short time. As a result, people are vulnerable to lightning while working in labor-intensive agriculture [1], occupying lightning-unsafe buildings such as schools [3] and dwellings [4], and tending animals and fishing for a livelihood [5]. Compounding this vulnerability is the lack of access to lightning-specific warnings, medical treatment, and the large number of beliefs surrounding lightning due to a lack of knowledge about the scientific nature of lightning as provided in the following chapters of this book. These beliefs and lack of knowledge may actually increase the injury risk by people either not taking action because they believe they are helpless to avert injury or believing they are safe when they are not.

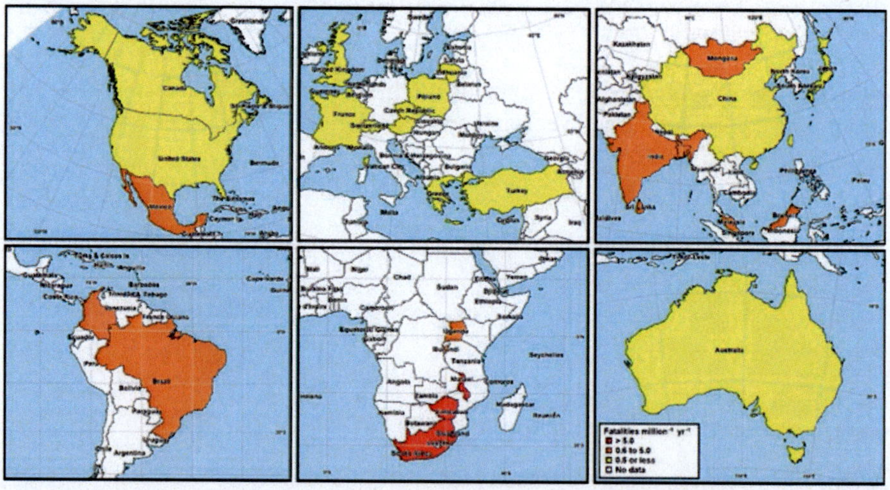

Fig. 1 Lightning fatality rate per million people per year by country. Red shading indicates a rate of more than 5.0 fatalities per million per year, orange is 0.6–5.0, and yellow is 0.5 or less. White indicates no national summaries have been published for datasets ending in 1979 or later (updated from Holle [1])

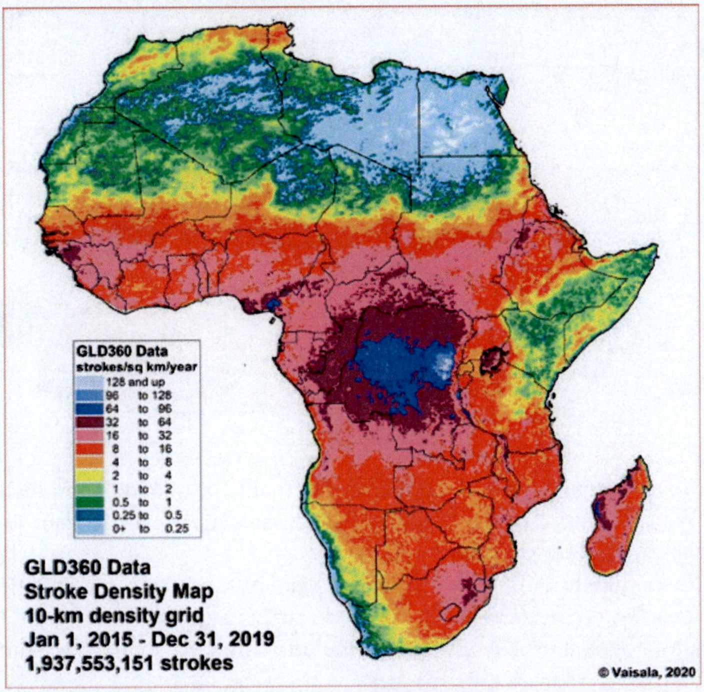

Fig. 2 Lightning stroke density rate per square km per year over Africa, based on 1, 937, 553, 151 detections from 2015 through 2019 by the Global Lightning Dataset GLD360 network [2]

2 Injury Mechanisms and Medical Treatment

i. Mechanisms of injury

Although the pathophysiology of lightning injury could be studied from several approaches, if we want to prevent injury, knowing the mechanisms of injury is the most useful place to start. In this way, we can help construct warnings and lightning safety education that includes the actions individuals can use to avoid injury [5–8].

The five commonly accepted mechanisms of injury have to do with how lightning energy reaches a person [7]. Figure 3 shows the relative distribution of these mechanisms in developed countries. The distribution is not known in developing countries, but ground current is suspected to play an even larger role [5]. A sixth mechanism, blunt or barotraumatic injury from a person being thrown either by concussive force or by induced muscle contraction can be a part of any of the first five as well [9].

1. **Ground current (also called earth potential rise and step voltage)**: The mechanism that kills the most people is ground current, where lightning strikes the surface of the earth and spreads through the earth to injure nearby people. It

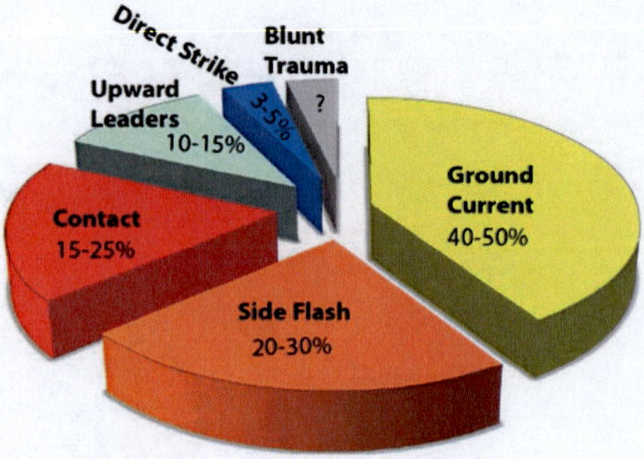

Fig. 3 Relative frequencies of the primary lightning fatality mechanisms [7]

can affect a large number of people, either inside or outside unprotected buildings. Workers in rice paddies, children in classrooms or worshippers at open-air churches are good examples.

2. **Side flash/splash**: This occurs when trees, poles, towers, and many other objects that need not necessarily be very tall are struck and a portion of the lightning jumps to a person nearby. Examples include someone sheltering from the rain under a tree.
3. **Contact**: This occurs when the person is in contact with conducting paths such as plumbing, corded telephones or appliances, headsets, or wiring, either outdoors or inside structures. Contact injury may also occur as animals gather next to long ungrounded wire fences (Fig. 4).
4. **Upward streamer (upward leader)**: Thunderstorms contain strong electrical fields. Whenever a thunderstorm moves across an area, opposite charges are induced on objects on the ground near the cloud including trees, towers, people, and animals. Upward leaders, not usually visible from these objects, will reach up seeking to connect with the downward-moving lightning channel. Even if the lightning attachment (completion of the channel) does not occur, as the upward leader collapses, it contains enough energy to cause injury which has been documented both theoretically and clinically [10–12].
5. **Direct strike**: Contrary to public belief, direct strike is the least common mechanism and causes only perhaps 3–5% of deaths [7]. A direct injury occurs when the lightning stroke attaches directly to the victim and is most likely to occur in the open. While one might intuit that a direct strike is more likely to cause fatalities than the other mechanisms, this has not been shown in any clinical studies.
6. **Blunt trauma (concussive/explosive trauma, barotrauma)**: Blunt trauma has long been suggested as a mechanism in lightning injury. As lightning passes through the air, rapid heating and expansion of the air occurs so that those nearby may experience a concussive force similar to being near an explosion. Blumenthal investigated barotrauma and likened it to being near a blast of 5 kg

Fig. 4 Cows killed by lightning as they gathered by an ungrounded wire fence. This is a common occurrence and can be from contact injury as lightning energy is conducted from a distance or from ground current. Side-flash from the fence is less likely. ©MA Cooper

of TNT [9]. Barotrauma is independent of the other electrical mechanisms of injury but may potentially overlay any of them [5, 7, 8]. Shrapnel from trees or other struck objects exploding can cause injury as well [13]. People who are thrown by lightning experience musculoskeletal injuries, as would be expected [6, 14].

ii. Lightning injury and clinical manifestations

Although nearly 90% of those injured by lightning in developed countries survive, many are left with disabling sequelae [6, 8, 15–19]. The percentage of survivors in developing countries is unknown but may be considerably smaller where people cannot access high quality medical and rehabilitative care [5, 20]. Lightning, despite its extremely high-voltage current, causes substantially different injuries from those caused by high-voltage alternating current electricity as shown in Table 1 [8].

Lightning discharge exerts an electrical effect at the multisystemic level [8]. A wide range of immediate clinical manifestations ranging from minor injuries to serious complications such as cardiorespiratory arrest can occur. Even with early apparently minor injuries, later complications mainly affecting the nervous system are frequent [21, 22]. Although many symptoms occur simultaneously, the clinical manifestations are presented separately in this discussion for academic purposes.

a. Cardiac injuries: The most severe early injury and the most common cause of death is cardiopulmonary arrest [18]. It occurs immediately after the lightning

Table 1 Lightning injuries compared with high-voltage electrical injuries [8]

Factor	Lightning	High voltage
Energy level	30 million volts (V), 50,000 amperes (A)	Usually much lower
Time of exposure	Brief, instantaneous	Seconds
Pathway	Flashover	Deep, internal
Burns	Superficial, minor	Deep, major injury
Renal	Rare myoglobinuria	Myoglobinuric renal failure common
Blunt injury	Explosive concussive effect	Falls, being thrown

discharge and is manifested as absence of cardiac electrical activity (asystole) and respiratory standstill. This primary cardiac arrest may be transient because of the heart's automaticity. This intrinsic property of the heart allows it to restart the cardiac electrical activity and myocardial contraction within a short time. Unfortunately, respiratory arrest may persist, and unless the victim receives immediate ventilatory assistance, attendant hypoxia may induce impaired or irregular heart rhythm (arrhythmia) and secondary hypoxic cardiac arrest [8, 23].

Multiple mechanisms, such as direct thermal damage, coronary artery spasm, increased circulating catecholamine levels, myocardial ischemia secondary to arrhythmia, autonomic nervous system injury and coronary artery ischemia as part of a generalized vascular injury, have been suggested to explain the cardiovascular events following lightning strike [24, 25]. In addition to sudden death, the victim may manifest alterations in the cardiac electrical conduction system (prolonged QTc, dysrhythmias, bundle branch block), myocardial ischemia (chest pain, ST-T segment abnormalities, myocardial infarction without coronary artery disease), cardiogenic shock (Takotsubo-shaped hypokinesis with aneurysmal dilation, abnormal contractility) or hypertension (catecholamine release) [26, 27].

b. Pulmonary injuries: Pulmonary contusion and pulmonary hemorrhage may result from blunt injury or direct lung damage. Other complications include pulmonary edema and aspiration pneumonia secondary to altered mental status [8].

c. Neurologic injuries: Lightning injury is primarily neurologic with damage possible to central, peripheral, and sympathetic nervous systems. Injury to the nervous system causes the greatest number of long-term problems for survivors. The victims can present central nervous system injuries (cerebral edema, intracranial hemorrhages and hematomas, anoxic brain injury, spinal cord injury), peripheral nerve injury and autonomic nervous system injury. Clinical manifestations are altered mental status, coma, agitation, seizures, headache, chronic pain, aphasia, weakness, partial or complete paralysis of one or more extremities, spinal cord dysfunction, and peripheral neuropathy [8, 16, 21].

d. Burns: Less than one-third of lightning survivors have any signs of burns or skin marks. Unfortunately, the lack of burns can result in physician skepticism and legal disputes with workers insurance denial [28–32]. Since the majority of injuries are from mechanisms where the strength of the strike has dissipated due to ground current, contact injury, side flash or upward streamer, and the short duration of the discharge, the flashover phenomenon and lower skin resistance from rain or sweat allow the absence of burns as a reasonable finding [8].
In developed countries, lightning burns tend to be superficial and insignificant compared to the neurologic injuries that are suffered. Burn location provides a prognostic indicator. Cranial and lower-extremity burns are associated with a fourfold and fivefold increase in mortality, respectively, compared to burns in other locations [18]. An occasional, but pathognomonic, finding is the Lichtenberg figure. There are not true burns and usually disappear within hours of the injury [8].
In developing countries, reports often describe lightning victims as "charred" or "burned beyond recognition. In developing countries, where mud brick walls, thatched roofs, and insubstantial buildings are the norm, the possibility of a fire increases (Fig. 5). Keraunoparalysis, a usually temporary paralysis or severe weakness lasting a few minutes to hours, may explain why victims cannot escape from burning buildings [8, 16, 20, 33]. This is present in about one half to two thirds of reported lightning cases [18].

e. Eye and ear injuries: These may be caused by direct thermal or electrical damage, intense light, the shock wave, or combinations of these factors. Clinical manifestations are corneal burns, intraocular hemorrhage or thrombosis,

Fig. 5 Eleven tribal leaders were killed when the thatch building where they were meeting caught fire from a lightning strike (used with tribal permission)

uveitis, macular damage, retinal detachment, optic neuritis, delayed cataract, tympanic membrane rupture, hearing loss, and vertigo [34, 35].

f. Musculoskeletal injuries: Spinal and other injuries may be suffered if the person is thrown by the concussive effect of lightning, by the intense muscle contraction that lightning can induce, or by falls [6, 14].

g. Miscellaneous: Deep muscle damage is rare and may be caused by burns, blunt injury or arterial spasm and secondary ischemia. Compartment syndrome, rhabdomyolysis, myoglobinuria and renal failure are rare. Disseminated intravascular coagulation, sexual dysfunction, menstrual irregularities and other endocrine dysfunction have been reported [8].

h. Psychological and neurocognitive problems: Lightning survivors may suffer temporary or permanent neurological sequalae similar to post-concussive syndrome. Post-traumatic stress disorder, cognitive impairment, severe short-term memory difficulty, attention and concentration deficit, difficulty with learning new information, phobias, emotional lability and irritability, insomnia, decreased exercise tolerance, personality changes and depression are common [8, 36, 37].

iii. Medical treatment

In the event of multiple casualties, victims who appear to be dead should be treated first and aggressively, since these patients may have a good prognosis. Ensuring scene safety is paramount. Rescuers are at risk if thunderstorms are in the area, and, if possible, the victim should be moved to the nearest safe area. Intense vasospasm may prevent discovery of a pulse. If the victim is unresponsive with no pulse or no normal breathing, the victim may have suffered a cardiac arrest and the rescuers should immediately activate the emergency response system and start CPR [23]. CPR may be continued until spontaneous adequate respirations resume, the victim is pronounced dead, the rescuer is exhausted, or there is danger to rescuers' survival [8]. Note that 77% of victims do not respond to CPR [18].

Lightning injury victims should be approached as blunt multiple trauma patients with spinal immobilization and the victim transferred to the nearest health center to perform the medical assessment. Patients with persistent musculoskeletal symptoms, neurologic, cardiac rhythm or vascular abnormalities, or significant burns require admission to a critical care unit. Vital signs are usually stable, but victims may demonstrate acute transient mild hypertension and tachycardia caused by sympathetic activation which usually does not need pharmacological treatment. Hypotension is rare and should prompt investigation for non-visible hemorrhage. Early seizures are probably caused by hypoxia. Any action that improves breathing in addition to administering oxygen can help the victim. If repetitive or recurrent seizures occur or mental status deteriorates, it is necessary to rule out an injury at the brain level [8].

Spinal alignment and immobilization are always required with spinal cord injury-like symptoms such as paraplegia. Keraunoparalysis should be a diagnosis of exclusion. Mottled, pulseless extremities associated with lightning injury often improve over several hours so that fasciotomy is rarely indicated. Lichtenberg figures commonly disappear spontaneously within hours of the injury. Eye and ear examinations should not be overlooked. Cataracts may develop either immediately or over a prolonged period.

Neurocognitive function and behavior disturbances commonly occur but may not be recognized until a victim returns to work or school. As with other brain-injured people, frustration, impatience, instant rage, and other personality changes may drive away family members, further compromising the survivor's recovery [28, 29, 32]. Unfortunately, there are reports of suicides as survivors become despondent when they cannot find help for their brain injury and other sequelae, may not be able to return to work and lose their homes, or lose the support of their friends and family due to personality changes and other stressors [28, 29, 32]. Some self-medicate with alcohol, drugs, or herbs for their post-injury chronic pain. Family and work dynamics can be difficult.

A further setback to the victim's family, particularly in developing countries such as those in Africa, is a common belief that a family affected by lightning injury is 'cursed' or was punished for bad behavior such as beating their wives or children. The community may shun the entire family so that they have little choice but to leave their community, home and employment to start over in a new community where their tragedy is unknown [38–41].

Even when the family supports the victims, sequelae may exceed their knowledge and ability to cope. This is one of the many reasons the victims often require interdisciplinary assistance services, social, family and community support to avoid isolation and frustration in the face of their new reality.

References

1. Holle RL (2016a) A summary of recent national-scale lightning fatality studies, weather, climate, and society 8(1):35–42
2. Said R (2017) Towards a global lightning locating system. Weather 72(2):36–40
3. Holle RL, Cooper MA (2016) Lightning-caused deaths and injuries at schools (preprints). In: 33rd international conference on lightning protection, Estoril, p 5, Sept 25–30
4. Holle RL (2010) Lightning-caused casualties in and near dwellings and other buildings (preprints). In: International lightning meteorology conference, Vaisala, Orlando, Florida, p 19, April 21–22
5. Cooper MA, Holle RL (2018) Reducing lightning injuries worldwide. Springer Natural Hazards, New York, p 233
6. Cooper MA, Holle RL (2011) Mechanisms of lightning injury should affect lightning safety messages, National Weather Association, Newsletter, pp 2–3
7. Cooper MA, Holle RL, Andrews CJ (2008) Distributions of lightning injury mechanisms (preprints). In: 20th international lightning detection conference, Vaisala, April 21–23, Tucson, Arizona, p 4
8. Cooper MA, Andrews CJ, Holle RL, Blumenthal R, Navarrete N (2017) Lightning-related injuries and safety. In: Auerbach P (ed) Wilderness Medicine, 7th edn. Elsevier, Philadelphia, Pennsylvania, pp 71–117
9. Blumenthal R, West NJ (2015) Investigating the risk of lightning's pressure blast wave. South African J Sci 3(3/4):5

10. Anderson RB (2001) Does a fifth mechanism exist to explain lightning injuries? IEEE Eng Med Biol 105–113
11. Anderson RB, Jandrell IR, Nematswerani HE (2002) The upward streamer mechanism versus step potential as a cause of injuries from close lightning discharges. Trans South Africa Inst Electr Eng 33–43
12. Cooper MA (2002) A fifth mechanism of lightning injury, Society for Academic Emergency Medicine 9:172–174
13. Blumenthal R (2012) Secondary missile injury from lightning strike. Am J Forensic Med Pathol 33(1):83–85
14. Hendler N (2005) Overlooked diagnoses in chronic pain: analysis of survivors of electric shock and lightning strike. J Occup Environ Med 47:796–805. https://doi.org/10.1097/01.jom.0000165753.52977.ab
15. Andrews CJ, Darveniza M (1989) Telephone mediated lightning injury: an Australian survey. J Trauma 29(5):665–671
16. Cherington M (2005) Spectrum of neurologic complications of lightning injuries. NeuroRehabilitation 20:3–8
17. Cherington M, Walker J, Boyson M, Glancy R, Hedegaard H, Clark S (1999) Closing the gap on the actual numbers of lightning casualties and deaths (preprints). In: 11th conference on applied climatology, American Meteorological Society, Dallas, Texas, pp 379–380, Jan 10–15
18. Cooper MA (1980) Lightning injuries: prognostic signs for death. Ann Emerg Med 9(3):134–138
19. Cooper MA (2001) Disability, not death, is the main problem with lightning injury. Nat Weather Dig 25:43–47
20. Cooper MA (2012) Whether the medical aspects of lightning injury are different in developing countries (preprints). In: 33rd international conference on lightning protection, Vienna, p 6, Sept 2–7
21. Cherington M, Yarnell PR, London SF (1995) Neurologic complications of lightning injuries. West J Med 162(5):413–417
22. Silva LM, Cooper MA, Blumenthal R, Pliskin N (2016) A follow-up study of a large group of children struck by lightning. S Afr Med J 106(9):929–932
23. Cooper MA, Holle RL, Andrews CJ (2012) Electrical current and lightning injury. In: Field J (ed) The textbook of emergency cardiovascular care and CPR: ACLS for the experienced provider, Williams & Wilkins, AHA/ACEP, Lippincott, pp 498–511
24. Christophides T, Khan S, Ahmad M, Fayed H, Bogle R (2017) Cardiac effects of lightning strikes. Arrhythmia Electrophysiol Rev 6(3):114–117
25. Lichtenberg R, Dries D, Ward K, Marshall W, Scanlon P (1993) Cardiovascular effects of lightning strikes. J Am Coll Cardiologists 21(2):531–536
26. Dundon BK, Puri R, Leong DP, Worthley MI (2008) Takotsubo cardiomyopathy following lightning strike. J Emerg Med 25(7):460–461
27. Rivera J, Romero KA, Gonzalez-Chon O, Uruchurtu E, Márquez MF, Guevara M (2007) Severe stunned myocardium after lightning strike. Crit Care Med 35(1):280–285
28. Andrews CJ, Reisner AD, Cooper MA (2017) Post electrical or lightning injury syndrome: a proposal for an American Psychiatric Association's diagnostic and statistical manual formulation with implications for treatment. Neural Regen Res 12:1405–1412
29. Cooper MA, Marshburn S, Marshburn J (2001) Lightning strike and electric shock survivors, international. Natl Weather Digest 25:48–50
30. Cooper MA (1995) Emergent care of lightning and electrical injuries. Semin Neurol 15(3):268–278
31. Cooper MA (1995) Myths, miracles, and mirages. Semin Neurol 15(4):358–361
32. Cooper MA, Marshburn S (2005) Lightning strike and electric shock survivors, international. In: Cherington M (ed) NeuroRehabilitation 20(1):43–47
33. Villamil DE, Navarrete N, Cooper MA (2019) Keraunoparalysis—an explanation for the more severe lightning injuries reported in developing countries. International symposium on lightning protection (XV SIPDA), September 30–October 4, São Paulo, p 5

34. Norman ME, Albertson D, Younge BR (2001) Ophthalmic manifestations of lightning strike. Surv Ophthalmol 46(1):19–24
35. Toquica JE, Gomez HF (2016) Ocular injuries caused by lightning strikes: review of the literature and presentation of two clinical cases. Pan-American J Ophthalmol 15(3):84–86
36. Cherington M (2003) Neurologic manifestations of lightning strikes. Neurology 60(2):182–185
37. Van Zomeren AH, ten Duis HJ, Minderhoud JM, Sipma M (1998) Lightning stroke and neuropsychological impairment: cases and questions. J Neurol Neurosurg Psychiatry 64(6):763–769
38. Cooper MA, Tushemereirwe R, Holle RL, Villamil DE (2019) African centres for lightning and electromagnetics network (ACLENet)—application to South America? In: International symposium on lightning protection (XV SIPDA), September 30–October 4, São Paulo, p 6
39. Mulder MB, Msalu L, Caro T, Salerno J (2012) Remarkable rates of lightning strike mortality in Malawi. PLoS One 7(1):e29281. https://doi.org/10.1371/journal.pone.0029281.2012
40. Phiri GR (2017) Enhancing lightning hazard mitigation through traditional customs and religion in Malawi. In: Holle R, Ataremwa E (eds) Lightning impacts in developing countries of Africa and Asia, Centre for science and technology of the non-aligned and other developing countries, New Delhi, pp 16–27
41. Trengove E (2013) Lightning myths and beliefs in South Africa: their effect on personal safety. Ph.D. dissertation, University of the Witwatersrand, Johannesburg

Lightning: Public Concepts and Safety Education

Chandima Gomes and Ashen Gomes

Abstract Being a spectacular atmospheric phenomenon, lightning could induce significant interest in the human mind since the beginning of history. Most often the public perceptions of many communities about lightning were marked by divinity, power, fear and punishment. The belief systems extend up to the present time, despite the development of modern science and technology. In the last two centuries, the scientific understanding of the thunderstorm and lightning phenomena gradually improved and at present we have a significant awareness of the nature of lightning, injury mechanisms and lightning-related medicine. The modern safety guidelines and safety modules for communities in lightning-dense geographical areas are based on proven scientific facts. In several developed countries, a marked decrease in the number of lightning casualties could be observed during the past century, due to the continuous safety awareness programs. However, lightning safety modules which are highly successful in one part of the world may not be that successful in another part of the world.

Keywords Hierarchical order · Lightning myths · Safe structures · Belief system · Underprivileged communities

1 Public Concepts

Lightning has most often been treated as two combined phenomena, the flash of light and the thunder, in the public perceptions. Due to its destructive effects, and high sound intensity, the ancient belief systems attributed lightning to either a powerful deity or wrath or weapon of such divinity. Thus, the God of lightning was often treated as a holy entity who punishes human beings or even animals with evil qualities or

C. Gomes (✉)
School of Electrical and Information Engineering, University of Witwatersrand, Johannesburg, South Africa

A. Gomes
School of Electrical Engineering and Computer Science, KTH, Royal Institute of Technology, Stockholm, Sweden

C. Gomes (ed.), *Lightning*, Lecture Notes in Electrical Engineering 780,
https://doi.org/10.1007/978-981-16-3440-6_9

spirits. This perception of the people of ancient civilizations has been depicted in the sculptures, paintings and symbols of the lightning God [1, 2].

In the development of community-specific lightning safety modules, it is of utmost importance to understand the beliefs of the public on the lightning phenomenon. Failure to do so, may results in the safety modules, fail miserably in the attempts of applying to target communities, as it has been witnessed in South and South East Asia in introducing several modules which were successful in other regions. The investigations on public beliefs includes analysis of the divinity induced on lighting, the mythical powers attributed to lightning or followers of lightning deities, beliefs on the lightning and its abilities, at various societal levels and up to what extent the modern scientific knowledge is compatible with such belief systems. Consequently, it is only through the smooth integration of accepted lightning safety guidelines with the deeply rooted beliefs of the societies on natural phenomena, one can meaningfully introduce safety programs to a given society [3, 4]. Such investigations, could also produce comprehensive databases on various concepts of the lightning phenomenon that will be useful in the applications of various non-technical sciences such as sociology, fine arts, anthropology, psychology and behavioural sciences etc.

2 Ancient Belief Systems

A striking point of the beliefs of ancient civilizations on lightning as divinity is their awareness of the massive power that lightning possesses. However despite this kmowledge they could not understand the form of energy dissipation in the event of a lightning strike. It is common to all these beliefs that the thrower of the lightning bolt is a divine entity while the receiver of the thunderbolt is a sinful human being, animal or an evil entity. Thus naturally, a person struck by lightning was branded as a sinner and a structure hit by lightning was treated as a place not suitable for residence. The people treated lightning striking an important structure such as a religious place or a palace as a very bad omen. In several communities in Africa, such beliefs are deeply rooted in the public of many tribes even today [1, 2, 5–7].

The earliest form of the lightning-thunder god was the 'thunderbird', who created lightning and thunder either from its beak or from its wings. Engravings of the thunderbird have been found in the archaeological sites of the Bronze Age in Dodona and Minussinsk in Siberia, Dong Son in Vietnam and on pots in north Peru [8, 9].

Several native tribes who lived in North America also believed that lightning is due to the flashing of feathers of a mystical thunderbird. Thunder is the sound of the flapping wings, according to their belief [8].

As it is found in the mythology, Zeus of Greece, Jupiter of Rome and Typhon of Egypt, send lightning bolts from heaven. The Greek legends state that the thunderbolt was invented by Minerva the goddess of wisdom and gifted to Lord Zeus to punish the bad entities. According to one of the Greek legendary stories, a Cretan called Iasios was struck by lightning hurled by Zeus for attempting to ravish Demeter, the goddess of corn [10]. The ancient Roman empires practised a cult of deducing the powers of god by visualizing and observing lightning and thunder. Since lightning was a manifestation of the gods, a location struck by lightning was regarded as sacred. Greek and Roman temples were often erected at such sites, where the gods were worshipped in an attempt to appease them [11, 12].

According to the Vedic mythology of India, Indra, the god of earth, conquered innumerable human and demon enemies and killed the dragon Vrtra, who had prevented the monsoon from breaking, by means of the power of lightning and thunder [13]. The weapon of Indra, Vajrayudh, is the lightning strike and symbolized the supreme power against evil. In many parts of India and Nepal, monuments of Vajrayudha could be seen at both Hindu and Buddhist temples. Figure 1 shows the Vajrayudha made of bronze at Swayambhunath Buddhist temple in Kathmandu, Nepal. Alongside the monument is the world-renowned lightning expert Prof. Vladimir Rakov of the USA. Interestingly this Vajrayudh itself has been struck by lightning around the year 2010, of which the damaged sign could clearly be seen. Agni, the god of fire, also uses lightning as a major weapon to demolish the enemies of divinity. Several sectors of Mythologists argue that the God Agni and God Indra may be the same entities.

The power of controlling lightning energy to harm enemies is also a popular voodoo cult in the African Continent [5, 7]. According to these studies, there are also records of people who claim that they can create lightning bolts at their will and direct towards the desired target. Practices and beliefs of such powers exist even today, in many parts of Africa.

Fig. 1 Vajrayudh, made of bronze at Swayambhunath Buddhist temple in Kathmandu, Nepal

The Holy Quran of the Islamic religion also mentioned God's power on lightning at five places. Surah 24: 43; "… And He sends down hail from the sky hail mountains (or there are in the heaven mountains of hail from where He sends down hail), and strikes therewith whom He wills and averts it from whom he wills. The vivid flash of its lightning nearly blinds the sight." (Tafsir At-Tabari).

At 46 places of the Holly Bible of Christian religion, the thunder and lightning are spoken of as tokens of God's wrath or a representation of God's glorious and awful majesty or some judgment of God on the world. Eg. Psalm 18:14; "He shot his arrows and scattered the enemies, great bolts of lightning and routed them".

The lightning and thunder gods still continue in the popular beliefs of several communities in the world. Some Eastern Europeans believe that St. Elijah is the controller of lightning while some Latin Americans treat Santiago as the saint of lightning.

The following list provides the names of the divine powers treated as the God/Goddess of lightning or the power bearer of lightning [1, 2].

Tien Mu (goddess), Lei Tsu and Lei Kung in China.

Thor of Scandinavians whose name is the origin of Thursday.

Tlaloc of Aztecs in Central America (now Mexico).

Aktzin of Totonacs in Central America (now Mexico).

Jasso of Mesoamericans in Central America (Now Honduras and Nicaragua).

Chaac of Mayans in Central America (now Guatemala).

Apocatequil of Incas in South America (now Peru).

Haokah of Lakotas in North America (Now Dakota).

Perun in Slavia and Bulgaria.

Raijin and Ajisukitakahikone in Japan.

Perkunas in Latvia and Lithuania.

Teshab of Hurrians in North Mesopotamia.

Taru of Hattians in Anatolia and Turkey.

Ishkur of Sumerians in Babylonia.

Adad of Akkadian in Babylonia.

Haddad in Middle East and Minor Asia.

Taranis in Gaul and Britain.

Perendi in Albania.

Ukko in Finland and Uku in Estonia.

Oya (goddess) and Shango of Youruba tribe in Nigeria.

Azaka-Tonnerre of Voodoo in Haiti.

Haikili and Kaha'i of Hawaiians in Polynesia.

Tāwhaki and Uira of Maoris in Polynesia.

3 Recent-Past and Current Belief Systems

One of the present communities in which, the belief of lightning god (or the lightning ghost or spirit) is deep-rooted is the Nigerian Yoruba tribe. They call the lightning spirit Shango, the thrower of thunderbolts. Shango or Chango is portrayed as a fierce god having a human-like figure whose axe could throw fire (Fig. 2). Shango is most often related to fire, perhaps, as the lightning strikes are often observed to ignite houses, and trees. In this Nigerian society, a person getting a lightning strike is treated as a sinful offender of god and nobody dares to touch his body except for the closest relations [14–16]. Usually, the entire family of the lightning target is extradited from the society. Unfortunately, Nigeria is a region of very high lightning density thus the probability of a person receiving a lightning strike is higher than the world average.

The lightning-bird-god is still practised in the Bantu tribe in Africa by the name Umpundulo. The indigenous doctors of the Bantu tribe, even at present, go out in storms and command the lightning to strike far away [17].

Interestingly, man-made lightning or more accurately witch-made lightning is very much a part of African life so that the subject has already come to the scientific

Fig. 2 Shango (or Chango) the lightning god of Youruba Tribe in Nigeria

fora. At International Colloquium on Overvoltages and Insulation Coordination, held in Harare, Zimbabwe in 1995, a paper has been presented on the observations of witch-created lightning in Zimbabwe [18]. As per this study, there are several ways of “manufacturing” lightning by the witch-doctors in the country. These methods are as follows,

- Two specifically made stones, by means of traditional magic, are charmed with ritual prayers until they are ready to generate lightning. Once that part is completed another mantra is enchanted stating the name and offence of the target who should be victimized. As the chanting is over, the lightning will reach its intended target.
- The witch doctor keeps the lightning in a calabash, a vessel made of seasoned bottle gourd. As it is required, the calabash is hit more than five times while the name of the target person and his offence (for which the punishment is given by directing the lightning) is chanted. With a loud bang of thunder, one can see the lightning emanate from the calabash. Once it hits the target, the lightning is supposed to return to the calabash.
- The witch doctor makes a fire with some magic sticks while chanting mantras. The smoke that emerges from the fire makes artificial clouds, which starts lightning and thunder. The generated lightning is dictated to reach a certain target determined by the creator.
- The witch doctor climbs a tree and smears a skin of goat, buck, cattle or any other animal with herbs. When the skin is dried, it is hung on a tree branch and struck several times with a stick. The lightning emanates from the skin and goes after the target. This type of lightning is considered to be uncontrolled once you dispatch it to find the predetermined target. If the lightning bolt fails to locate the target it will come back to the sender and attack him.

Obviously, these descriptions are highly unrealistic in the views of present-day accepted science. However, the author of [18], a Ph.D. holder in science, has written the paper in a way that he himself firmly believes these lightning-manufacturing powers of the people in Zimbabwe. Many recent news reports state that even very evidently natural lightning strikes, often associated with witchcraft in Zimbabwe, do spark acrimonious witch-hunts in rural areas. Many studies show that such lightning generation by witch-craft is not confined only to Zimbabwe. It is spread over many parts of Africa [7].

The above observations regarding the mindset of the African public have been re-confirmed by the interviews conducted in the studies presented in Trengove and Jandrell [6] and Trengove [19]. As per the outcomes of their work, 40% of the interviewees have given the response “yes” to the question of whether they believe that witchcraft can control lightning. Out of the sample, 18% were not certain about such ability. In response to another query, an overwhelming 91% of the sample has said that they believe that mirrors can attract lightning. This outcome shows that even after more than three decades in modern times the beliefs hardly change in the region. The information in Trengove and Jandrell [6] and Trengove [19] reveals that Zulus (a tribe in South Africa) believe that lightning is attracted to shiny objects.

Sri Lanka, an island in the Indian Ocean, which is located south of India, it is a long-standing tradition to curse on the most hated enemies with the phrase "may lightning-without-rain fall upon you". A close look at the characteristics of positive lightning shows the scientific significance of this curse as far as its intended purpose is concerned: Most often lightning that reaches the ground in the absence of lightning is positive in polarity, as they most often emerged from the leading or trailing edge of the tower [20]. Positive lightning contains much larger peak impulse current, longer time duration of both impulse and continuing current components thus, carry enormous energy [21]. Such characteristics will fit into a curse that is meant for spelling maximum disaster to the enemy.

A well-documented incident, from the nineteenth century in Sri Lanka provides, substantial support to the public belief of "a bad person is punished by lightning" [22]. Major Thomas William Rogers was a British National who was appointed in the 1840s as the Assistant Government Agent and the District Judge of Badulla, a mountainous district of then Ceylon, a Colony of the British Empire (Now Sri Lanka). He was an elephant hunter and is credited with killing over one thousand five hundred elephants within a short span of 4 years. One of the elephants that Major Roger's killed was from the sacred area of Kataragama, a sacred city. An old patriarch had warned him that he had done wrong in killing this elephant within the sacred boundaries of one of Lanka's holiest citadels and that he should beware of a tragic death. On the 7th of June 1845, when Roger's stepped outside from his shelter on a stormy night, onlookers suddenly witnessed a flash of lightning and saw the elephant hunter fall face forwards. In the government cemetery of Nuwara Eliya lies the tomb of Rogers, which, according to the records, have been struck by lightning twice since it was erected. The authors have visited the burial place of the subject, in 2008. The incident is still highlighting a riddle to be solved in lightning science.

4 Lightning Safety Education

There were several efforts taken at the international level during the last few years to address lightning safety issues in the developing world, especially in countries with a high risk of lightning accidents. The International Roundtable on Lightning Protection, which was held in Colombo, Sri Lanka in 2007 where the formulation and endorsement of "Colombo Declaration on Lightning Safety took place, International Symposium on Lightning Protection which was held in Kathmandu, Nepal in 2011, African Regional Conferences on Lightning Protection held in Entebbe, Uganda in 2013 and 2014 and Lusaka, Zambia in 2015, World Meeting on Lightning, Cartagena, Colombia, in 2016, Roundtable Meeting on Lightning and Thunderstorms, Agartala, India in 2019 are few such events. Several positive steps for the way forward have been taken in these programs; however, during the discussion sessions, many stakeholders cited that the lack of compiled information on lightning safety awareness programs conducted in developing regions as the main obstacle in optimizing new programs.

In the above backdrop, a comprehensive summary on the success and failure of lightning safety in regions with high lightning risk is a need at present for the benefit of lightning safety promoters, especially in the developing world. Furthermore, such information will be very useful for the safety module developers and funding agencies in developed countries, to strategize their road maps, preparation of work plans and decision making on prioritizing fund allocations. This study is done with the view of filling this void in the field of lightning safety.

Educators and community workers in developing countries are at a distinct disadvantage as they do not get opportunities to access up-to-date knowledge or training in lightning safety measures. Furthermore, the awareness promotion methodologies and techniques of imparting knowledge, practised in developed countries, may not be applicable directly in developing countries. For example, the web-based lightning safety guidance and training which has shown fruitful results in developed countries such as the USA and Australia, has not been a very successful technique of educating the public in many of the third world countries up to now [23]. However, such conditions may change in the next 5–10 years, as computer literacy among the common people is raised to a higher level.

Lightning safety programs developed for any region should consider minimizing the effects of each of these mechanisms of injury. However, in different parts of the world different types of injuries dominate as per the analysis of injury statistics [24–26].

People in the following environments should be given priority in the process of developing lightning safety promotion modules.

1. Live in areas of high lightning ground flash density
2. Permanently reside at elevated locations (hilltops, plateaus etc.), exposed areas (large landscapes with low-grown or no vegetation, rivers (boat people) etc.
3. Permanently reside in unprotected wooden, thatched and clay huts, small shelters with metal roofing on non-metal structures and canvas/polythene tents etc.
4. Often involved with outdoor activities for employment (farmers, fishermen, power and communication line repairers, outdoor labourers etc.)
5. Often involved with outdoor recreational activities (cycling, hiking, golfing, boating, adventure walking etc.)
6. Reside or work at locations (indoor or outdoor) close to metal transmission or communication towers.

5 Sheltering Under Thunderstorm Conditions

In the event of a natural random atmospheric phenomenon such as lightning, no place is 100% safe or having zero risks, however, some places are safer than others. Therefore, in the event of an approaching thunderstorm, one should seek shelter in a low-risk location that is reachable within a reasonable time. It is always recommended that one should be inside a safe shelter or safe structure during a thunderstorm period.

It is essential to define what is meant by a lightning safe shelter (safe structure). The best definition, in this case, is "A structure that protects the occupants against the five primary lightning injury mechanisms, namely direct strike, side flash, step potential, touch potential and upward leader". These mechanisms have been described in the previous chapter. As per the definition, the safe structure may not safeguard the occupants from secondary effects such as barotrauma (due to shockwave), vision imparity (due to intense light), splinters and fragments from exploding objects, falling objects (tree branches, masonry etc.), choking (due to smoke from fires ignited by lightning) etc. Hence, the occupant may pay attention to the possibilities of such secondary hazards.

One of the safest locations during a thunderstorm is inside a substantially constructed building, preferably with steel reinforcement (concrete slabs and pillars reinforced with steel), plumbing and electrical wiring with a sound grounding system. Such structures are residential complexes, fully enclosed factories, shopping malls, cinema halls, schools, office buildings, and private residences made with brick, concrete etc. If lightning strikes the building, the steel bars, plumbing and wiring will conduct the electricity more efficiently than a human body. Therefore, the chances of lightning current entering the human body through an electric spark from the roof or walls is negligibly small. The risk is further reduced if the building is installed with a properly designed structural protection system. The design and installation of a lightning structural protection system should be done by a competent engineer specialized in the subject. Such design descriptions are given in national and international standards (IEC 62305 2010; NFPA 780 2008; AS/NZS 1768 2007 etc.)

When one is inside a building, he should stay in the middle of a room or a hall. It is advisable to sit on a chair or bed and keep the feet up. If one is in a standing position, he should keep his feet close together. One should never sleep on the floor, especially inside a risky building, when thunder is roaring around. One should stay inside for at least 30 min after hearing the last thunder. Once lightning strikes a structure, the current is most likely to flow along with metal parts such as railings, fences etc. Therefore, touching or staying very close to such components should be avoided.

A structure made of non-metallic materials or having large exposed areas is not safe during a thunderstorm. The risk of injuries and death will greatly be increased if such structures are covered with combustible material (eg. wood, paper pulp, thatch, polymeric materials such as PVC or rubber, fabrics etc.) The following structures fall into the above categories, thus offer no safety from lightning. One must refrain from seeking shelter in such structures under thunderstorm conditions.

- Thatched roofed houses or temporary shelters
- Wooden or non-metallic structures with metallic roofs
- Beach shacks and cabanas
- Camping tents and picnic huts (irrespective of the material)
- Sports pavilions and open stages
- Carports (especially the ones having no walls)

- Rooftop terraces (even when the terrace is covered with glass or transparent polymeric materials)
- Structures with no walls or half walls (Dharma-Shala of most of the temples, most of the schools in rural areas (even in urban areas), and public gathering places such as Praja-Shala etc.)

One should not stay inside a building (even if it does not fall into the above categories), which stores (or manufactures) fireworks, gun powder, explosives, volatile fluids, poisonous or compressed gases, petrochemicals etc., if the building is not installed with a structural protection system that complies with national or international standards. The relevant government authorities should take strict measures to ensure that such structures are comprehensively protected against lightning, to safeguard the occupants and neighbourhood.

It should also be emphasized that structures with metal roofs are very much likely to attract lightning. If the roof is fixed on a structure that is not properly earthed, the occupants will be at a very high risk of getting side flashes if the structure is struck by lightning.

If no proper building is available for sheltering under lightning conditions, then an enclosed sturdy metallic vehicle such as train, car, van, bus or large ship makes a good alternative. However, convertible vehicles offer no safety from lightning, even if the top is covered with the foldable flap. Other unsafe vehicles during lightning storms are those which have exposed parts such as open cabs, golf carts, tractors, trailers, three-wheelers, motorcycles and bicycles, agricultural vehicles, construction equipment such as cranes and elevators, canoes, and open boats etc. Inside a ship, one should refrain from staying in open decks.

Inside a vehicle, one should keep the windows up, and avoid contact with any conducting paths leading to the outside or connected to the body of the vehicle (e.g. radios, body-fixed telephones and keys in keyholes etc.). One should also avoid leaning against the metal parts of the vehicles. If lightning strikes in the close vicinity, one should cover his ears with hands if a suitable ear protector (earphones, cotton buds etc.) is not around.

In recent years, several studies have been done on designing purpose made safety structures. These structures have been designed for both group protection and individual protection. The most popular group protection structure is simply the abandoned cargo containers modified to facilitate short-duration human occupancy. Such containers, being fully metallic enclosures provide total protection to the occupants even if the structure is not properly grounded. However, such ungrounded or poorly grounded structures could pose a significant threat to human beings and animals who are outside the structure but at proximity. Thus, if the structure is not properly grounded, it is strongly advised to display warning signs emphasizing the danger of reaching the structure under thunderstorm conditions. These container-type safety structures are highly practical at remote construction sites, mining fields, agro-fields, and large gathering sites etc.

Several purpose-made individual safety structures have also been tested in a few studies in the recent past. These are most often tripod or pyramid skeletal type structures where one or two people may occupy. These structures could be used in golf courses, remote hiking and camping sites, fishing and bathing locations (at the beach, lakeshores or river banks) etc. Figure 3 shows such a safety structure being tested under high current impulses at the University of Witwatersrand, South Africa (Constructed and tested by Mr. Musa T. Mukansi and Mr. Mathew L. Woodhead).

Small boats should also be given protection in the form of a series of flexible cables starting from a purpose-installed or existing metal final and ending up at least one meter inside the water [27]. Figure 4 depicts a diagram of such a protection system. Note that these protection systems, including personal safety structures, are yet to be tested against real lightning. The most feasible way of testing them is to apply triggered lightning currents to grounded structures.

Fig. 3 Lightning safety structre for individual occupation (Constructed and tested by Mr. Musa T. Mukansi and Mr. Mathew L. Woodhead, School of Electrical and Information Engineering, University of Witwatersrand, South Africa)

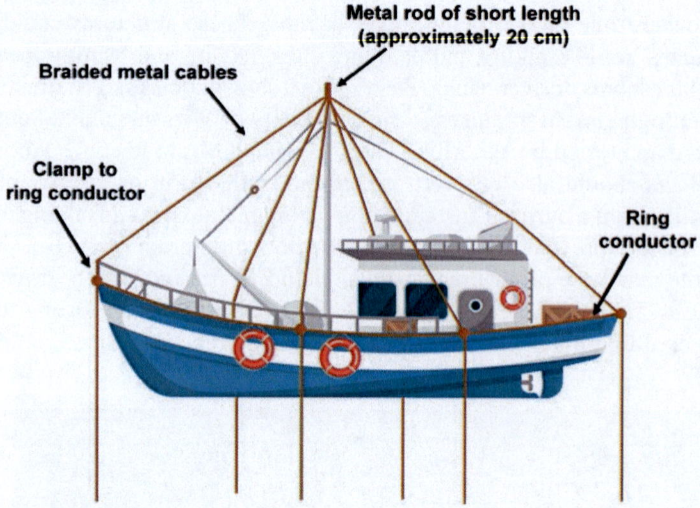

Fig. 4 Lightning protection system for a small boat. Adopted from Gomes [28]

6 Lightning Threat from Equipment

Power and communication lines are frequently struck by lightning due to their exposure to electricity from the sky. When such a service line is subjected to lightning, the current may travel along the wires and enter nearby buildings. Therefore, under thunderstorm conditions, electrical appliances should not be handled if they are connected to the power supply or communication line. For the safety of the equipment, they should be kept plugged off from the service lines. It is also advisable to remove the external antenna jack of the Television and place it outside the building. However, it should be emphasized that the unplugging of the TV antenna jack, power connection, telecommunication connection etc. should be done well in advance. Such removal should not be done after the arrival of the thunderstorm.

Corded telephones and wired microphones should not be used unless it is an emergency. However, there is no additional lightning threat of using mobile phones, cordless phones or FM microphones. Nevertheless, it should be noted that the person who handles the electronics of the public addressing system is at risk of getting a shock if the system is connected to the electricity service. Working on computers is also dangerous if they are connected to communication and electrical services. If the trip-switch (RCD) or other circuit breakers get switched off under thunderstorm conditions they should be kept at off-position until the storm is over. One should also not attend to the rectification of faulty conditions in the electrical wiring system or corded telephone systems during the thunderstorm period.

If the budget permits the building should be fitted with a system of coordinated surge protective devices. Selection and installation criteria are given in many international standards including IEC, IEEE and ITU, and in some literature on easy guidance [29].

7 Dangerous Acts Indoors

There are many domestic activities that one needs to suspend in the event of an approaching thunderstorm. A number of reported lightning accidents show that hazards could have been avoided if victims have suspended activities that they were involved with. Most often, people are reluctant to give up their activities either due to ignorance/stubbornness or financial/opportunity cost.

The repairing of leakage in the roof under overcast conditions should strictly be avoided. One should not take a shower or bath or use a hot tub during an intense thunderstorm. Using the swimming pools (both indoor and outdoor) should also be avoided during the entire thunderstorm period even if the building is installed with a structural protection system. The shock wave and the intense light generated by a close-by lightning strike and the potential gradient in the water that can be developed by the lightning current, may temporarily paralyze the person who uses the swimming pool, thus drowning him to death.

8 Outdoor Safety Measures

It is important to plan the outdoor activities during the lightning season to avoid being caught up in a thunderstorm before reaching a safe shelter. The lightning season or seasons of a country depends on its geographic location. For example, in Sri Lanka, the acute lightning seasons are the inter-monsoon periods; March–April and September–October. During the acute lightning seasons, most of the thunderstorm activities take place in the evening. Therefore, one should keep an eye on the weather forecast and plan outdoor activities accordingly.

If a person becomes a tall protrusion in a certain landscape his body may be the unfortunate object that sends the first upward channel that meets the downward stream of charge from the cloud. Therefore, to avoid being subjected to a direct lightning strike one should not expose himself to the down coming stepped leader.

Under thunderstorm conditions people should not stay at high risk areas such as.

- Playgrounds, racing tracks and other outdoor recreational areas
- Paddy fields and other agricultural landscapes including gardens with low growth
- Beaches, riverbanks, open wells, bridges and open roads
- Open construction sites, worksites and aerodromes etc.
- Higher elevations such as mountain tops, and building tops etc.

- Close to isolated trees and other tall isolated objects.

To avoid such places, one should refrain from playing outdoor games and doing recreation activities, farming, boating, cycling and riding, hiking, gathering for open rallies, repairing power and other service lines etc.

One of the most important rules of outdoor lightning safety is to avoid seeking shelter under large isolated trees during thunderstorm periods. The electrical resistance of a human body; about 300 Ω is much less than that of a tree which is in the order of mega Ohms. Therefore, once a tree is subjected to a lightning strike the large current that is flowing along the tree trunk may jump to the bodies of the people who gather around the tree and passes into the earth in a low resistive path. This side flashing may kill even 5–6 people, according to the records that we have from Sri Lanka, Bangladesh and Pakistan etc. [30]. Although sheltering under isolated trees are very risky under thunderstorm conditions, in comparison with open terrains or mountain tops, seeking shelter in a uniformly grown forest patch or clumps of shrubs may be less dangerous.

When one is in contact with an object which will be subjected to a lightning strike a part of the current may flow across his body as well. This has been described earlier as the touch potential. To prevent the body from being subjected to touch potential one should keep away from flag poles, metallic masts, wire fences, metallic walls and doors, metal railings, etc.

One should also avoid taking a bath or swim in open pools, streams, rivers, lakes, sea etc., under thunderstorm conditions. A person may be drowned to death if he falls unconscious in an unattended environment while he is taking a bath or swim in such water masses (even if the water is only a couple of feet deep). One should also discontinue fishing, water skiing, scuba diving, swimming or other water activities when there is lightning or even when weather conditions look threatening.

If a person is in a small watercraft such as a boat, canoe, raft etc., move fast as possible to the land and seek a proper shelter. If such movement is not possible, try to take shelter under a bridge. In the worst scenario, be inside the cabin or any other enclosure if such location is available and take the safety position that will be described later. It is highly recommended that those who regularly use small to medium-sized boats should adopt proper lightning protection systems in the watercraft. A low-cost protection system for small boats is given in [27].

If one stays close to a tall communication or broadcasting tower he has to take extra measures in protecting himself and his equipment. This is due to the high chance of lightning current flowing near to his house or factory. In case of poor earthing at the tower base there can be a so-called "earth potential rise" in the nearby area, so that a person outside may be subjected to a "step potential" [31]. As a result, he may be injured or temporarily paralyzed. Such paralysis may lead to severe injuries and even death if he is standing close to a pit or unprotected well or taking a bath in a water pool. Thus, those who have such towers in the neighbourhood should strictly be adhere to the safety guidelines described in this paper. In addition to human and livestock injuries, there is a high probability of equipment damage in buildings in the neighbourhood such as towers both due to ground potential rise and induced voltages [32].

9 Estimation of Timing

In many countries such as the USA, Canada and Australia lightning safety plans essentially include the so-called 30/30 rule (30/30-R). As per the 30/30-R people should get into a protective shelter (sturdily built building or an all-metal vehicle) if the illumination-to-thunder time delay (duration of time in seconds between the vision of the lightning flash and the subsequent hearing of thunder) is 30 s or less and that they should not leave their shelter of protection until 30 min after the final sound of thunder.

As light travels almost instantly compared to the speed of sound (approximately 330 m/s), a 30 s time-to-thunder corresponds to a lightning strike about 10 km away. The analysis of lightning detection data in several countries shows that at the beginning of the lighting activity, strikes can be scattered within a space of about 10–15 km [33]. Hence, at least 30 s time-to-thunder lead is necessary prior to the arrival of the thunderstorm as there is a possibility of distant strikes. A 30-min time delay, after the sound of the last thunder, is required as the trailing part of thunderstorms may carry a net residual charge in either the negative charge centre or the positive charge centres. This charge may produce lightning on the passing edge of a storm, tens of minutes after the rain has ended. Note that in a thunderstorm, rain is produced typically from the cloud base, which may be quite small in coverage compared to that of the upper parts. However, there is no solid scientific evidence to justify the validity of the 30-min delay from the last sound of thunder to restart the normal activities.

Several studies have revealed that most people affected by lightning are struck not at the most active stage of a thunderstorm but before and after the storm peak. This can be explained scientifically as in many cumulonimbus clouds that produce lightning, the anvil of the cloud from which lightning can be emanated, is several tens of kilometres shifted from the rain base due to the wind shear. Most importantly this part of the thundercloud houses the positive charge that drives positive cloud to ground lightning. As per the literature [20] such positive lightning may drive much larger impulse currents (in the order of 500 kA) and long continuing currents (currents in the order of about 1 kA flowing for a considerably longer period). Furthermore, if the lightning strikes before the rain the chances of triggering fire is also larger due to the dry conditions that may prevail. Therefore, such lightning poses a much higher threat to human beings, animals and property than their negative counterparts.

The above facts show that many people are unaware of how far lightning can strike from its parent thunderstorm. Therefore, one should not wait for the rain to start seeking shelter and should not leave shelter just because the rain has ended.

Although, application of 30/30-R is successful in developed countries such as the USA, Canada and Australia, the same may not be the case in many developing countries, especially in communities that work on a daily wage basis. Interviews conducted by authors in Bangladesh and Sri Lanka reveal that a majority of low-income societies are not ready to give up or delay their professional activities for more than 5–10 min, even though they have an understanding of the risk they pose (experience of authors in the two countries during the awareness programs conducted). Hence, a suitable rule or guideline should be adopted region-wise to replace 30/30-R, if it is practically nonviable in a given region.

10 Safety Position

If one cannot go elsewhere and is compelled to stay outdoors in a severe thunderstorm (as he may be far away from a proper shelter), he should move to the safest location available (away from open fields, higher elevations, water etc.) and adopt the safety position described below.

The person should crouch down, put the feet together and place hands over ears to minimize hearing damage from thunder and duck the head as much as possible (Fig. 5). One should make sure that he does not take the safety position at a place that has a chance of falling material (very close to a large wall or underneath an overhanging roof), flooding (dry river beds, floodplains, pits etc.), land sliding (eroded slopes, newly filled lands and close to wells etc.) or explosion (close to underground ammunition dumps, minefields etc.).

Each person in a group, in a safety position, should at least be 2-3 m away from one another, thus if one unlucky person is struck, the others are protected and can provide first aid to the victim.

In the event of very close thunder activities, one better not use earphones and headsets. All removable metallic parts on the body such as backpacks, caps with metal

Fig. 5 Lightning safety position

tips, wristwatches, metallic badges etc. and any metallic items such as golf clubs, fishing rods, agricultural tools, tennis rackets, umbrellas etc. should be removed or dropped aside. The reason for getting rid of such metal objects is to avoid getting side flashes and also to prevent heat from being trapped into a single point in the event of a lightning strike to an unfortunate person [34]. There are several records where people have been severely injured as metal parts on the body garments were melted due to the heat of the lightning current and stuck into the body. However, one should note that there are no scientific evidence to conclude that metal parts attached to the body have any influence on the probability of a direct strike to a human being [34].

It will be advantageous to wear shoes or slippers made of insulation material (such as rubber, clothes, leather, plastic, etc.), as that will minimize the effects of being subjected to step potential. Studies that have been done in Bangladesh reveal that step potential may lead to the death of people more often than one would expect [24]. It should be repeated that such footwear also does not influence the probability of the person being subjected to direct lightning strikes.

11 Safety at Workplace

Lightning safety should be an integral part of the safety plan of workplaces in areas of high lightning occurrence density. This is specifically important in the industrial and service sectors where.

- considerable outdoor activities are involved; power distribution, communication (tower related sites, line maintenance etc.), building construction, road and other civil constructions, defence, police, dockyards, transportation, airport and aviation, hydro projects, fisheries, plantations, metal crushers, playgrounds, Golf courses, swimming pools etc.,
- large masses of employees are engaged; garment industry, hotel industry, hospitals etc.
- a high-risk environment exist; firework industry, explosive manufacturing, petrochemical industry, compressed gas distribution etc.

The employees of such sectors should be given a mandatory short training program together with demonstrations on lightning safety and protection on annual basis. Typically a three-hour program will be sufficient to enlighten the awareness of workers. Such training program should include.

- basic concepts of lightning
- human safety concerns
- techniques of lightning protection of equipment and properties
- training on first aid
- maintenance and record-keeping, troubleshooting and regular inspection.

The following measures can be taken to improve the lightning safety environment of the workplace.

- Installing of proper structural and surge protection systems to the buildings.
- Displaying of "do"s and "should-not-do"s under thunderstorm conditions, at frequently-visited places of employees; restaurants or lunch/refreshment rooms, reception, restrooms, recreation centres etc.
- Installing of lightning warning systems at vulnerable places.
- Displaying of warning signs at dangerous locations such as playgrounds, swimming pools, outdoor recreation centres, beaches, isolated trees, open spaces, flag poles, close to down conductors of the structural protection system, etc. Few such warning signs are
  "Do not use XX under the lightning conditions". XX: Playground, swimming pool etc.
  "Keep away from this XX under the lightning conditions". XX: flagpole, down conductor, tree etc.
  "Don't go out of the building under the lightning conditions". In beaches, gardens, hotels etc.
  It is highly meaningful to incorporate these warning signs (displayed in both English and native languages) with a lightning warning system.
- Covering of the locations of the earthing pits (of down conductors or power) with a few-centimeter layer of gravel or crushed rock (area of radius about 2 m around the pit).
- Planning of outdoor events such as repairing of power and communication lines, plantation activities, construction work etc., according to the weather forecast or information obtained from a lightning detection system. This is specifically important in the case of repairing power systems where a lineman is lifted by an insulated-boom crane to be in contact with low voltage or high voltage overhead lines. As far as the bucket is insulated from the body of the crane (and in most cases, the bucket is temporarily bonded to the line as well) the lineman is safe from electrocution due to power frequency currents. However, in the event of a lightning strike to the line, the bucket will become a floating electrode that facilitates the lightning current to flow into the ground in the form of an aerial spark & a resistive flow combination. In other words, a lightning generated spark may leap through the insulation of the bucket (bridging the gap electrically) so that the lightning current may pass through the body of the lineman into the ground (killing or injuring him).

12 Organized Lightning Safety Promotion

Lightning safety has been promoted at various levels by individuals and organizations in South Asia for several decades. However, due to the disorganized manner that such programs have been conducted, the maximum benefits of the investment and efforts could not be harvested [23]. Observations of authors in Bangladesh, Sri Lanka and India reveal the following shortfalls of disorganized and unplanned lightning awareness campaigns.

a. Overlapping of target groups: There were occasions that the same village or same school has been approached by multiple organizations for the lightning safety programs. This is often observed following an occurrence of a catastrophic incident in a given area. Repeated programs of a similar type may wear out the audience and deprive other parts of the community of acquiring awareness.
b. Overlapping of safety promotion modes: Similar media programs, quiz competitions and seminar series etc. at the same time in a given region may be less effective.
c. Lack of opportunities for background studies: It is very important to do the success-failure analysis of previous programs prior to the launching of similar programs in a given region.
d. Difficulties in validating program outcomes: It is much easier to evaluate success-failure rates of programs when data can be shared through collaboration.
e. Lack of confidence and trust of the public: It will be quite an uphill task to build up public trust and confidence when promoters reach masses at the individual or solitary organizational level. On the other hand, the same task may be quite viable by approaching at the organized institutional level.
f. Promotion of inadvertent misinformation: Lightning protection is plagued with many products and technologies that are rejected by the international scientific community and many reputed standards due to their lack of scientific acceptability. However, vendors of these products may infiltrate unsuspecting safety promoters and include misinformation into the safety programs with the view of boosting their fraudulent products [35].

Therefore, it is proposed to establish lightning centres (LCs) in each developing country to address the needs and issues of the respective regions. The LC may serve the public in the region in many ways.

LCs need not be charity organizations. They may generate income and achieve personal/team goals. However, it is strongly advised to decide the financial model of the LC from the right beginning and get the consent of all the stakeholders. The objectives of the LC may be;

To reduce lightning accidents in a target region.

To reduce property damage in a target region.

To increase the scientific knowledge among both the scholars and the public.

The decision-makers of each centre should plan out the best ways of disseminating knowledge and promoting awareness in each region. It is also important to share the experience of each centre with others. The following general activities have been recommended to be conducted by the regional lightning awareness centres; however, the activities should not be restricted to the given list.

- Publishing awareness material in local languages (with diagrams and pictures) and distributing them among schools, public service sectors etc.
- Conducting lightning safety seminars and demonstrations/training on first aid for school children, social workers etc.
- Educate the private and government sector in lightning protection techniques and the importance of industrial lightning safety.

- Displaying of banners, posters, cut-outs on lightning safety in highly lightning prone areas.
- Conducting awareness programs for community leaders such as religious heads, doctors (both western medicinal and indigenous), public servants of local authorities and officers of the police.
- Training youths in the region to practice lightning protection as self-employment (especially on low-cost protection measures).

The above points are categorized in concise form in Fig. 6. Note that all LCs may not be able to put all these points into practice in their roadmap due to various financial and strategic constraints. Similarly, depending on the public needs, some LCs may include region-specific additional objectives and activities.

Modes of promoting lightning safety are strongly dependent on the cultural, social and economic backgrounds of a given region. Hence, it is not a good practice to use everywhere a generalized formula in developing lightning safety modules. As it is described in the literature [23] lightning safety modules used in the USA were not very successful in South Asia. Even within South Asia, modules successfully practised in one country was not that fruitful in another country in the region [23].

Another point of significance is the attitudinal trends of certain nations. For example, as it was reported by [23], in Sri Lanka, the level of awareness on lightning safety and protection among the public is quite high due to various programs conducted over the years, however, as a majority simply neglect or overlook the safety advice due to ignorance or stubbornness, lightning accidents in the country is in the increasing trend for the last decade. Jayaratne and Gomes [23] attribute this to the decade long attitudinal practices of the island nation regarding hazard safety.

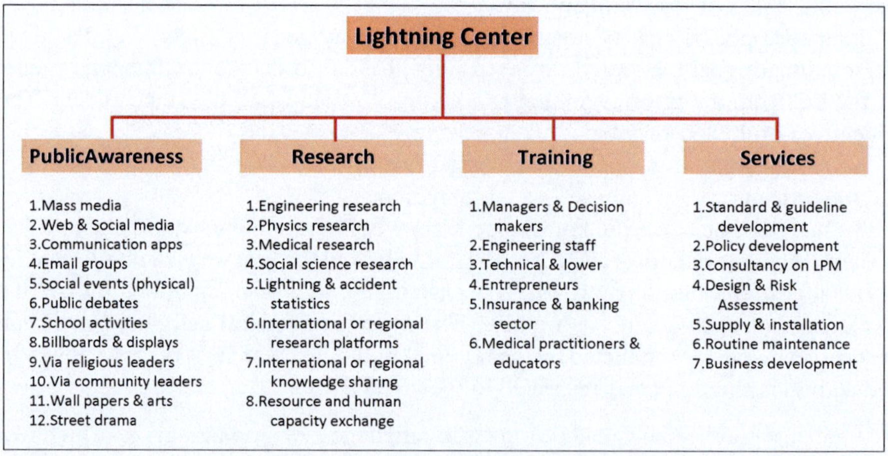

Fig. 6 Major goals and activities that an LC may include in their roadmap

The LAC should consider these regional and local factors into account in developing awareness programs for the people within their territorial coverage.

Table 1 depicts the timeline and related concerns of developing an LC. Note that, usually the concept of LC is developed by a single person, who may be most suitable as the driver (or leader) of the LC. However, there were cases where the concept developer preferred to be sideline giving the driving seat to abler individuals. Therefore, there is no firm rule that the concept developer should be the leader. Once the road map is developed, it is very advantageous to stick to the timeline of the roadmap as much as possible. It is advisable to develop both a short term (say one year) and a long term (say five years) roadmap where the former is more detailed

Table 1 The typical timeline and relevant concerns that an LC may follow

Timeline activities	Concerns
A driver (LC Leader) by volunteering	• Key person that initiates and drags the cart forward • Should be multiplied at least to two with time. Very risky to move on with one driver
Decision on the LC platform	• The institute under which the LC is established • Usually, the institute where the driver is affiliated • Could be standalone but usually as a business entity
Selection of the core team	• Select a team that has the least personal conflicts • Better have a couple of people at decision-making level • Could be standalone but usually as a business entity
Agreement on the LC model	• Business Entity, University/institute funded, NOGO/INGO/GO funded, not for profit earning but income-generating, totally charity based etc • Sometimes it may be a combination, also could be changed with time
Development of the roadmap and time plan	• Prepare six-month, one-year and three-year roadmaps and time plans • It's advisable that the driver prepares it and discuss it with the core team • Be realistic in both activities and time frames
Selection of the advisory team	• First, decide the roles of the advisers (it should be solid and useful) • Select a list of advisers and ask for their consent providing them with the document regarding the expectations from them
Formal establishment of the LC	• It is very important that the LC is officiated • It is good to inaugurate the LC with a regional or international event • Even a nominal physical location makes a big impact on the success of LC

and specific. The LC can develop short-term roadmaps periodically following the long-term roadmap as the guideline.

13 Hierarchy of Hazard Control

A low-income society with below-par literacy rate is much tougher to be mechanized for adopting lightning safety measures compared with the same operation in developed countries [23]. However, the interviews conducted by the authors in several South Asian countries with a number of potential victim communities, revealed that many social and religious leaders are concerned about human safety against lightning and they are willing to be educated. Such observations prompted us to develop a hierarchy of hazard control mechanisms that may successfully be applied to the communities in high lightning risk regions. Although such mechanisms are employed in enclosed work environments (factories, harbours, cargo control divisions, outdoor sites with task boundaries etc.), any community with reasonable size and common interests (fisher communities. farmers, livestock-based communities, highlanders etc.) may provide the operational feasibility for such mechanism.

A group of people, even very large in number that engaged with a similar type of employment or routinely practices can be treated as a bound community. Such a community is often composed of many interacting subsystems and sub-processes. Thus, the safety of such a social system concerning any natural hazard cannot be easily ensured either by centralized control alone or individual control alone. However, the bound nature of the community either by profession or by other mass activities makes it viable for the implementation of safety measures to the community through distributed responsibility of control. A hierarchal hazard control approach is needed for the lightning safety of such a community under this backdrop.

The first attempt at formulating a hierarchy of hazard control was done by Gomes and Gomes [1, 2], where they have applied the concept to the fisheries community along the shores of Lake Victoria in Uganda. In this study, we expand this concept in a broader perspective to make the applications more generalized. Based on the inferences and recommendations given in previous sections, the following hierarchy of control map is proposed for the lightning safety of a bound community, as shown in Fig. 7. Such bound communities should have a common parameter that integrates them into similar practices or activities; e.g. by profession (fisheries, farming, outdoor construction, mining, highway cleaning etc.), by social and religious norms (congregation, pilgrims, outdoor rituals, mass rallying etc.) and by recreational activities (group hiking, outdoor sporting, amusement and adventurous activities etc.).

Forecasting: The government (through the Department of Meteorology) or a relevant private sector that owns a region-wide lightning detection system should provide thunderstorm forecasting and lightning nowcasting information to the concerned community. This should be done in collaboration with mass media, especially audio-visual media such as radio and television. Even electronic media such as the internet

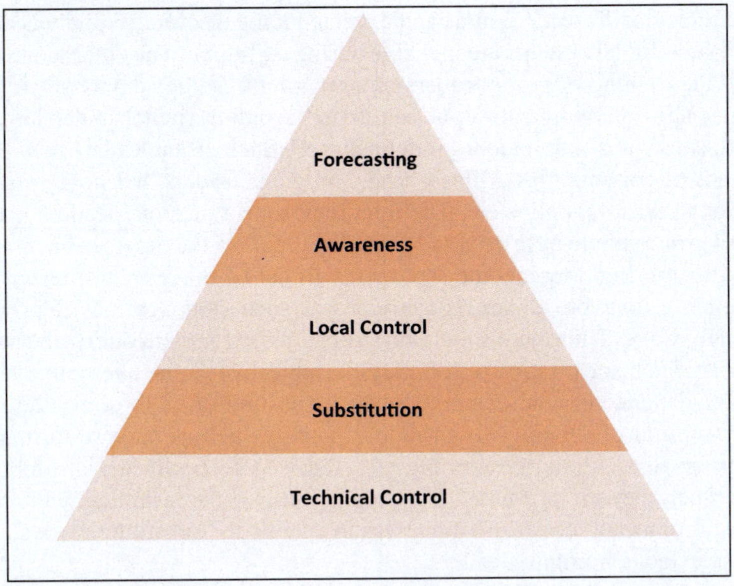

Fig. 7 Hierarchy of hazard control mechanism (adopted from [1] with modifications)

is fast reaching even remote communities. The need of providing accurate information in local languages is a key factor to the successful adoption of safety measures following such news broadcasts.

Awareness: The experience in Uganda [36], Zambia [3], Bangladesh and Sri Lanka [24], shows that thunderstorm forecasting, safety guidelines, protective structures etc. have no impact on community protection unless the society is well aware of the danger of lightning and safety measures that should be taken. The promotion of awareness, even for a single community is not a once and for all process. Such promotion should be done on a periodic basis. Local authorities, governmental institutions (police, educational institutes, hospitals etc.) and non-governmental organizations may take part in this process with the help of local community leaders.

Local Control: Although a general consensus can be reached among the community to act on the thunderstorm forecasting information, in most of the cases of regular non-dramatic natural hazards, the public needs local directives in starting safety procedures. Such directives or leadership are more prominently felt in low-educated societies than in their opposite counterpart. During floods in Thailand and Malaysia, and debris flow in Pakistan and Iran it has been observed that a majority of victims haven't followed even simple safety guidelines due to the lack of initiatives by local leadership. In lightning safety in a bound-community, such local control can be achieved by lightning warning systems located at regular intervals in a way that they can be seen at distance. The most appropriate location for such a warning system is the lakeshore. These warning systems may preferably be in the form of coloured lights (Green-Red or Green-Orange-Red sequences). Alternatively, large signal systems in different colours can be erected if the electricity supply is an issue. However, such signal systems are invisible during nighttime. The other mode of local control is the training of group leaders on executing rules on activity stop/start (eg. 30/30 rule) and following safety measures (eg. avoiding shelter under large trees, going into safety position, indoor guidelines etc.). Such group leaders may be landlords, heads of communities, village-heads, religious leaders, teachers, responsible civil servants, doctors, police etc. The important aspect of group leaders is that the concerned group should have a natural tendency to follow the orders of such a leader.

Substitution: In a low-income society it will not be that easy to prevent people from attending their bread-earning activities as such stoppage may deprive them of their daily wage. Thus there should be a substitution for them during the stoppage of the work. Such substitution will highly be subjective as the alternative tasks are community-dependent. One example of such substitution is to direct the farmers in an agricultural community to an indoor activity such as harvest sorting, stock taking, group discussions on weeding etc., that can be conducted inside a sturdy structure when they are prevented from going out into the farm fields. Planning of such substitutions and providing directives to take up the substitute work should be done by selected community leaders.

Technical Control: As a standard solution for those who seek shelter in places of low risk and the last resort solution for those who do not willing to give up their outdoor activities under any cost, lightning protection systems can be implemented and viable protection measures can be adopted appropriately. These can be implemented at the community level, most probably with the help of external experts. These may include low-cost structural protection systems [27] at all buildings in the community (if possible) or at least at several selected structures where the mass gathering takes place, less complicated protection system for small water vessels such as fishing boats [27, 28], insisting on wearing rubber sole boots to minimize step potentials etc. Placement of metal structures specially made for lightning protection in the farming fields, worksites etc. at regular intervals is strongly recommended as such structures could be developed at quite a low cost. Properly designed such structures can be placed at several locations even at offshore locations with the aid of anchors, thus, fishermen or workers at water-based employment in unsafe boats can get inside such in the event of acute thunderstorms.

References

1. Gomes A, Gomes C (2014) Hierarchy of hazard control to minimize lightning risk. In: 32nd international conference on lightning protection-2014, Shanghai, Oct 2014, pp 1405–1414
2. Gomes C, Gomes A (2014) Lightning: gods and sciences. In: International conference on lightning protection (ICLP), Shanghai, pp 1909–1918
3. Lubasi FC, Gomes C, Ab Kadir MZA, Cooper MA (2012) Lightning related injuries and property damage in Zambia. In: 31st international conference on lightning protection, Vienna
4. Mary AK, Gomes C, Gomes A, Ahmad WFW (2014) Lightning hazard mitigation in Uganda. In: 32nd international conference on lightning protectio, Shanghai, pp 1770–1779

5. Kizito N, Phéneas N (2019) Lightning myths versus science facts: traditional beliefs on thunderstorm among Rwandans. Int J Arts and Humanit (IJAH) 8(2): 1–10, S/No 29. https://doi.org/10.4314/ijah.v8i2.1
6. Trengove E, Jandrell IR (2010) Strategies for understanding lightning myths and beliefs. In: 30th international conference on lightning protection (ICLP), Cagliari
7. Trengove E, Jandrell IR (2015) Lightning myths in southern Africa. Nat Hazards. https://doi.org/10.1007/s11069-014-1579-4
8. Lankford GE (2011) Native American legends of the Southeast: tales from the Natchez, Caddo, Biloxi, Chickasaw, and other nations. University of Alabama Press, Tuscaloosa
9. Zolotarjov AM (1980) Társadalomszervezet és dualisztikus teremtésmítoszok Szibériában. In: M. Hoppál: A Tejút fiai. Tanulmányok a finnugor népek (Social structure and dualistic creation myths in Siberia. In: The Book of M. Hoppál: sons of Milky Way. Studies on the belief systems of Finno-Ugric peoples). Európa Könyvkiadó, Budapest, Hungaria, pp 29–58
10. Burkert W (1985) Greek religion. Harvard University Press, Cambridge
11. Bremmer JN, Horsfall NM (1987) Roman myth and mythography. University of London Institute of Classical Studies, London, pp 49–62
12. Grandazzi A (1997) The foundation of rome: myth and history. Cornell University Press, New York, pp 45–46
13. Smith S (2008) Indra (God of thunder and lightning). B. Vedic deities, deities and entities, Mahavidya, June 2008
14. Bewaji JAI (1998) Olodumare: god in Yoruba belief and the theistic problem of evil. Afr Stud Q 2(1)
15. Idowu EB (1973) African traditional religion: a definition. S.C.M. Press
16. Lucas JO (1996) The religion of the Yorubas. Athelia Henrietta PR
17. Lynch PA (2004) African mythology A to Z. Facts on File (J)
18. Sibanda PM (1995) The traditional views of lightning in Zimbabwe. International colloquium on overvoltages and insulation coordination. Holiday Inn Hotel, Harare, May 1995
19. Trengove E (2013) Lightning myths and beliefs in South Africa: their effect on personal safety. Ph.D. dissertation, University of the Witwatersrand, Johannesburg
20. Cooray V (2003) The lightning flash. IET, London
21. Visacro S, Soares A Jr, Aurélio M, Schroeder O, Cherchiglia LCL, De Sousa VJ (2004) Statistical analysis of lightning current parameters: measurements at Morro do Cachimbo Station. J Geophy Res Atmos 109(D1):16
22. Santiapillai C, Silva A, Karyawasam C, Esufali S, Jayaniththi S, Basnayake M, Unantenne V, Wijeyamohan S (1999) Trade in Asian elephant ivory in Sri Lanka. Fauna and Flora International 33(2):176–180
23. Jayaratne C, Gomes C (2012) Public perceptions and lightning safety education in Sri Lanka. In: 31st international conference on lightning protection, Vienna, Sept 2012
24. Gomes C, Ahmed M, Abeysinghe KR, Hussain F (2006) Lightning accidents and awareness in South Asia: experience in Sri Lanka and Bangladesh. In: Proceedings of the 28th international conference on lightning protection (ICLP), Kanasawa, Sept 2006
25. Mary AK, Gomes C (2012) Lightning accidents in Uganda, 31st International Conference on Lightning Protection, Vienna, Austria, September 2012
26. Cardoso I, Pinto OJr, Pinto IRCA, Holle RL (2014) Lightning casualty demographics in Brazil and their implications for safety rules. 135–136, 374–379
27. Gomes C, Ab Kadir MZA, Cooper MA (2012) Lightning safety scheme for sheltering structures in low-income societies and problematic environments. In: 31st international conference on lightning protection, Vienna, Sept 2012
28. Gomes C (2019) Lightning safety structures for applications in the industrial sector and underprivileged communities in Africa, Wattnow, Aug 2019, pp 18–26
29. Gomes C (2011) On the selection and installation of surge protection devices in a TT wiring system for equipment and human safety. Saf Sci 49:861–870
30. Gomes C, Ab Kadir MZA (2011) A theoretical approach to estimate the annual lightning hazards on human beings. Atmos Res 101:719–725. https://doi.org/10.1016/j.atmosres.2011.04.020

31. Gomes C, Diego AG (2011) Lightning protection scenarios of communication tower sites; human hazards and equipment damage. Saf Sci 49:1355–1364. https://doi.org/10.1016/j.ssci.2011.05.006
32. Chandimal APL, Gomes C (2012) Lightning related effects to the neighborhood due to the presence of telecommunication towers. In: 31st international conference on lightning protection, Vienna, Sept 2012
33. Christian HJ et al (2003) Global frequency and distribution of lightning as observed from space by the optical transient detector. J Geophys Res Atmos 108(D1), ACL 4–1-ACL 4–15. https://doi.org/10.1029/2002JD002347
34. Gomes C, Khurshid ZM (2021) Do metal objects such as mobile phones increase lightning risk? Geomat Nat Haz Risk 12(1):1819–1836. https://doi.org/10.1080/19475705.2021.1946172
35. Gomes C (2020) Lightning protection of structures: how to do it wrong. Wattnow, 50–59, Aug 2020
36. Mary AK, Gomes C (2015) Lightning safety of under-privileged communities around Lake Victoria. Geomat Nat Haz Risk 6(8):669–685

Economic, Technical and Human Implications of Lightning Protection

Chandima Gomes and Ashen Gomes

Abstract Despite the significantly large number of deaths and injuries, the property loses and service downtime, lightning is not treated as a serious natural hazard in many countries of which lightning ground flash density is notably high. In this chapter, we highlight this lack of attention from both government and non-governmental sectors as a substantial barrier to curb lightning related losses. The ignorance of experts and statutory bodies that control the implementation of standards and guidelines have paved the way to the flooding of fraudulent products and technologies into the respective countries. The attitudinal issues and negligence of responsibilities of engineering and managerial capacities of both private and government sectors contributes to the mishaps and losses due to lightning-related incidents. The chapter finally discusses possible mechanisms of promoting lightning protection as business ventures in less-privileged communities by developing entrepreneurship among people having low to medium levels of technical know-how.

Keywords Entrepreneurship · Business model · Public perceptions · Economic implications · Protection measures

1 Lightning Protection as a Safety Concern

Lightning protection (LP) is a safety concern at both public and professional levels, as per the Extremely low number of papers published in safety science journals with Scopus or Web of Science database inclusion indicates that lightning has not been treated as a serious safety concern by the safety research community, both academic and professional levels, so far [1].

C. Gomes (✉)
School of Electrical and Information Engineering, University of Witwatersrand, Johannesburg, South Africa

A. Gomes
School of Electrical Engineering and Computer Science, KTH, Royal Institute of Technology, Stockholm, Sweden

C. Gomes (ed.), *Lightning*, Lecture Notes in Electrical Engineering 780,
https://doi.org/10.1007/978-981-16-3440-6_10

Even at policy-making level, lightning has been overlooked in many countries having high lightning ground flash density. A few examples can be given in this regard from several tropical and subtropical regions;

a. The Occupational Safety and Health Act, The Government of Uganda, a comprehensive document of 89 pages released in 2006, covers almost every aspect of occupational safety risks, except lightning, a term which has not been even included once in the document. Note that Uganda has one of the highest lightning ground flash densities in the world and recorded the worst single-event lightning accident in recent history where 18 students were killed while they were inside their school building [9].
b. The Occupational Safety and Health Act, The Government of Kenya, published in 2007, a 128-page document contains the term 'chemical', 'fire' and 'lightning' at 38, 28 and zero times respectively.
c. Similar observations were made in the developed countries as well. The 416-page document, Occupational Safety and Health Regulations of Western Australia published in 1996 does not contain the word 'lightning' at least once.

The above observations depict that both political and professional/academic entities, either due to ignorance or lack of awareness, treat lightning as an insignificant safety concern at the national level. This inert and non-proactive attitude of the statutory bodies towards lightning results lack risk assessment and threat level identification concerning thunderstorm related hazards, which in turn prevents the relevant parties from taking suitable measures to curb injuries, property damage and service interruption in the respective countries. Several investigations that have been done in the last decade [2, 3, 8–11] attributed the overwhelming number of lightning-related deaths and financial losses due to the lack of awareness and educational programs on lightning safety/protection among the public.

It is advantageous to investigate the causes of the attitudes of the authorities and other concerned parties on lightning safety and protection. Intuitively, one can suggest that the sporadic nature of lightning accidents and a small number of affected people per incident may be the main reason for overlooking lightning as a serious safety threat. However, collective annual statistics, clearly delineate that lightning stands as the second or third largest threat with respect to human and animal life, properties (direct and indirect effects) and information security (power downtime, data losses, signal distortion and loss of signals etc.). It is of interest to note that countries of relatively low lightning ground flash density, but having "developed country" status (eg. Europe, and North America) pay high attention to curbing lightning threats through, standards and guidelines, safety programs, government or state legislations, and inclusion of lightning safety in school curricula, compared with the attention of less developed countries in the lightning dense geographical locations.

2 Decision Making in Adopting LP

The lack of interest in adopting appropriate lightning safety and protection measures, in many countries, is not only a drawback among the statutory bodies, policy-making institutes etc. alone. Even LP seekers at the industrial level and among the general public have similar perceptions. This fact is well supported by the research done by Gomes and Gomes [4]. As per their study, in South Asia, out of those who have adopted LP measures for their buildings, less than 1% have conducted proper risk assessment before making the decision on the type of LP system to be installed. The same percentage has been found to be 50% in South Asia and 70% in Middle East Asia, despite the fact that in these regions the reported lightning hazards is almost 20 times less than that in South Asia.

As per the same study, the questionnaire surveys reveal that the decision of adopting LP measures by the commercial and industrial sectors is based on several factors in the Asian region. The list given below depicts these reasons in descending order of the level of prominence given to each reason by the respondents.

- The lightning accident has happened at the own premises
- A marketing representative from an LP vendor has visited and persuaded them to install the LP system
- The lightning accident has happened in the neighbourhood
- Insurance companies have insisted (imposed a higher premium for not having LP measures)
- A high ranked company representative has participated in an LP training program
- The maintenance engineer or any other senior engineer has anticipated lightning threats.

A similar survey that has been carried out by the team lead by the author, in 2019, reveals that the situation is not very much changed in South and South East Asia. The latest survey also shows that a risk assessment has been demanded by the LP seeker (from the LP provider) only in the case of the last two reasons listed above. Such cases were a clear minority among the total list of respondents. The awareness of a majority of the respondents on the globally or nationally accepted LP standards, was disappointingly low, despite the majority of the respondents were either engineers or engineering managers. Two relatively small countries, Sri Lanka and Singapore showed marked difference in the knowledge and awareness of their engineers with respect to the LP measures, standards and good practices. A large number of training programs conducted by several expert groups in these countries could be the reason for this exception.

As it has been observed in the surveys done by Gomes and Gomes [4] and also that by the authors' team in the recent past, the lack of correct motivation in the decision making of installing LP system opens an opportunity for the vendor to dictate the selection of LP scheme to the client. The lack of paying any significant attention to the risk assessment by the client gives the advantage to the vendor to decide the level of protection and the number of items, where and what to be installed etc.

Most often it has been noticed that the opportunistic vendors have made unnecessary or sometimes even hazardous selections. In recommending these LP measures, one could clearly notice the ulterior motives of the vendor, which is far from technically and scientifically sound facts and figures.

The most common unscrupulous practice among many LP vendors was to provide a structural protection system to a building that has an extremely low risk of strike as per a proper risk assessment. These buildings include low rise buildings in areas of low lightning occurrence density, totally metallic structures made of suitably thick materials where only a simple earthing system is the only requirement, and buildings protected by high rise structures in the near vicinity (e.g. base stations underneath tall and well-grounded metallic towers). The vast majority of installations in such cases are non-conventional LP systems, predominantly, early streamer emission (ESE) devices. The ESE technology has not been adopted by the International Standards (IEC) or most other national standards due to its lack of scientific grounds or technical justification. As the installations mentioned above are on low or no risk buildings they do not pose any significant probability of strike even without the LP system. Thus, the vendors who provided LP is at almost zero risks of failure. Such cases also contribute immensely to the no-accident statistics of installations with non-conventional devices, which is a false indication of the success of the technology.

Implementing LP systems, especially those consist of copper components on all-metal structures, have another adverse effect. Copper down conductors on buildings made of steel metal structure and corrugated steel roofing strongly promote the corrosion of the steel due to the galvanic effect. In most of the cases, the dimensions of the roofing and supporting materials are well above the minimum values specified in IEC 62305-3 [7] for being self-sufficient air-termination and down conductor system. However, the unnecessary installation of the LP system causes serious corrosion problems which are detrimental to the building structure, especially in areas with high salinity and acidity in rainwater.

Another financially and technically erroneous LP practice that can be noticed in many countries irrespective of the region (as per the informal investigations done in the Middle-East, South and South-East Asia, and Africa), is the installation of an unnecessary number of surge protective devices (SPDs), without a proper coordination plan. It is not uncommon that commercial and industrial sites, where the authorities have requested LP vendors to provide protection to their equipment, SPDs have been installed without any justification with respect to the selection of location and specifications. In these cases, it has frequently been noticed that some robust equipment and components are over-protected whereas some sophisticated equipment and systems are under-protected or not protected at all.

The issue of the suitability and scientific background of the non-conventional air terminations has been debated for almost two decades now. The percentage of LP systems installed according to IEC 62305-3 [7] or similar is less than 5% in Sri Lanka, about 20% in Malaysia and about 10% in Indonesia. The rest are partial protection systems (usually single copper rod grounded by a single down conductor). Pakistan, Bangladesh, Nepal, and the Middle-East is yet infiltrated by the vendors of non-conventional technology. In Singapore, such technology is forcefully suppressed as a result of the dominance of few individuals.

It is of interest to analyse the human psyche that tempts to adopt technologies that have not been accepted by international and their own national standards and rejected by a clear majority of the scientific community. Irrespective of the fact that the performance of such technologies is unproven, theoretically, experimentally or statistically, it seems that the products based on such technology are successfully marketed in South and South-East Asia, and even in several developed countries in Europe. In some countries, such as Malaysia, despite government warning not to adopt non-conventional LP systems, the visual observations could make a rough estimation that over 75% of lightning structural protection systems installed are non-conventional. The following reasons could be figured out for such an overwhelming number of non-conventional systems.

- Inclusion of the predominant non-conventional LP system, ESE technology has been included in French and Spanish Standards, despite they are not recommended by many other national standards in addition to the IEC Standards. Being well-developed influential European countries, France and Spain make a considerable psychological impact on developing nations. Once said that such countries have included ESE technology in their standards the product gets an automatic endorsement to convince the general public regarding its efficiency. Even under a legal framework, the vendor is safe as his product is in compliance with a European Standard.
- Non-rejection of ESE technology by any standard is another Standard-related issue that could be noted. Although many standards have not included ESE technology in their recommendations, none of the standards has a vehement rejection of the ESE technology. Under such circumstances, there are simply no grounds to persuade an ordinary engineer to reject a product based on ESE technology. One cannot expect a field engineer to read research papers or scientific documents as they are burdened with routinely work. The situation becomes even tougher for an anti-ESE campaigner when it comes to the convincing of decision-makers who are most often non-technical personnel.
- The introduction of ESE and other non-conventional technology as a new technology is another marketing strategy adopted by such vendors. It is very natural that the general public loves modern technology. The ESE proponents use this human thinking pattern much to their advantage.
- The lucrative appearance of many non-conventional air-termination systems compared to simple metal rods is a much attractive object to the human eye and mind. In contrast, ESE air-terminations come with various shapes and chrome plated surfaces. To compete with such advantages acquired by non-conventional LP vendors several manufacturers in some countries promote conventional systems.
- In contrast to an LP system designed according to IEC 62305-3 [7] or similar, an ESE technology-based LP system, designed according to a standard such as NFC-17-102 [12] or UNE 21186 [14] is less laborious and more convenient to

be installed. Especially in the case of buildings that have decorated and complex shaped roofing, non-conventional technologies are much convenient to be installed and also be hidden from the public view (to preserve the aesthetic appearance).

- The marketing promotional campaigns of vendors that sell non-conventional LP systems are much more rigorous and aggressive than those of other companies. The major reason for such affordability for marketing campaigns is the large profit margins that have been enjoyed by the ESE device vendors.
- Another, unchallengeable strategy adopted by non-conventional system vendors is to reward heavily the consultants that recommend and promote such technologies. The large profits gained by the vendors make it permissible for them to offer lavish rewards to the consultants, thus, in turn, they get more business; hence more profits. This positive feedback loop gradually adds more consultants into the loop and destroys the companies that are reluctant to stay away from the loop. During the last five years of this investigation, most of the companies that were previously reluctant to market ESE technology were sucked into the loop as they could not survive in the business outside the loop.
- The non-conventional system manufacturers and marketers have created an atmosphere in many countries that the total solution of LP depends on the efficiency of non-conventional technology. Hence, there are numerous cases where the LP seekers demand non-conventional systems, despite the attempts made by the scientific community to make them aware of the reality. In such cases, the LP assignment consists of both structural and surge protection, thus refusal of providing a non-conventional structural protection system results in the rejection of the total solution by the client. Such a situation prompts even those who are against non-conventional systems to include such systems in their business plan to ensure that they will not lose opportunities to market their surge protective devices.

3 Issues Due to Erroneous Engineering Practices

Similar to any other technical solution in power systems, LP should also be considered as an integral part of the entire electrical network. Thus its success strongly depends on the overall correctness and appropriateness of the engineering practices adopted in the electrical system. Several investigations have reported that in a number of cases the losses and damages are attributed to are wrongly lightning; whereas the real cause is the erroneous electrical installation and maintenance. The LP vendors are also responsible (partly) for such mishaps. As per the surveys conducted by Gomes and Kadir [5] and Gomes and Gomes [4], LP vendors often forward quotations, without getting the service of a consultant, or even visiting the premises to find the actual situation. Thus, the recommendations are solely depending on a line diagram in most cases. Even if a team visits the premises either in advance or during the installation of LP systems, the vendor hardly requests the customer to rectify the drawbacks of the electrical system, when such are noticed. This is done either due to the lack of knowledge on the problem or the fear of losing the contract. The result will be the failure of equipment even after the installation of the LP scheme. Several of such problems in the electrical system are listed below.

- Grounding at various points of the wiring system: This is one of the commonest problems in the subcontinent. About 20% of the engineers interviewed in the study by Gomes and Gomes [4] have expressed the view that greater the number of leads from the wiring system to the mother earth better the safety. Most of the engineers who had this view are electronic/communication engineers. As we suspect the recommendation of the manufacturers of communication equipment to have separate (or dedicated) ground may have prompted the engineers to have such view. There were also many installations where the SPDs are connected to a grounding system different from the power ground.
- Wiring system defects: There are many malpractices in the installation and maintenance of the wiring system, as we have observed. A few of them are stated below.
  - Selection of wrong colour code: We have come across few sites where even green/yellow grounding wires are used for live or neutral at few locations of the wiring system, which may be extremely hazardous.
  - Damaged wires due to mechanical mishaps and rat/squirrel bites: Ina addition to the safety threat, such damaged points give rise to regular arcing, generating transients in the system. In the Middle-East and some parts of South and South-East Asia, we frequently observe damage to outdoor cable insulation due to extreme weather conditions.
  - Birds/squirrels make nests in panel boxes: Such animals and parts of their nests may cause sporadic arcing between bus bars.
  - Unplanned power feeds routed outdoor: One of the biggest challenges of rectifying the wiring systems is to figure out and remove/re-route power lines that extend to outdoor feeds from points within the building. To make the situation worse, the extensions are most often taken from unsuspecting points; plug points, lamp holders, or even splitting the insulation of the wires at any place convenient to the technician. Typically, these extensions are done on a temporary basis and after the purpose is served the extension is left unattended.
- Absence of electrical safety devices: In the entire sub-continent it is only in Sri Lanka the installation of both earth fault tripping and over current tripping devices are compulsory. In the Middle-East and South East Asia most of the industrial sites are installed with such devices but not in small-scale out-of-city industries.
- Unattended defective electrical appliances: Flickering fluorescent lamps, noise generating old UPSs, defective capacitor banks and inductive loads etc. are few examples of transient generators within premises.
- Irresponsible switching operations: We have come across on several occasions the on–off operations of some sophisticated loads have been done by inappropriate means. One common example is the switching on–off large number of computers by on circuit breaker, to save time and labour.

4 Unethical Marketing Practises

There are two unethical marketing strategies used by a few sectors of the LP service providers, mostly in Asia and Africa. One of them is the testing of their products for a certain quality of their product and interpret the certification from a totally different perspective. A common example is the wrong interpretation of the impulse current withstanding capacity test outlined in IEC 62305-3 [7] for structural protection components, as an endorsement of ESE and other non-conventional technologies. The IEC and several other reputed standards specify a certain lightning current that all protection components should withstand. The test results reflect the ability of the component to withstand up to a certain maximum impulse current once it is injected into the component. This withstanding value depends on the material and dimension of the component and waveform characteristics of the injected current. The test results do not indicate or even imply at all the ability of the component, especially the air-terminations to attract or repel lightning step leaders. However, pathetically, the ESE vendors interpret such test results as the endorsement of their product by the relevant international or national standard in effectively attracting lightning, whereas proponents of lightning repelling systems (another non-conventional technology) interpret the same certification as an endorsement of their product's ability to diverge lightning away from the structure.

Another very problematic issue in many countries (both developed and developing) is the marketing of fake LP devices, especially SPDs. The issue is much serious in developing countries compared to that in developed countries due to the below standard practices of policy implementations and import regulations. There are many electrical items in these countries, especially those related to LP systems, that have been counterfeited by copying popular product brands. In many other electrical products, the counterfeit has some value, although the quality is most often less than that of the genuine item. Hence if the price is proportionately low, people buy them, sometimes knowingly, although such purchase is not ethical in the strict sense. On the contrary, the counterfeit SPDs have zero value (except for the plastic casing and the material filled inside to increase the weight).

In one such case that has been come across during an investigation, Gomes and Kadir [5] report the following observation. The SPDs showed in Fig. 1 bear a reputed international brand name, and are installed in the premises of a financial institute in South Asia. The investigators have noticed that one of the SPDs in the panel is discoloured with smoke/burn signs in the panel and panel cover. The fault indicator of the SPD (the red colour button that should pop out in the event of SPD failure) shows that it is in good condition. The investigation team has requested the technical personnel of the institution to remove the SPD from the panel and dissected it to inspect inside. The team has found a partially melt lump of PVC-like material inside. There were no signs of any surge protective component such as MOVs, GDTs, Zener diodes etc. The fake product designers most often become very successful in their endeavours so that their products look identical in appearance to the genuine component so that even the experts in the technical field could not identify the

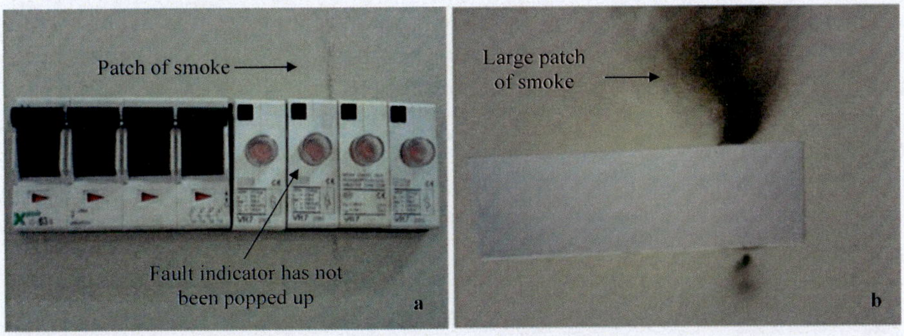

Fig. 1 **a** Damaged SPD with a faked brand name. Note that the fault indicator has not been popped up irrespective of arc signs. The defected SPD has been discoloured showing signs of internal heating, which was verified on opening the device. **b** Cover of the panel with a large patch of black smoke. Adopted from Gomes and Kadir [5]

counterfeit without conducting a proper test. Therefore, it is a tough task to provide public guidance on identifying fake devices with simple observation, in the case of LP items. The national standard institutions and such committees should bring up strict policies and guidelines to monitor the quality of products that are imported to the country and also produced within the country.

5 Engineering and Management Attitudes

The experience of authors in the South and South-East Asia, and Africa reveals that in most cases of electrical system failures, sociological and psychological causes are responsible for the mishaps than the technical causes. These causes play a significant role in the prevention of lightning-related hazards in many industrial and commercial premises. An investigation that was done by Gomes and Gomes [4] show that administrative barriers and rigid attitudes at managerial capacities could hinder the performance of subordinate technical staff in taking the right decision at the right time. A few of these cases are discussed below.

Several field-level technical personal who have been interviewed stated that they are highly reluctant to work beyond their routine workload in rectifying any technical issues unless there are catastrophic cases. In a majority of institutions, the decisions to invest in infrastructure development are taken at the managerial or directorial capacity of which the composition has no technical or scientific background. Hence, to convince such capacity the technical staff needs to quantify the safety and protection in terms of monitory values. Such practise is tedious and the ordinary engineer/technician is hardly rewarded for such efforts. Thus, once lightning-related losses and damages are observed, instead of rectifying ground-level issues (such as wiring defects, power/signal line routing errors, placement errors of equipment, general grounding faults etc.) the engineer or technical in charge tends to make a recommendation to the management asking for installing an LP system. On this recommendation, the authorities call quotations from LP suppliers. As there are, most often, no knowledgeable personnel in the staff to evaluate the quotations, the contract is given to the lowest bid, unless there are some other non-technical reasons to award it to a higher bid. Sometimes, the management rejects the request to install a surge protection system after installing the structural protection system (or vice versa), stating that an LP system has already been installed.

Gomes and Gomes [4] has also reported that at several institutions the management has asked the engineering staff to refer the request for an LP system to the financial departments to check the possibility of obtaining insurance coverage against the lightning hazards in the place of an LP scheme. Due to the high competition among the insurance companies and also due to the lack of knowledge on LP, most of the insurance companies cover the risk against lightning damage (and even the losses due to downtime) without demanding a proper protection scheme.

In most of the industrial and service sectors, the administers are very reluctant to shut down the power supply for the requirements of rectification or replacement. However, the management overlooks the fact that in the event of transient damage the most probable outcome is the unexpected and uncontrolled power outing, which may cause an extensive downtime loss.

Sometimes the installation of the LP system needs structural or electrical network alteration which is a burden for the maintenance engineers as they need to plan and design modifications and relocations that are outside their routinely work. Apart from this attitudinal barrier from the maintenance sector, some psychological inertia could also be displayed at the administrative or managerial level as such additional infrastructure modification needs various financial and regulatory approvals. Thus, in such situations, taking an insurance policy to cover potential losses becomes most often preferred to the adaptation of comprehensive LP measures.

6 Lightning Protection as a Business Model

Being a relatively focussed field of engineering sciences, LP is treated as a business sector that requires highly specialized technical expertise. Such conceptual norms at the societal level place LP consultancy, design, supply and commissioning, a commercial venture that is operated by a narrow layer of the business community. Since such expertise is not very common even among electrical engineering and technical sectors, only a limited number of private and public institutes could provide the service to the protection seekers. This is one of the prime reasons that in many countries LP is treated as a practice of affordable society. In reality, the research done in many developing countries reveals that lightning-related deaths and injuries are more prevalent in underprivileged communities than that in the affluent society [6, 13]. In this backdrop, it is highly recommended to popularize LP as a business venture among industrially and economically backwords societies in developing countries. Such expansion of business activities will serve two burning issues of many countries, especially those in landscapes of high lightning density; LP measures will be more affordable to many underprivileged societies that badly need such, and a significant number of employment opportunities will be opened to the job market at the lower-technical end.

LP as a business model yields several expertise and skills at a few levels of education. Thus, if LP is introduced as a viable business at lower-economic layers, one should develop a practical model to address these requirements. On top of the business model, there should be an entity that provides a mechanism for guiding the stakeholders.

- Role of the lightning centre

As it is described in the previous chapter, a lightning centre (LC) in a given region has to play a key role in developing entrepreneurship among suitable candidates. The service required from such LC includes developing financial model, selection and training of entrepreneurs, updating the entrepreneurs with the latest knowledge, and information on various components and materials available at the local level, supporting them in addressing technical issues and providing the trained personal the social recognition in the target communities. The LC is also expected to ensure that the trained entrepreneurs do not fall into prays of the vendors that market fraudulent and unscientific products/technologies.

- Basic training and skills development

The LC may select 5–10 people with medium financial strength and preferably some fundamental technical knowledge and educate them in LP measures. LP education should include, basic concepts of lightning, human safety, risk assessment, structural protection, surge protection, installation concerns, costing and budgeting and engineering, technical and business ethics. As it was discussed in the previous chapter, this knowledge should come from the international/national level to the LC level. The selected entrepreneur should be a trainer for his employees thus, the LC needs to conduct regular programs of training of trainers. Once the candidate successfully completes the training program, the LC could issue a certificate and an identification card to the entrepreneur recognizing his skills and abilities. Such recognition is essential for a fresh technical entrepreneur to build up the confidence and trust of society. The LC can also publish a list of authorised technical personal (by the LC) on their website so that protection seekers can check for the validity of the entrepreneur's claims.

- Financial model

One of the toughest hurdles for a low or medium-income entrepreneur in developing countries is to find the capital and initial operating investment. A majority of financing institutes such as banks are reluctant to provide loans for such individuals due to the lack of confidence. The LC can be the mediator in such a case as they get a better impression of the ability of the entrepreneurs by the end of the training program. There is also a possibility that the LC itself could be able to win funding from various institutions (both governmental and non-governmental) to support the entrepreneurs.

In such cases, the LC may develop their own business-monitoring mechanism to make sure that the ventures move in the right direction.

- Risk assessment and design

The biggest challenge in venturing into LP as a viable business is the requirement of conducting risk assessment and consequent design of the LP system for a complex installation, according to the outcomes of the risk assessment.

In the modern world, computer literacy is fast improving among even the lower economic layers of many developing countries. Therefore, it may not be difficult for the business owner (henceforth termed as the entrepreneur) to get the service of a person to use risk assessment and design software to produce the required LP layout. As such computation and simulations need occasionally, the entrepreneur may come to an agreement with the IT specialist for getting his service on a hire-on-demand basis.

The available software of risk assessment and LP measure design in the market are relatively expensive and the usage of such software by one entrepreneur is occasional. Hence, it is greatly advantageous if the LC purchases the software and hire it at an affordable fee to the entrepreneur. In this case, based on the demand, the LC may hire an IT expert or train its own staff member to run the computation.

- The responsibility of the company (entrepreneur)

The LC has the prime responsibility of educating the entrepreneurs on engineering and business ethics as an essential part of the training program. In contrast to marketing consumables or other luxuries, LP is a business that involves the safety of human life. Thus, the businessperson should have deep knowledge on his/her responsibilities and commitments to the protection seeker and the society as a whole. The lack of knowledge on a particular technology is not a justification for failures or lapses in the products/services provided. Both the LC and the companies are urged to look for scientifically accepted low-cost solutions that can be given to the shelters of underprivileged societies, who could not afford standard LP measures even at a significantly reduced cost. The international research frontiers are regularly updated at present with information of the availability of such measures.

References

1. Gomes A, Gomes C (2014) Hierarchy of hazard control to minimize lightning risk. In: 32nd international conference on lightning protection, pp 1405–1414, Shanghai, China, October 2014
2. Cardoso I, Pinto O, Pinto IRCA, Holle R (2014) Lightning casualty demographics in Brazil and their implications for safety rules. Atmospheric Res 135–136:374–379
3. Doljinsuren M, Gomes C (2015) Lightning incidents in Mongolia. Geomat Nat Haz Risk 6(8):686–701
4. Gomes C, Gomes A (2016) Lightning safety psyche. In: 33rd international conference on lightning protection, Estoril, Sept 2016

5. Gomes C, Kadir MZAA (2011) Lightning protection: getting it wrong. IEEE Technol Soc Mag 30(2):13–21
6. Gomes C, Izadi M (2019) Lightning caused multiple deaths: lethality of taking shelter in unprotected buildings. XI SIPDA—international symposium on lightning protection, Sao Paulo
7. IEC 62305-3 (2010) Protection against lightning: physical damage to structures and life
8. Lubasi FC, Gomes C, Ab Kadir MZA, Cooper MA (2012) Lightning related injuries and property damage in Zambia. In: 31st international conference on lightning protection, Vienna, Sept 2012
9. Mary AK, Gomes C (2012) Lightning accidents in Uganda. In: 31st international conference on lightning protection, Vienna, Sept 2012
10. Mary AK, Gomes C (2015) Lightning safety of under-privileged communities around lake Victoria. Geomat Nat Haz Risk 6(8):669–685
11. Mulder MB, Msalu L, Caro T, Salerno J (2012) Remarkable rates of lightning strike mortality in Malawi. PLoS ONE 7:1–4
12. NFC-17-102:2011 (2011) Protection against lightning—early streamer emission lightning protection system
13. Syakura AR, Izadi M, Osman M, Ab Kadir MZA, Elistina AB, Gomes C, Jasni J (2019) On the comparison of lightning fatality rates between states in Malaysia from 2008–2019. In: 11th IEEE Asia-Pacific international conference on lightning (APL), Hong Kong
14. UNE 21186:2011 (2011) Protection against lighting: surge arresters using early streamer emission air terminals

Frontiers in Lightning Research and Opportunities for Scientists from Developing Countries

Adonis F. R. Leal

Abstract Lightning research has been playing an important role for decades in the scientific community. Lightning researchers have been putting a huge effort into solving issues in electromagnetic modeling, instrumentation, and the analysis of large datasets. Scientists from developing countries are in constant improvement of their techniques for lightning research because they need to do more with fewer resources. In this chapter, we discuss a brief history of lightning research to understand at which point we are now. After that, it is shown the main topics of lighting research that, the most relevant journals published at the beginning of the XXI century. Later, it is discussed some studies that were carried in developing countries including Brazil, Colombia, India, Malaysia, South Africa, and Sri-Lanka, that are the more prominent developing countries in lightning research. In the end, it is discussed the frontiers in lightning research for developing countries, which it is addressed the use of artificial intelligence in lightning research.

Keywords Developing countries · Lightning research · Main topics of lighting research · Artifitial inteteligence

1 A Brief History of Lightning Research

Lightning research has been paying attention to scientists for a long time. The scientific research of lightning may have begun in the middle of the eighteenth century. Between 1747 and 1950, Benjamin Franklin proposed an experiment to prove and determine that lightning is electricity [1]. In his proposal, he said "On top of some high tower or steeple place a kind of sentry-box, big enough to contain a man and an electrical stand (insulating stool or table). From the middle of the stand let an iron rod rise and pass bending out of the door, and then upright 20 or 40 ft, pointed very sharp at the end. If the electrical stand is kept clean and dry, a man standing on it when such clouds are passing low might be electrified and afford sparks, the rod drawing fire to

A. F. R. Leal (✉)
Department of Electrical and Biomedical Engineering, Federal University of Para, Belem, Brazil
e-mail: adonisleal@ufpa.br

C. Gomes (ed.), *Lightning*, Lecture Notes in Electrical Engineering 780,
https://doi.org/10.1007/978-981-16-3440-6_11

him from a cloud." Later, he realized that the experiment could be made by flying a kite. In this second approach one needed to fly a kite in a thunderstorm, a conductor wire would be attached to the kite and a grounded metal should be placed close to the metal wire in order to see small sparks. This experiment was the so-called kite experiment. Curiously, it was not Benjamin Franklin the first one to perform this kind of experiment. On 10 May 1752 in the village of Marly-la Ville, near Paris, Mr. Thomas-François Dalibard and helpers performed an experiment based on the kite experiment. They drew sparks from a tall iron rod that was insulated from the ground by a wine bottle. Those sparks were the first direct proof that thunderclouds contain electricity. After Mr. Thomas-François Dalibard's experiment, many other scientists conducted systematic experiments in order to investigate the electrical properties of lightning [2].

Only about one month later of Mr. Thomas-François Dalibard's experiment, in June 1752 Benjamin Franklin performed the kite experiment in Philadelphia and drew sparks himself from a key attached to the conducting string of the metal wire of his kite that was insulated from the ground by a silk ribbon. In 1753, he received the prestigious Copley Medal from the Royal Society, in recognition of his "curious experiments and observations on electricity."

Benjamin Franklin, Thomas-François Dalibard, and other scientists of the eighteenth century finally proved that somehow clouds store electricity, and lightning is a manifestation of that electricity, and so it is not a manifestation of Gods as ancient myths explained the nature of lightning. That was the beginning of the systematic study and research of lightning nature.

2 Main Topics of Lighting Research at the Beginning of the XXI Century

In this section, we discuss the main topics in lightning research at the beginning of the XXI century. The most relevant papers, from the most important Journals, that were published from 2000 to 2019 were analyzed regarding their titles and abstracts. They were classified according to different topics in lightning research. Six journal platforms were used for searching, including IEEE Explorer, Nature, Journal of Geophysical Research: Atmospheres, Geophysical Research Letters, Electric Power Systems Research, and Atmospheric Research.

In all platforms, it was applied the following filters during the search process. The title of the paper should have the word "lightning". The period of publication should be since 2000. In all searches, the papers were sorted by relevance, and only the first 20 in each platform were considered.

In the IEEE Explorer and in Nature platforms that have different journals in their dataset, the journals that appeared in the searching (the first 20 sorted by relevance) were; IEEE Transactions on Electromagnetic Compatibility, IEEE Transactions on Power Delivery, IEEE Transactions on Sustainable Energy, IEEE Transactions on Plasma Science, IEEE Transactions on Magnetics, IEEE Transactions on Dielectrics and Electrical Insulation, Nature, Nature Chemistry, Nature Astronomy, and Scientific Reports.

The elected topics in lightning research that the papers were classified are **(1) Lightning related to renewable energy**, such as lightning interaction with PV (photovoltaic) systems and lightning incidence in wind farms. **(2) Upward Lightning**. **(3) Lightning protection**, including devices for lightning protection, standards, and lightning protection systems—LPS. **(4) Lightning attachment process**. **(5) Lightning in transmission and distribution lines** that include measurements and calculation of overvoltages, specific arrangements for lightning protection of distribution and transmission lines. **(6) Lightning measurements** that include measurements of E-fields, B-fields, direct measurements of currents, measurements of sound and infrasound generated by lightning, and measurements of X-ray produced during lightning processes. **(7) Lightning tests** that include the analyses of the performance of methods and equipment using lightning impulse voltages and currents. **(8) Lightning location systems**, including different techniques for lightning location. However, in most of the cases, it noticed the newest technologies such as interferometry, lightning mapping arrays—LMA and the new Geostationary Lightning Mapper (GLM) placed on board of the GOES-16 satellite in a geostationary orbit. **(9) FDTD simulations related to lightning**. **(10) Lightning types**, including studies of negative and positive cloud-to-ground lightning, compact intracloud discharges (CID), and ball lightning. **(11) Triggered lightning**, in this topic, it is included works that used artificial initiated lightning experiments as the main source for investigating parameters of lightning or investigate techniques and methodologies related to lightning research. **(12) Lightning initiation**. **(13) Lightning occurrence**, in this topic it is included works that show the lightning occurrence in different places of the globe, and lightning density, maps, nowcast, forecast, and the relation between lightning and rainfall. **(14) Lightning Physics/Chemistry**, including the study of plasma, lightning channel structure, and etc. **(15) Lightning in other planets**. **(16) Volcanic lightning**. **(17) Modeling and equations**, and **(18) Lightning fatalities**.

As shown above, there are many topics regarding lightning research that were addressed since 2000. Now is given the information regarding which topic was the most addressed in each of the 6 platforms. It is given the percentage for the topic with the most occurrence considering the 20 most relevant papers for each platform. In IEEE Explorer journals, 45% (9 out 20) of the papers were regarding **Lightning in transmission and distribution lines**. In the Nature platform, 30% (6 out 20) were regarding **Lightning Physics/Chemistry**. In the Journal of Geophysical Research: Atmospheres, 30% (6 out of 20) of the papers were related to **Lightning location systems**. In Geophysical Research Letters 35% (7 out 20) of the works were regarding **Lightning measurements**. 30% (6 out 20) of the papers in the Electric Power Systems Research journal were regarding **Lightning related to renewable**

Table 1 Percentage of topics for all papers analyzed in this work

1	2	3	4	5	6	7	8	9	10	11	12	13	14	15	16	17	18
8%	3%	3%	3%	11%	13%	3%	13%	2%	3%	3%	4%	17%	6%	3%	3%	2%	1%

[a]Sample size, 120 papers
1. Lightning related to renewable energy, 2. Upward Lightning, 3. Lightning protection, 4. Lightning attachment process, 5. Lightning in transmission and distribution lines, 6. Lightning measurements, 7. Lightning tests, 8. Lightning location systems, 9. FDTD simulations related to lightning, 10. Lightning types, 11. Triggered lightning, 12. Lightning initiation, 13. Lightning occurrence, 14. Lightning Physics/Chemistry, 15. Lightning in other planets, 16. Volcanic lightning, 17. Modeling and equations, 18. Lightning fatalities

energy. In the Atmospheric Research journal, 50% (10 out of 20) of the works were about **Lightning occurrence**.

The results showed in the last paragraph are in agreement with the scope of each journal, or group of journals (IEEE Explorer, Nature).

Table 1 shows the overall results taking into account all the 120 papers found in the 6 platforms. According to Table 1, the topic with the highest percentage of occurrence is the "**Lightning occurrence**" in which, researchers investigate what is the incidence of lightning in different parts of the globe, and generate lightning density maps. In addition, this topic includes how lightning occurrence is related to climate or non-climate parameters such as temperature, rainfall, aerosols among others. Among the 18 topics two of them were tied in second place with 13% each "**Lightning measurements**" and "**Lightning location systems**". The third place in the ranking of the most relevant topic went to "**Lightning related to renewable energy**". The growing of papers related to lightning and renewable energy are directly related to the fast increase of renewable energy in the XXI century society.

3 Lightning Research in Developing Countries

Developing countries are characterized by being less developed industrially and have a lower Human Development Index (HDI) when compared to other countries, and have in most cases, a medium to a low standard of living. The lack of funding in education is one of the most difficult challenges to be overcome in most developing countries. Lightning research relies on observations of lightning phenomena and modeling. Both observations/measurements and modeling are together in new findings in lightning research.

In this section, it is discussed some studies that were carried in developing countries including Brazil, Colombia, India, Malaysia, South Africa, and Sri-Lanka, that are the more prominent developing countries in lightning research. All the countries are in tropical or subtropical regions. Regions in the globe in which lightning incidence are more likely to occur.

i. Brazil

Brazil is a continental country, in which most of its territory has a tropical climate, with the south being relatively temperate. Most of the lightning researcher centers in Brazil are located in the southeast region of the country in which most Brazilian people live and where is located the largest cities. Perhaps, for these reasons, lightning research in Brazil started in São Paulo. The history of lightning detection in Brazil is well addressed in [3]. In Brazil, the reference center for lightning research is the National Institute for Space Research (INPE). In 1987 the Atmospheric Electricity Group (ELAT) was created at INPE, and since then has performing different types of lightning research.

In the past decades, most of the lightning research in Brazil was restricted to the southeastern region of the country. For instance, Pinto et al. [4] analyzed the characteristics of cloud-to-ground lightning flashes that occurred in the summer of 1992–1993 in the state of Minas Gerais (see Fig. 1). The data was provided by the first Lightning Location System (LLS) in operation in Brazil. They found, that 35% of cloud-to-ground lightning flashes were positive, and for negative cloud-to-ground lightning flashes, the average number of strokes per flash was 2.9. Later, other different studies were conducted in the same region, such as the seven-year study of negative cloud-to-ground lightning characteristics in Southeastern Brazil [5], climatology study of large peak current from cloud-to-ground lightning flashes in southeastern Brazil [6], analyses of anomalous lightning activity over the metropolitan region of São Paulo due to urban effects [7], among others.

On the other hand, lightning research in other regions of Brazil is relatively scarce. It is known that most of the lightning activity in Brazil is in the Amazon region of the country (see Fig. 2). According to the national ranking of lightning incidence in

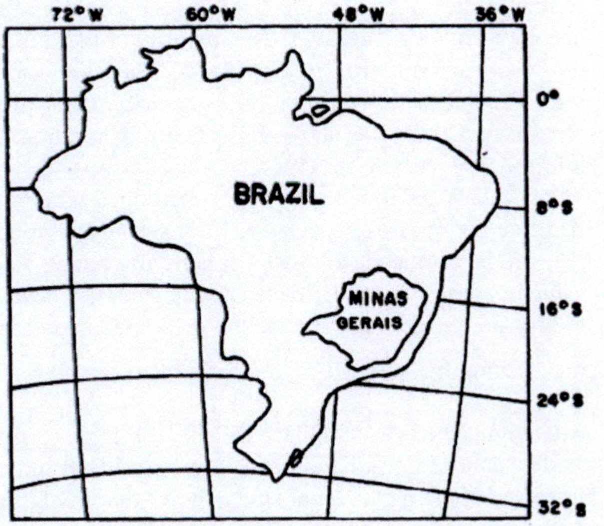

Fig. 1 Map of Brazil, indicating the state of Minas Gerais where was installed the firs lightning location system in Brazil [4]

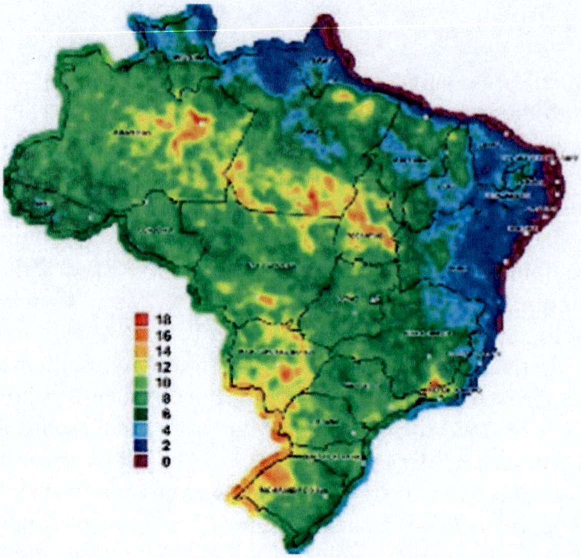

Fig. 2 Map of lightning flash density in Brazil according to the Brazilian standard for lightning protection (NBR-5419)

Brazil provided by INPE/ELAT, all of the 10 top cities regarding lightning density are located in the Amazon region. Additionally, according to INPE/ELAT, the state with the highest lightning density is Tocantins, with 17.1 lightning/km^2/year, followed by Amazonas (15.8), Acre (15.8), Maranhão (13.3), Pará (12.4), Rondônia (11.4), Mato Grosso (11.1), Roraima (7.9), Piauí (7.7) and São Paulo (5.2), being the top 10 states with the highest lightning density per square kilometer per year.

ii. **Amazon region of Brazil**

A few studies related to the lightning incidence in the Amazon region have been developed in recent decades. The majority of the studies were dedicated to the lightning characteristics in the region [8–10]. The latter studies on lightning in the Amazon region found that a considerable percentage (7%) of lightning return strokes have reported peak currents above 100 kA [8]. In addition, they found that the median of inferred peak current for first return strokes was 42 kA, and the average was 49 kA. These values are considered larger than ones found in the literature, that is, in the Amazon region, it is expected to have strong lightning events. Recently, Leal et al. [11, 12] investigated some atmospheric discharges in the Amazon region of Brazil using lightning electric field measurements. They developed low-cost E-field sensors to be used for lightning research in the Amazon region of Brazil.

iii. **Colombia**

In Colombia, the Ground Flash Density (GFD) is very high in some places. Different studies conducted in Colombia have found GDF higher than 40 flashes/km^2/year. In addition, the most recent study shows that lightning activity in different regions of the country reaches values up to 60 flashes/km^2/year. In South America, 7 out of the10 top hotspots of lightning occurrence are in Colombian territory. All of them are located near or at the valleys or foothills of the northern Andes Mountains [13]. Figure 3 shows the GFD map in Colombia using LINET network.

Because of the high lightning incidence rate in the mountains in Colombia, lightning warning systems for those regions are well investigated. In the study of Lopez et al. [15], they presented a methodology to set thresholds in lightning warning systems for preventing human accidents and sensitive systems damages. According to Tovar et al. [16], lightning risk evaluation at countries with high ground flash densities and with industries involving hazardous environments commonly doesn't satisfy the tolerable limits (IEC62305-2), especially for human losses. This situation should be well investigated in developing countries.

In 2015, a Lightning Mapping Array-LMA network was installed in Colombia (COLMA). With the LMA network (COLMA) it was possible to have 3-dimensional VHF lightning measurements in tropic areas for the first time [17]. They found that the electrical charge distribution in tropical thunderstorms shows higher vertical development reaching higher altitudes compared to thunderstorms in temperate latitudes.

iv. India

In India, between the years 1979 and 2011 about 5259 persons have been killed by lightning strikes [18]. Lightning is a real threat to human life in India, and due to its high population, lightning researchers are paying attention to study lightning fatalities in India. At the beginning of the XXI century, the number of fatalities in India increased, as seen in Fig. 4.

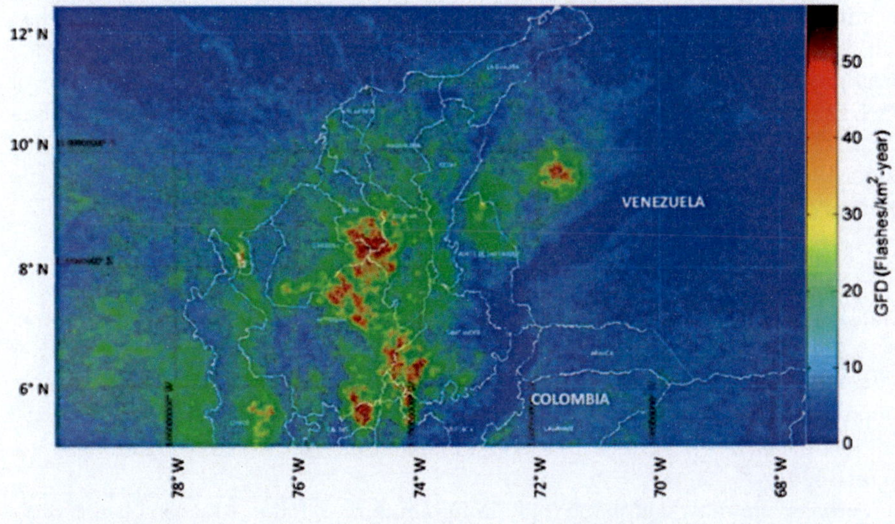

Fig. 3 Ground Flash Density in Colombia according to the LINET network. Adapted from [14]

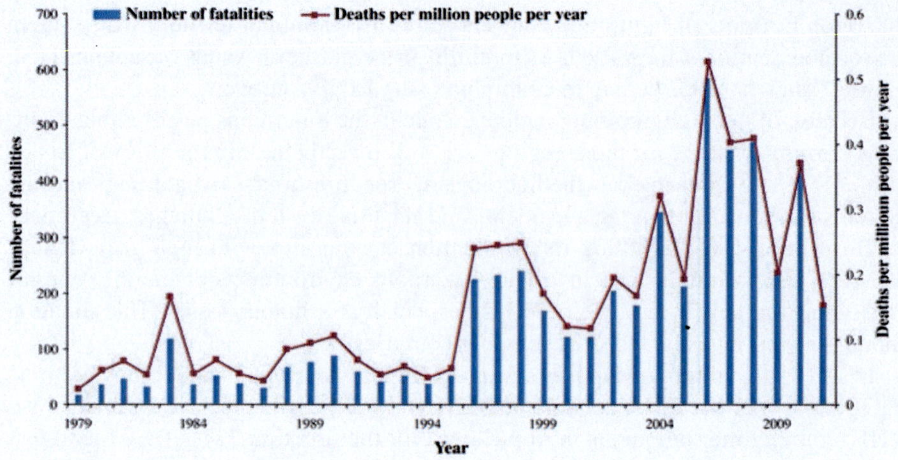

Fig. 4 Annual number and rate per million people per year of fatalities in India during 1979–2011. Adapted from [18]

According to Singh and Singh [18] males are more likely to be killed by lightning strikes in India than females, possibly due to the agricultural economy and higher work participation by males in labor-intensive practices. In addition, they observed that on the seasonal timescale, the monsoon season has the maximum percentage of lightning fatality frequencies, which suggests that major activities need to be avoided during a thunderstorm in this season. People need to know that they should to adjust their work during this time of the year.

Using the Global Lightning Dataset GLD360, which is a long-range lightning location system, Nag et al. [19] studied cloud-to-ground lightning over the Indian subcontinent. They observed data from 2012 to 2016 (see Fig. 5). For this period and location, cloud-to-ground flash and cloud-to-ground stroke detection efficiencies of the GLD360 are expected to be around 45–60% and 30–40%, respectively. According to GLD360 data, about 64,566,091 strokes occur in India per year, on average. Strong regional variations indicated that strokes were most frequent near the Himalayas and oceanic coastlines.

v. Malaysia

Malaysia is among the top three in the world in terms of lightning density, more than any other country in the Asia region. In Malaysia, more than 70% of power outages are due to lightning. Furthermore, the direct and indirect effects of lightning on power systems, communication networks, and structures may cost over RM 250 million for Malaysia, besides thousands of humans that are killed by lightning or got injured [20].

Perhaps, the most addressed topic in lightning research in Malaysia is the study of electric fields generated by lightning flashes (i.e. Arshad et al [21]; Azlinda et al. [22]; Baharudin et al. [23]; Mehranzamir et al. [24]; Wooi et al. [25]).

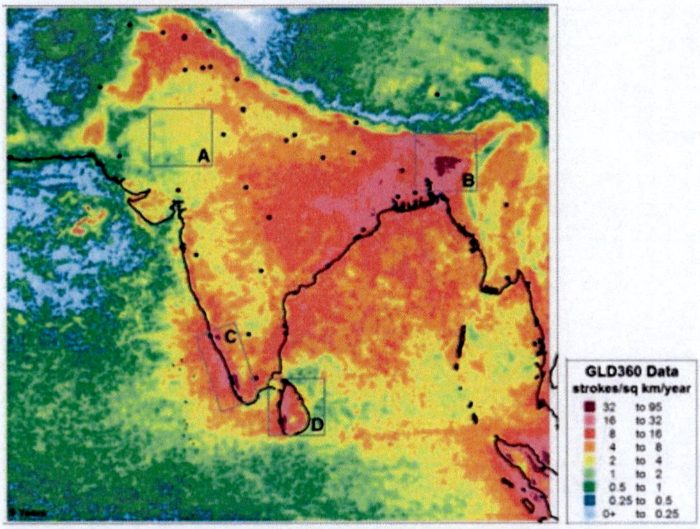

Fig. 5 Average annual stroke density (strokes km-2 yr-1) from 2012–2016. Adapted from [19]

Using parallel flat plate antennas to record lightning electric fields, Baharudin et al. [23] analyzed 100 negative cloud-to-ground lightning flashes containing 405 return-strokes. They found that the inter-stroke intervals have an arithmetic mean value of 86 ms, a geometric mean value of 67 ms, and do not depend on the return stroke order. In addition, they found that 38 flashes (38%) have at least one subsequent return-stroke (SRS) whose electric field peak was greater than that of the first return-stroke (RS). The percentage of single-stroke flashes was 16%, and the mean number of strokes per flash was 4.

A few years later, Wooi et al. [25] performed a similar study in which they analyzed statistics of electric field parameters for negative lightning in Malaysia. In their study, 104 negative lightning flashes containing 277 negative return strokes occurring within 10–100 km from the measuring station were analyzed. Differently from Baharudin et al. [23], they found that 27% of the recorded flashes were single-stroke flashes, and the average number of strokes per flashes was 2.6.

vi. South Africa

In South Africa, lightning researchers have an important tool for lightning research since 2005 when the South African Lightning Detection Network (SALDN) started its operation. The performance of the SALDN regarding flash detection efficiency and location accuracy was evaluated by Hunt et al. [26]. After the upgrade of the network in 2011 the flash detection efficiency was 76% and the median location accuracy was found to be 280 m.

According to lightning climatology in South Africa for the period between 2006 and 2010, the cloud-to-ground lightning flash density is higher along the windward

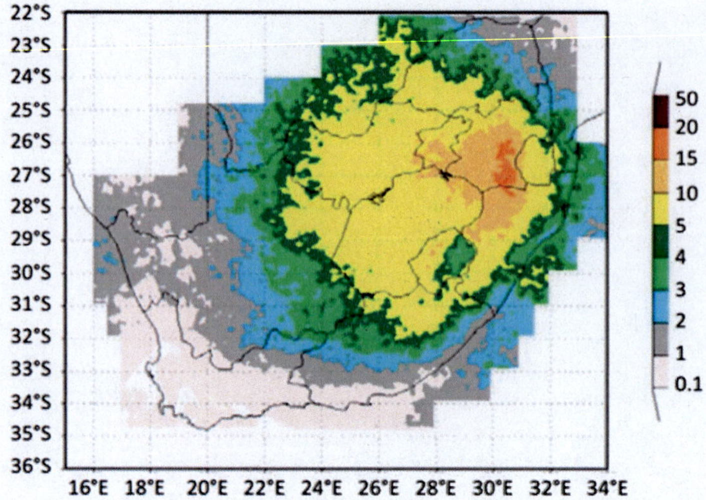

Fig. 6 Average annual lightning ground flash densities per square kilometer for positive and negative lightning flashes. Adapted from [27]

slopes of the northern parts of the eastern escarpment, extending from the northern parts of KwaZulu-Natal into the Mpumalanga (see Fig. 6) [27].

vii. **Sri Lanka**

Sri Lanka is an island country in South Asia, located in the tropics. One of the pioneer lightning research conducted in Sri Lanka was performed by Cooray and Lundquist [28]. They characterized radiation fields from lightning in Sri Lanka for the first time. They found an average number of strokes per flash of 3.2 with a standard deviation of 2.3 for a sample size of 100 lightning flashes. They also investigated the radiation field characteristics of stepping leader pulses and cloud flashes. A few years later, with a more sophisticated measuring system, Cooray and Jayaratne [29] studied lightning grounded flashes in Sri Lanka. In this other study, they found that the mean number of strokes per flash was 4.5, being 21% of the flashes single-stroke flashes. In addition, they measured the time interval of subsequent strokes and they found the geometric and arithmetic mean being 57 ms and 83 ms, respectively.

More recently, Weerasekera et al. [30], using a commercial LLS, observed cloud to ground lightning in Sri Lanka. They found that the time of the day with more lightning occurrence is 16:00 h (Local Time) (see Fig. 7), which is normal for tropic regions. They also found that 52% of negative flashes are single stroke flashes and 95% of positive flashes are single stroke flashes.

Currently, lightning researchers in Sri Lanka are investigating specific lightning discharges and lightning processes that occur in the country, such as narrow bipolar pulses (NBP) or compact intracloud discharges (CID), isolated breakdown pulses (IBPs), and the first electric field pulse of cloud and cloud-to-ground flashes [31–34].

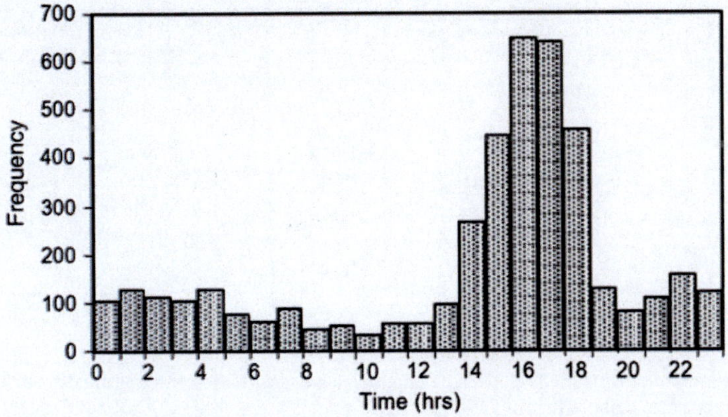

Fig. 7 Frequency of cloud-to-ground lightning flashes occurrence vs. time of the day (Local Time) in Sri Lanka. The data were recorded during the months of February, May, June, August, September and October in 1999. Adapted from [30]

4 Frontiers in Lightning Research for Developing Countries

As shown in the last section, most of the lightning research in developing countries relies on electromagnetic field measurements or analyses of data from LLS. Only specific studies use sophisticated instrumentations such as LMA or high-speed cameras. Such sophisticated instrumentations are sometimes over budget for research centers in developing countries. These limitations for lightning research in developing countries are overcome through cooperation with lightning research centers in developed countries or by developing their own low-cost instrumentation.

Disregarding the budget for lightning research, the other frontier in lightning research is the same for all researchers around the world, lightning modeling, lightning physics, and analyses of large datasets.

One trend in different fields of scientific research is the use of Artificial Intelligence (AI) to help in solving different types of problems. The use of AI techniques can help lightning researchers solve issues in lightning modeling and analyses of large lightning datasets. Scientists in developing countries can use AI to overcome the issues with a low budget and can perform more efficient lightning research.

Yang et al. [35] for instance, were able to retrieve the current waveform of a lightning return-stroke using time series neural network (TSNN), backpropagation neural network (BPNN), and measurements of lightning electromagnetic fields. They found a very good agreement between the waveforms obtained from the electromagnetic fields and the one measured at the lightning channel-base (see Fig. 8). Measurements of current waveforms at the lightning channel-base are not an easy task and applying this methodology one can have useful data for studies on lightning protection in the power system, as well as lightning characteristics.

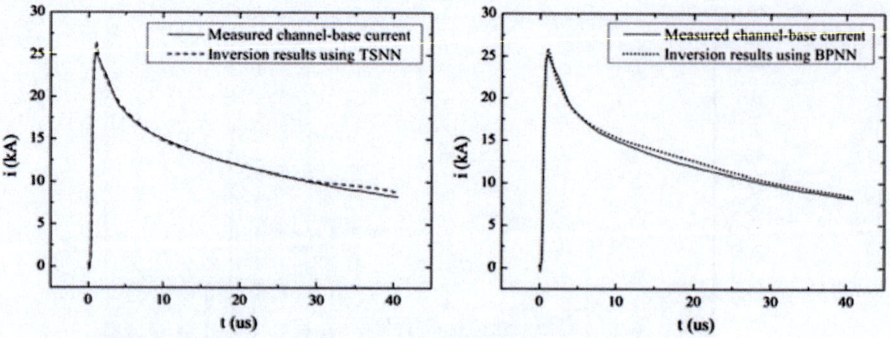

Fig. 8 Current waveforms measured at the channel-base (solid line) and obtained from TSNN and BPNN, respectively. Adapted from [35]

AI is also being used for the classification of lightning electromagnetic field signatures. Currently, many lightning researchers are collecting lightning electromagnetic fields around the world. The amount of raw lightning data is becoming so huge that is impossible to classify lightning waveforms manually. Peng et al. [36] used a convolutional neural network for the classification of lightning waveforms in Low Frequency/Very Low Frequency (LF/VLF) bands. They trained the network using a dataset including over 50,000 lightning waveforms. They used five categories to classify the waveforms, including negative and positive cloud-to-ground (N/P CG) flash, ordinary intracloud (IC) flash, preliminary breakdown pulse (PB), and narrow bipolar event (NBE). Their classifier had an average accuracy higher than 95%. Leal et al. [37] used different Machine Learning (ML) techniques to estimate lightning distance for return strokes (RS) in negative cloud-to-ground (-CG) flashes. They were able to estimate the lightning distance using a single-station E-field sensor. They splinted their lightning waveform dataset in day-time and night-time records because lightning E-field waveforms are different depending on the time in which they were recorded due to the ionosphere reflection (see examples in Fig. 9).

The summary of the results found by Leal et al. [37] is in Table 2.

Other very important applications of AI in lightning research are the Nowcast and Forecast of lightning occurrence. Wang et al. [38] used data from an electric field mill (EFM) and two orthogonal loop antennas to predict lightning occurrence. They applied artificial neural network algorithms in the data to achieve a prediction accuracy of 93.9%. In another study, Mostajabi et al. [39] used machine learning techniques to nowcast lightning occurrence using commonly available meteorological parameters (air pressure at station level (QFE), air temperature, relative humidity, and wind speed). They were able to predict lightning occurrence up to 30 min before lightning occurred. In the study of Alves et al. [40], they used data from the vertical profile of air temperature obtained from satellite NOAA-19 to predict lightning occurrence from one to five hours in advance. They used Artificial Neural Network techniques in order to build their prediction algorithms. They obtained 95.6% accuracy in predict lightning with five hours of advance.

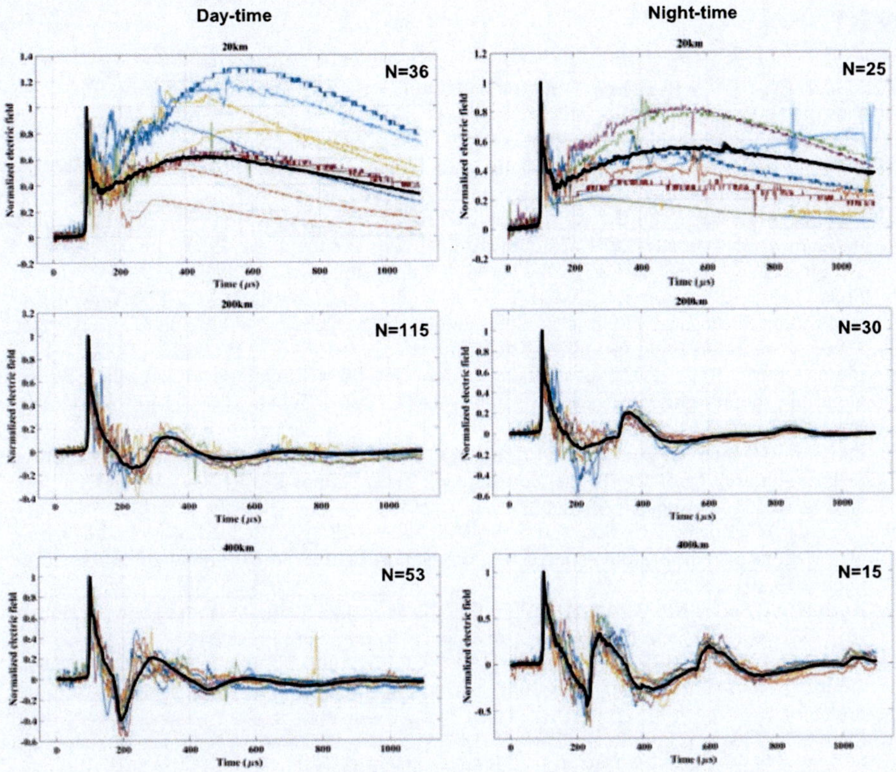

Fig. 9 Examples of electric field waveforms for ranges of 20, 200, and 400 km for day-time and night-time conditions. The fine colored lines are the individual waveforms. The thick black line corresponds to the mean waveform that was obtained by averaging individual waveforms. N is the sample size. Adapted from [37]

Table 2 Best classifier for each machine learning model technique used in Leal et al. [37]

	Day-time		Night-time	
	Best classifier	Accuracy (%)	Best classifier	Accuracy (%)
Decision tree	Fine tree	58	Fine tree	69
	Medium tree		Medium tree	
KNN	Weighed KNN	68	Fine KNN	82
SVM	Quadratic SVM	80	Linear SVM	88

Overall, AI is an important tool that lightning researchers in developing countries can use to improve their performance in finding new and important discoveries about lightning.

References

1. Krider EP (2010) Benjamin Franklin and lightning rods. In: Lightning protection. IET Publishers, London
2. Cohen IB (1990) Benjamin Franklin's science. Harvard University Press
3. Pinto Junior O (2017) The history of lightning detection and location in Brazil. In: International symposium on lightning protection (XIV SIPDA). Natal, Br
4. Pinto O, Gin RBB, Pinto IRCA, Mendes O, Diniz JH, Carvalho AM (1996) Cloud-to-ground lightning flash characteristics in Southeastern Brazil for the 1992–1993 summer season. J Geophys Res: Atmos 101(D23):29627–29635. https://doi.org/10.1029/96JD01865
5. Pinto O, Pinto IRCA, Diniz JH, Filho AC, Cherchiglia LCL, Carvalho AM (2003) A seven-year study about the negative cloud-to-ground lightning flash characteristics in Southeastern Brazil. J Atmos Solar-Terr Phys 65(6):739–748. https://doi.org/10.1016/S1364-6826(03)00077-4
6. Pinto O, Pinto IRCA, de Campos DR, Naccarato KP (2009) Climatology of large peak current cloud-to-ground lightning flashes in southeastern Brazil. J Geophys Res 114(D16):D16105. https://doi.org/10.1029/2009JD012029
7. Farias WRG, Pinto O, Naccarato KP, Pinto IRC (2009) Anomalous lightning activity over the metropolitan region of São Paulo due to urban effects. Atmos Res 91(2–4):485–490. https://doi.org/10.1016/j.atmosres.2008.06.009
8. Almeida AC, Rocha BRP, Souza JRS, Sá JAS, Filho JAP (2012) Cloud-to-ground lightning observations over the eastern Amazon Region. Atmos Res 117:86–90. https://doi.org/10.1016/j.atmosres.2011.08.015
9. Rocha BRP, Souza JRS, Costa MJ (1996) Electric and magnetic fields from lighting in Belem. In: Congresso Brasileiro de Eletromagnetismo, pp 55–58. Ouro Preto, MG
10. Souza JRS, Rocha BRP, Carreira GT (1999) CG-lightning observation (and applications) around Belém, during the 1995–1998 period. In: International symposium on lighting protection, pp 17–21. São Paulo
11. Leal AFR, Rakov VA, da Rocha BRP (2018) Upgrading the low-cost lightning detection and waveform storage system. IEEE transactions on electromagnetic compatibility.
12. Leal AFR, Shinkai R, Lopes MNG, Rocha BRP, Rakov VA, Lapierre J (2018) First lightning electric field waveform recorder permanently operating in the Eastern Amazon: preliminary results. In: Ground'2018 & 8th LPE, pp 55–60. Pirenopolis, Brazil
13. Albrecht RI, Goodman SJ, Buechler DE, Blakeslee RJ, Christian HJ (2016) Where are the lightning hotspots on earth? Bull Am Meteor Soc 97(11):2051–2068. https://doi.org/10.1175/BAMS-D-14-00193.1
14. Torres H, Perez E, Younes C, Aranguren D, Montana J, Herrera J (2015) Contribution to lightning parameters study based on some American tropical regions observations. IEEE J Sel Top Appl Earth Obs Remote Sens 8(8):4086–4093. https://doi.org/10.1109/JSTARS.2015.2428217
15. Lopez J, Perez E, Herrera J, Aranguren D, Porras L (2012) Thunderstorm warning alarms methodology using electric field mills and lightning location networks in mountainous regions. In: 2012 International conference on lightning protection (ICLP), pp 1–6. IEEE. https://doi.org/10.1109/ICLP.2012.6344397
16. Tovar C, Aranguren D, Lopez J, Inampues J, Torres H (2014) Lightning risk assessment and thunderstorm warning systems. In: 2014 International conference on lightning protection (ICLP), pp 1870–1874. IEEE. https://doi.org/10.1109/ICLP.2014.6973434
17. Lopez JA, Montanya J, van der Velde O, Romero D, Aranguren, D, Torres H, et al (2016) First data of the Colombia lightning mapping array—COLMA. In: 2016 33rd International conference on lightning protection (ICLP), pp 1–5. IEEE. https://doi.org/10.1109/ICLP.2016.7791436
18. Singh O, Singh J (2015) Lightning fatalities over India: 1979–2011. Meteorol Appl 22(4):770–778. https://doi.org/10.1002/met.1520
19. Nag A, Holle RL, Murphy MJ (2017) Cloud-to-ground lightning over the Indian Subcontinent. In: 8th Conference on the Meteorological. American Meteorological Society, Seattle

20. Ab-Kadir M (2016) Lightning severity in Malaysia and some parameters of interest for engineering applications. Therm Sci 20(suppl 2):437–450. https://doi.org/10.2298/TSCI151026028A
21. Arshad SNM, Ab Kadir MZ, Izadi M, Hamzah MN, Gomes C, Jasni J (2014) Characterization of measured lightning electric fields observed in Malaysia. In: 2014 International conference on lightning protection (ICLP), pp 1058–1063. IEEE. https://doi.org/10.1109/ICLP.2014.6973281
22. Ahmad NA, Fernando M, Baharudin ZA, Cooray V, Ahmad H, Malek ZA (2010) Characteristics of narrow bipolar pulses observed in Malaysia. J Atmos Solar-Terr Phys 72(5–6):534–540. https://doi.org/10.1016/j.jastp.2010.02.006
23. Baharudin ZA, Ahmad NA, Mäkelä JS, Fernando M, Cooray V (2014) Negative cloud-to-ground lightning flashes in Malaysia. J Atmos Solar Terr Phys 108:61–67. https://doi.org/10.1016/j.jastp.2013.12.001
24. Mehranzamir K, Abdul-Malek Z, Salimi B, Ahmad NA (2014) Return strokes measurements of electric field produced by lightning discharges in Malaysia. Appl Mech Mater 554:618–622 https://doi.org/10.4028/www.scientific.net/AMM.554.618
25. Wooi C-L, Abdul-Malek Z, Ahmad N-A, El Gayar AI (2016) Statistical analysis of electric field parameters for negative lightning in Malaysia. J Atmos Solar Terr Phys 146:69–80. https://doi.org/10.1016/j.jastp.2016.05.007
26. Hunt HGP, Liu YC, Nixon KJ (2014) Evaluation of the South African lightning detection network using photographed tall tower lightning events from 2009–2013. In: 2014 International conference on lightning protection (ICLP), pp 1746–1751. IEEE. https://doi.org/10.1109/ICLP.2014.6973411
27. Gijben M (2012) The lightning climatology of South Africa. S Afr J Sci 108(3/4). https://doi.org/10.4102/sajs.v108i3/4.740
28. Cooray V, Lundquist S (1985) Characteristics of the radiation fields from lightning in Sri Lanka in the tropics. J Geophys Res 90(D4):6099. https://doi.org/10.1029/JD090iD04p06099
29. Cooray V, Jayaratne KPSC (1994) Characteristics of lightning flashes observed in Sri Lanka in the tropics. J Geophys Res 99(D10):21051. https://doi.org/10.1029/94JD01519
30. Weerasekera A, Sonnadara D, Fernando I, Liyanage J, Lelwala R, Ariyaratne T (2001) Activity of cloud-to-ground lightning observed in Sri Lanka and in surrounding area of the Indian Ocean. Sri Lankan J Phys 2:21. https://doi.org/10.4038/sljp.v2i0.177
31. Ahmad NA, Fernando M, Baharudin ZA, Rahman M, Cooray V, Saleh Z, Dwyer JR, Rassoul HK (2010) The first electric field pulse of cloud and cloud-to-ground lightning discharges. J Atmos Solar-Terr Phys 72(2–3):143–150. https://doi.org/10.1016/j.jastp.2009.11.001
32. Gunasekara TALN, Fernando M, Sonnadara U, Cooray V (2016) Characteristics of narrow bipolar pulses observed from lightning in Sri Lanka. J Atmos Solar Terr Phys 138–139:66–73. https://doi.org/10.1016/j.jastp.2015.12.010
33. Sharma SR, Fernando M, Cooray V (2008) Narrow positive bipolar radiation from lightning observed in Sri Lanka. J Atmos Solar Terr Phys 70(10):1251–1260. https://doi.org/10.1016/j.jastp.2008.03.002
34. Sharma SR, Cooray V, Fernando M (2011) Unique lightning activities pertinent to tropical and temperate thunderstorms. J Atmos Solar Terr Phys 73(4):483–487. https://doi.org/10.1016/j.jastp.2010.11.006
35. Yang G, Chen K, Yu Z, Zhang Y, Zhou F, He J (2017) An inversion method for evaluating lightning current waveform based on time series neural network. IEEE Trans Electromagn Compat 59(3):887–893. https://doi.org/10.1109/TEMC.2016.2621139
36. Peng C, Liu F, Zhu B, Wang W (2019) A convolutional neural network for classification of lightning LF/VLF waveform. In: 2019 11th Asia-Pacific international conference on lightning (APL), pp 1–4. IEEE. https://doi.org/10.1109/APL.2019.8815977
37. Leal AFR, Rakov VA, Alves ER, Lopes MNG (2019) Estimation of –CG lightning distances using singlestation E-field measurements and machine learning techniques. In: 2019 International symposium on lightning protection (XV SIPDA). Sao Paulo, Br

38. Wang G, Kim W-H, Kil G-S, Park D-W, Kim S-W (2019) An intelligent lightning warning system based on electromagnetic field and neural network. Energies 12(7):1275. https://doi.org/10.3390/en12071275
39. Mostajabi A, Finney DL, Rubinstein M, Rachidi F (2019) Nowcasting lightning occurrence from commonly available meteorological parameters using machine learning techniques. npj Clim Atmos Sci 2(1):41. https://doi.org/10.1038/s41612-019-0098-0
40. Alves ER, da Costa CT, Lopes MNG, da Rocha BRP, de Sá JAS (2017) Lightning prediction using satellite atmospheric sounding data and feed-forward artificial neural network. J Intell Fuzzy Syst 33(1):79–92. https://doi.org/10.3233/JIFS-161152